Graphene-Based Materials as Adsorbents for Wastewater Decontamination

This book aims to provide a fundamental grasp of graphene-based materials (GAMs) and their adsorption process. The effect of diverse process parameters, including pH, temperature, agitation, competing ions, etc., on the adsorption performance of GAMs as well as their recent and relevant applications in biomedical fields, are discussed. The current challenges and future outlook have been addressed as an independent chapter, and the recyclability of these adsorbent materials has also been covered.

Features:

- Focuses on graphene-based materials as adsorbents to remove contaminants from wastewater.
- Includes detailed computational and statistical analyses and cost comparison points.
- Compares the performance of graphene-based materials as adsorbents in the context of various other reported adsorbents, including other 2D materials, such as WS2 and BN.
- Provides fundamental comprehension of the graphene-based materials' adsorption process.
- Discusses the recyclable nature of graphene-based materials, as well as approaches used.

This book has been aimed at graduate students and researchers in wastewater treatment, environmental, materials, and chemical engineering.

Graphene-Based Materials as Adsorbents for Wastewater Decontamination

Suprakas Sinha Ray,
Jonathan Tersur Orasugh,
Lesego Tabea Temane, and
Sarah Constance Motshekga

CRC Press
Taylor & Francis Group
Boca Raton London New York

CRC Press is an imprint of the
Taylor & Francis Group, an **informa** business

Designed cover image: www.shutterstock.com

First edition published 2025
by CRC Press
2385 NW Executive Center Drive, Suite 320, Boca Raton FL 33431

and by CRC Press
4 Park Square, Milton Park, Abingdon, Oxon, OX14 4RN

CRC Press is an imprint of Taylor & Francis Group, LLC

ISBN: 978-1-032-60309-4 (hbk)
ISBN: 978-1-032-62125-8 (pbk)
ISBN: 978-1-032-62130-2 (ebk)

DOI: 10.1201/9781032621302

Typeset in Times
by SPi Technologies India Pvt Ltd (Straive)

Dedication

Dedicated to Our Parents

Contents

Preface

There are several ways to purify water; however, adsorption is among the most straightforward, efficient, yet cost-effective ways to do it. Around the world, adsorption techniques have been used for pollution prevention and remediation. In addition to being the best options for high-standard adsorption systems, composite materials are excellent candidates for adsorbing environmental toxins found in water when combined with graphene or its derivatives. A significant contribution to the adsorption of heavy metals, toxic organic chemicals (colorants/dyes, diverse VOCs, pesticides, chemical fertilizer, drugs, etc.), as well as other suspended particle pollutants in water, particularly industrial effluents, has also been demonstrated for graphene oxide and its engineered material nanostructures (hybrids, composites, etc.). The broad surfaces of both graphene oxide's derivatives and nanocomposites are attached with reactionary oxygen-containing functional groups from graphene oxide, which gives them exceptional stability and adsorption efficiency in aqueous environments and enables them to be recycled for numerous adsorption-desorption cycles. It is preferred to employ graphene-based adsorbents due to their stability and sustainability. All of these graphene-based materials, their adsorption processes, and their use for cleaning and purifying water will be covered in this book.

Certain classes of polymeric material can be used to modify surfaces to create responsive interfaces, which may also exhibit significantly diverse characteristics in response to even minute changes in the environment. Their surfaces can change from hydrophobic to hydrophilic or both. If the surface is a membrane, the pore size may also change if the membrane is chemically changed.

Multiple versions of functional groups with easily modifiable features, such as charge, polarity, and solvency, can be combined along a graphene-polymer backbone to produce drastic variations in the macroscopic properties of the material.

Synthetic adsorbents, such as graphene-polymers systems or graphene-ceramics, etc., created to replicate these naturally occurring adsorbent materials with greater efficiency have been created into various functional varieties covered in this book to satisfy scientific and industrial (practical) uses. These synthetic systems could be classified into many types based on their chemical characteristics. Some adsorbent materials have developed into a particularly valuable class, each with unique chemical characteristics and possible applications in various industries. The components and architectures of those sophisticated adsorbent materials can vary.

Along with many other topics, this book will examine the fundamentals of the synthesis of graphene and its derivatives, changes made to them to use them as adsorbents, advancements made in graphene-based adsorbents, contaminant-targeted adsorption, adsorption kinetics, and isotherms, as well as the use of graphene-based adsorbents for the removal of contaminants from wastewater.

This book also discusses the effect of diverse process parameters, including pH, temperature, agitation, competing ions, etc., on the adsorption performance of these graphene-based adsorbent systems and their recent and relevant applications as biomaterials. The current challenges and the way forward have been discussed as an independent chapter. Though neglected by other authors, this book has also covered the recyclability of these adsorbent materials. For instance, many books have covered the common factors affecting the adsorption process and neglected the effect of adsorbent structural configurations, which should be the critical factor. Moreover, this book focuses on the latest literature, trends, and innovations in this field, including the latest trends in computational and statical analyses and cost comparison.

This book will be a fundamental comprehension of the graphene-based materials adsorption process, the introduction of various emerging pollutants that can be absorbed by GBAMs, knowledge of GBAMs applications as biomaterials, their recyclable nature as well as approaches used thereof, current understanding of the principal difficulties and prospects facing GBAMs, and knowledge of the GBAMs' potential for commercialization.

In summary, the following are the key insights of this book.

- Provides a state-of-the-art overview of fundamental aspects of graphene and its analogs for environmental remediation;
- Demonstrates challenges and prospects for advancement, experimental strategy, fabrication, and improvement in the overall performance of graphene-based materials for water purification;
- Recyclable nature of graphene-based adsorbents; and
- Focuses on the latest literature, trends, and innovation, including computational and statical analyses, along with cost comparison, in this field.

This book contains eight comprehensive chapters. The **First Chapter** introduces the entire book regarding graphene-based adsorbent materials for the remediation of water/wastewater in recent decades. The **Second Chapter** discusses various adsorbents used for adsorptive wastewater decontamination and their advantages and disadvantages compared to graphene-based materials within recent decades regarding available literature. The **Third Chapter** reports a holistic overview of graphene and various graphene-based materials, their synthesis via diverse approaches, and a summary of their application in wastewater/water. The discussion of adsorption, its kinetics, and the isotherms, which are crucial to researchers working in the field, is the primary focus of **Chapter Four**, which serves as this book's fundamental and central argument. In **Chapter Five**, we critically summarized and explained how various factors influence the performance of graphene-based materials as adsorbents to remove contaminants from wastewater. **Chapter Six** covers, to a large extent, the diverse application areas of these innovative materials for the adsorption of contaminants from water/wastewater. In context to Chapter Six, **Chapter Seven** discusses the various graphene-based adsorbent regeneration approaches that researchers have utilized globally as per available literature in recent decades. Finally, the latest trends in computational and statical analyses, along with cost comparison, have been included in **Chapter Eight**. Moreover, the principal difficulties and latest knowledge of graphene-based materials and their potential for commercialization are also covered in this chapter.

This book is ideal for water scientists, material scientists, researchers, chemical and civil engineers, and under- and postgraduate students interested in this exciting field of research.

Moreover, this book will help industrial researchers and R&D managers to bring advanced graphene nanomaterials-based environmental remediation solutions into the market.

We sincerely appreciate the reviewers' critical evaluation of the proposal and manuscripts. Our special thanks go to Dr. Gagandeep Singh at CRC Press for his encouragement, suggestions, cooperation, and advice during the various stages of manuscript preparation, organization, and production of this book. The financial support from the Council for Scientific and Industrial Research, the Department of Science and Innovation, and the University of Johannesburg is highly appreciated.

AIMS

This book aims to provide a fundamental grasp of graphene-based materials (GAMs) and their adsorption process. The effect of diverse process parameters, including pH, temperature, agitation, competing ions, etc., on the adsorption performance of GAMs as well as their recent and relevant applications in biomedical fields, are discussed. The current challenges and future outlook have been addressed as an independent chapter. For instance, many books have covered the common factors affecting the adsorption process and neglected the effect of adsorbent structural configurations, which should be the critical factor. Though overlooked by other authors, this book has also covered the recyclability of these adsorbent materials.

Moreover, this book will focus on the latest literature, trends, and innovations in this field. The latest trends in computational and statical analyses, along with cost comparison, will be included in this book, particularly in Chapter 8. Finally, the principal difficulties and knowledge of the GAMs' potential for commercialization are also covered in this book.

SCOPE

Adsorption techniques have been applied globally for both pollution control and cleanup. When coupled with graphene or its derivatives, composite materials are the most effective choices for high-standard adsorption systems and ideal possibilities for adsorbing environmental pollutants in water. It has been shown that graphene oxide and its engineered material nanostructures (hybrids, composites, etc.) have a significant contribution to the adsorption of heavy metals, toxic organic chemicals (colorants/dyes, diverse VOCs, pesticides, chemical fertilizer, drugs, etc.), as well as other suspended particle pollutants in water, particularly industrial effluents. The reactionary oxygen-containing functional groups from graphene oxide are attached to the broad surfaces of both graphene oxide derivatives and nanocomposites, giving them exceptional stability and adsorption efficiency in aqueous environments and allowing them to be recycled for numerous adsorption-desorption cycles. Due to their stability and sustainability, graphene-based adsorbents are favored. This book will address all of these graphene-based materials, their adsorption procedures, and their applications to the cleaning and purification of water.

The fundamentals of the synthesis of graphene and its derivatives, modifications made to them for use as adsorbents, improvements made in graphene-based adsorbents, contaminant-targeted adsorption, adsorption kinetics, and isotherms, as well as the use of graphene-based adsorbents for the removal of contaminants from wastewater, will all be covered in this book along with many other topics.

For instance, although the impact of adsorbent structural configurations, which ought to be the determining factor, has been largely ignored, numerous books have discussed the common aspects affecting the adsorption process. The impact of various process variables, such as pH, temperature, agitation, competing ions, etc., on the adsorption performance of these graphene-based adsorbent systems is also covered in this book, along with information on their current and pertinent biomedical uses. The current difficulties and the path forward will also be covered in a separate chapter. Although overlooked by other authors, this work has also explored the recyclability of certain adsorbent materials.

Suprakas Sinha Ray
Jonathan Tersur Orasugh
Lesego Maubane
Sarah Constance Motshekga

Pretoria and Johannesburg
March 2024

Acknowledgements

The authors would like to thank the Department of Science and Technology and the Council for Scientific and Industrial Research, South Africa, for financial support. We express our sincerest appreciation to all colleagues, postdoctoral fellows, and students for their valuable contributions, as well as the reviewers for their critical evaluation of the proposal and chapters. We also thank the authors and publishers for their permission to reproduce their published works.

Authors

Suprakas Sinha Ray is a Chief Research Scientist and Manager of the Centre for Nanostructures and Advanced Materials, DSI-CSIR Nanotechnology Innovation Centre, Council for Scientific and Industrial Research, Pretoria, South Africa. He received his Ph.D. degree in Physical Chemistry at the University of Calcutta, India, in 2001 and was a recipient of the "Sir P. C Ray Research Award" for the best Ph.D. work. Prof. Ray's current research focuses on the applications of advanced nanostructured & polymeric materials. He is one of the most active and highly cited authors in polymer nanocomposite materials. Thomson Reuters has recently rated him as one of the Top 1% of most impactful and influential scientists and the Top 50 high-impact chemists. Prof. Ray is the Author of 7 books, co-author of 11 edited books, 32 book chapters on various aspects of polymer-based nanostructured materials & their applications, and Author and co-author of 465 articles in high-impact international journals, and 30 articles in national and international conference proceedings. He also has 7 patents and 25 newly demonstrated technologies shared with colleagues, collaborators, and industrial partners to his name. So far, his team commercialized 19 different products. His honors and awards include South Africa's most *Prestigious 2016 National Science and Technology Award* (NSTF); Prestigious 2014 CSIR-wide Leadership Award; Prestigious 2014 CSIR Human Capital Development award; *Prestigious 2013 Morand Lambla Awardee* (top award in the field of polymer processing worldwide), International Polymer Processing Society, USA. He is also appointed as an Extraordinary Professor, University of Pretoria, and a Distinguished Professor of Chemistry, University of Johannesburg.

Jonathan Tersur Orasugh received his Ph.D. in Polymer Science and Engineering (Nanoscience and Nanotechnology) from the University of Calcutta, India. He is currently working as a Postdoctoral Fellow at the Department of Chemical Sciences, University of Johannesburg, and associated as a senior researcher at the Centre for Nanostructures and Advanced Materials, DSI-CSIR Nanotechnology Innovation Centre, Council for Scientific and Industrial Research, Pretoria, South Africa. His research focuses on processing and characterizing biopolymer-based immiscible polymer blends for biomedical applications, particularly drug delivery and tissue engineering.

Lesego Tabea Temane received her M-Tech in Chemical Engineering from the University of Johannesburg, South Africa. She is a Polymer Characterization Technologist at the Centre for Nanostructures and Advanced Materials, DSI-CSIR Nanotechnology Innovation Centre, Council for Scientific and Industrial Research, Pretoria, South Africa. Her research focuses on synthesizing, processing, and characterizing advanced nanostructured polymeric materials.

Sarah Constance Motshekga holds a D-Tech qualification in Chemical Engineering from Tshwane University of Technology, South Africa. She works as a Senior Lecturer at the University of South Africa, Florida Campus. Her experience spans diverse areas such as municipal water reclamation and biogas production, research and development in material science, nanocomposites, and wastewater treatment. She is passionate about sharing best practices in research and is involved in post-graduate student supervision in wastewater treatment. She takes exceptional pride in her work and always strives for excellence.

1 Introduction

1.1 INTRODUCTION

Major concerns are being raised about air, soil, and water quality due to the contamination brought on by the rapid increase of industrialization, human population, and environmental activities worldwide (Figure 1.1). In particular, efforts are being made to treat contaminated water and to avoid additional contamination of water sources because it is now widely acknowledged that water contamination poses a serious threat.[1–3] The industrial revolution, the advancement of science and technology, and the lack of adequate and effective wastewater treatment methods are all considered causes of water contamination. In addition to rendering water sources unsafe for drinking, wastewater effluent from households and industrial sources negatively impacts water availability for agriculture, healthcare, pharmaceutical uses, etc. Most contaminants interfere with the natural ability of water to purify itself and disrupt photosynthesis in aquatic systems; one of their severe effects is the extinction of some aquatic species.[4–6] For instance, heavy metal ions and organic dyes exhibit adverse effects that may result in death. Their acute exposure offers substantial health risks to humans. Similarly, industrial chemical and oil spills and leaks into water bodies deteriorate water quality and harm aquatic life.[7] Accidents involving large-scale oil spills are considered to be disastrous and harmful to marine life, negatively impacting the infrastructure, the underwater ecosystem, the tourism economy, fishing and seafood industries, as well as costs associated with their rehabilitation.[7]

As a result, there is a significant amount of research being done globally to create effective and efficient materials and novel technologies for the rehabilitation of wastewater sources that contain contaminants, including but not limited to dyes, emerging contaminants, heavy metals, and oils.[7–15] Due to specific benefits in terms of cost-effective, simple process operation, environmental friendliness, and adaptability to various contaminants; sorption techniques (adsorption or absorption) emerged as widely preferred and economically practical approaches among the biological, chemical, and mechanical wastewater treatment processes. These processes include advanced chemical oxidation, coagulation and flocculation, ozonation, membrane filtration, ion exchange, etc.[16,17] For instance, methods for treating water that use chemical or membrane technologies typically have high operational costs and occasionally result in additional secondary harmful contaminants. Since the water industry needs to supply potable water, there is a need to develop affordable and reliable solutions to deal with the daily decline in water quality. Adsorption techniques, in particular, provide a promising alternative to membrane filtration and photocatalysis in treating metal ions that are generally challenging to remove. The method has proved to be highly efficient in removing various contaminants from wastewater streams of diverse sectors.[18,19] In contrast, absorption techniques are frequently used to remove specific nonpolar (hydrophobic) contaminants from aqueous emulsions or

DOI: 10.1201/9781032621302-1

FIGURE 1.1   Overall industrialization contributes to water contamination.

floating layers over water, including typical organic solvents, crude oil, mineral or vegetable oils. Although sorption is a general term, it is crucial to understand that the underlying adsorption and absorption sub-processes are phenomena connected to the surface and the bulk, respectively, and their driving principles are entirely different.

Furthermore, the type of the adsorbent-adsorbate mixture, the surface area of the adsorbent that is accessible, the amount of contaminants present, contact time, pH, temperature, etc., all have a role in the adsorption process.[18,20,21] In contrast, the pore size or pore volume of the adsorbent, composition, and wettability, coupled with the physical properties of the adsorbate species, and the parameters of the interface between the absorbent and the absorbate, all control how much material is absorbed. Many adsorbing-absorbing materials, such as zeolites, silica gel, activated alumina, coal ash, clay minerals, carbon-based materials, and polymeric materials, have been used for wastewater treatment.[22–25] All these materials render quality water when used for treatment, with more continuous development being done to explore them thoroughly. Carbon-based materials, for instance, are continuously being explored for their exceptional properties, including compatibility with the environment, high porosity, and a high specific surface area. Graphene, which is the newest addition to the carbon-materials, is another attractive alternative due to its outstanding chemical and mechanical properties combined with high specific surface area, cost-effective, and simple synthesis process.[26,27] Considering this, extensive research is being done to develop highly effective graphene-based adsorbents, understand their contaminant-sorption mechanism and evaluate the efficiency of their treatment outcomes. This is driven by the quest to explore and develop highly efficient adsorbents and advance the exploration of carbon-based materials.

Moreover, effective and potent solutions for treating domestic and industrial wastewater are needed. This can be done by creating new approaches or developing the current ones by making certain interventions. This chapter serves as the book's introduction to using graphene-based adsorbent materials for water and wastewater cleanup in recent years.

1.2 WASTEWATER

Although some contaminants are released into the environment naturally due to geological and biological activity, human activities continue to contribute more. Many of these contaminants have well-established toxicology; industrial wastewater is a significant component of these contaminants. Most industries use a lot of water; as a result, a lot of effluents are produced and discharged with contaminants of either organic or inorganic compounds.[1,2,5,6] Due to depleting water sources, rising costs of wastewater disposal, and tougher discharge laws that have reduced the allowable quantities of contaminants in water; wastewater treatment is becoming increasingly imperative. Several efficient treatment techniques that use modern technology and are cost-effective are available to treat water contaminants. Heavy metals, microorganisms, organic and inorganic compounds, and emerging contaminants in trace amounts are the most common contaminants.[11,28,29] Figure 1.2 summarizes some of these contaminants.

1.2.1 WASTEWATER CONTAMINANTS

1.2.1.1 Heavy Metals

Heavy metals have significantly high densities and are hazardous even in small amounts. They tend to accumulate and infiltrate food chains, where they can potentially become a health threat because of their persistence, toxicity, and non-biodegradability. Figure 1.3 summarizes the harmful effects of various heavy metals on human health. Different industries, including the production of energy and fuel, mining, iron and steel production, fertilizers and pesticides, electroplating, etc., are sources of heavy metals or heavy metals that are often used. Most metal ions exist as cations, although some metals also exist as anions.[5,31] In the last decades, great concern has been addressed on the high concentrations of heavy metal ions contaminating the environment. Whether discharged in low or high concentrations, the contamination of water sources by heavy metals is significant and may cause permanent harm to the environment and human health. Because of their durability and environmental resistance, technological advancements in the treatment of heavy metal ions from contaminated water have become a significant challenge. In the study of Belova,[32] the author investigated the sorption properties of natural zeolite on nickel, copper, cobalt, iron and their mixture. The author pointed out that zeolite is a

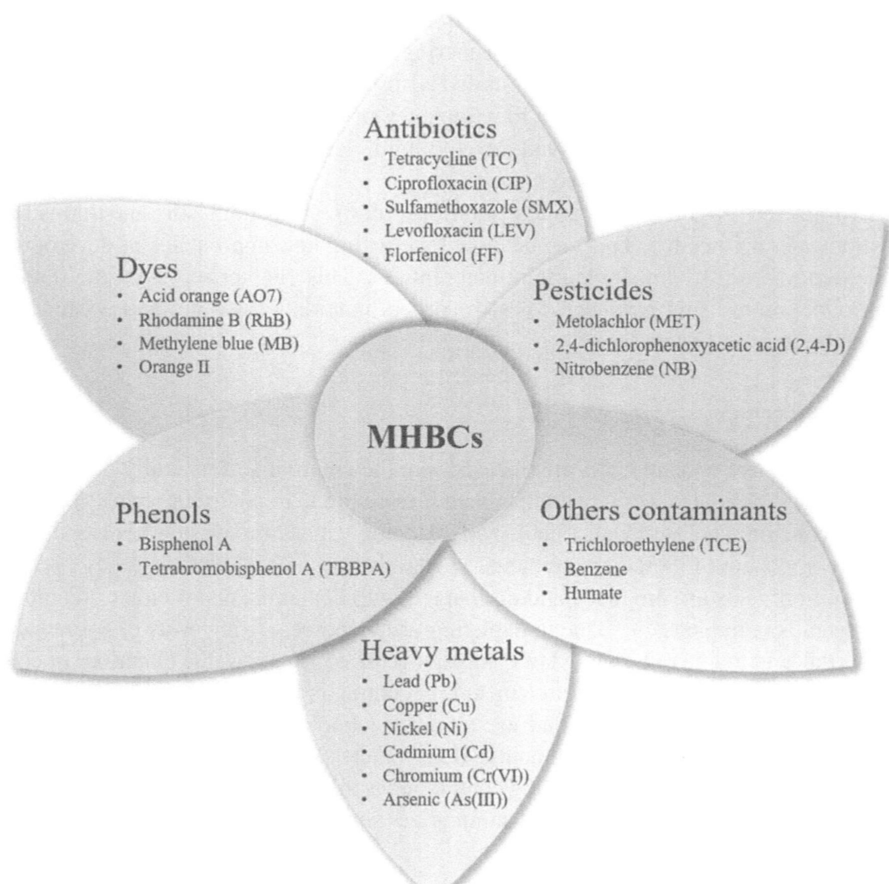

FIGURE 1.2   Summary of some of the contaminants found in wastewater. (Reproduced with permission from Liu et al.[30] Copyright 2022, Elsevier Science Ltd.)

potential material for extracting heavy metal ions from wastewater. This was because the zeolite's adsorption capacity (Q) increased with a concentration increase of the metal ions, where concentrations ranging from 0.5 to 3.5 mg-eq/L showed satisfactory results according to the Langmuir equation. Based on the results presented, the author concluded their study by promoting the use of natural zeolite as a practical adsorbent for removing contaminants in water. Another study by Taamneh and Sharadqah[33] showed the practicability of zeolite as a potential adsorbent for treating heavy metals, including copper and cadmium. The study employed batch studies to investigate the sorption properties of zeolite by varying factors such as adsorbent mass, contact time, initial concentration, etc. The authors concluded that these variables were important factors influencing adsorption. The results showed that zeolite is an effective material for treating copper and cadmium; the selected batch method showed significant advantages of both the material and the technique. Figure 1.4 briefly summarizes treatment technologies used to remove heavy metal ions.

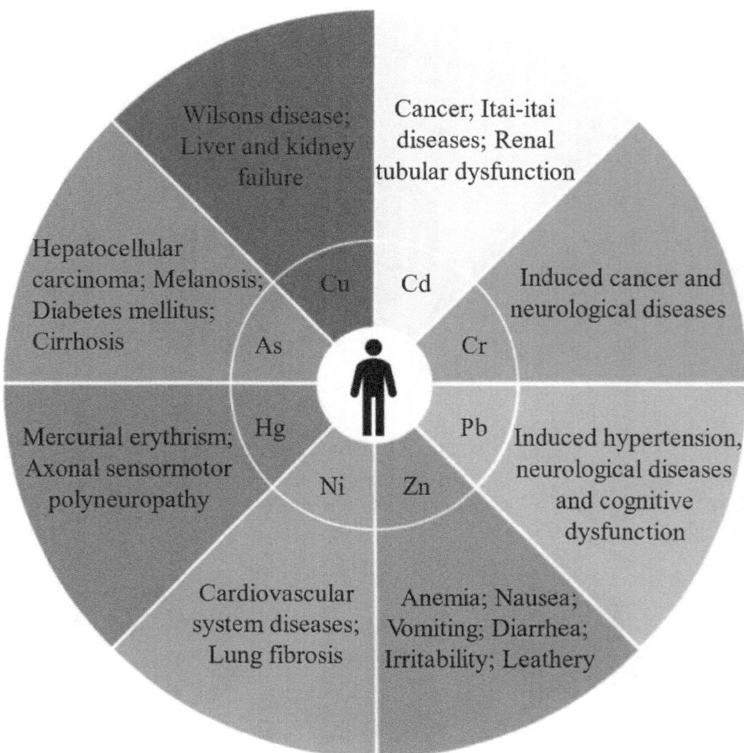

FIGURE 1.3 Various heavy metals' harmful effects on human health and ecosystems. Mercury (Hg), cadmium (Cd), arsenic (As), lead (Pb), chromium (Cr), nickel (Ni), Copper (Cu), and zinc (Zn) are heavy metals that cause a variety of hazards to human health and ecosystems. (Reproduced with permission from Jiang et al.[34] Copyright 2022, Elsevier Science Ltd.)

1.2.1.2 Emerging Contaminants

Emerging contaminants (ECs) are partially regulated or completely non-regulated contaminants which may pose a health threat to humans and the surroundings. Emerging contaminants include pharmaceuticals, personal care products, food additives, plasticizers and laundry detergents. The threat posed by emerging contaminants is unique to water sources, human health as well as the environment because of their limited toxicity data.[2,9,10,12,14] Due to the lack of the availability of additional data on emerging contaminants, there is uncertainty regarding the toxicological impact on human health. Moreover, they have been observed to exhibit substantial harm to the aquatic ecosystem with dysfunctional liver and lungs, impairment of reproduction, brain dysfunction, carcinogenic diseases, and the disruption of gene expression in the marine species, which results in the feminization of some of the aquatic species.[2,6] Furthermore, emerging contaminants have been associated with developing resistance towards the microbial community and genotoxicity in aquatic species. Water sources, especially drinking water, have been discovered to contain

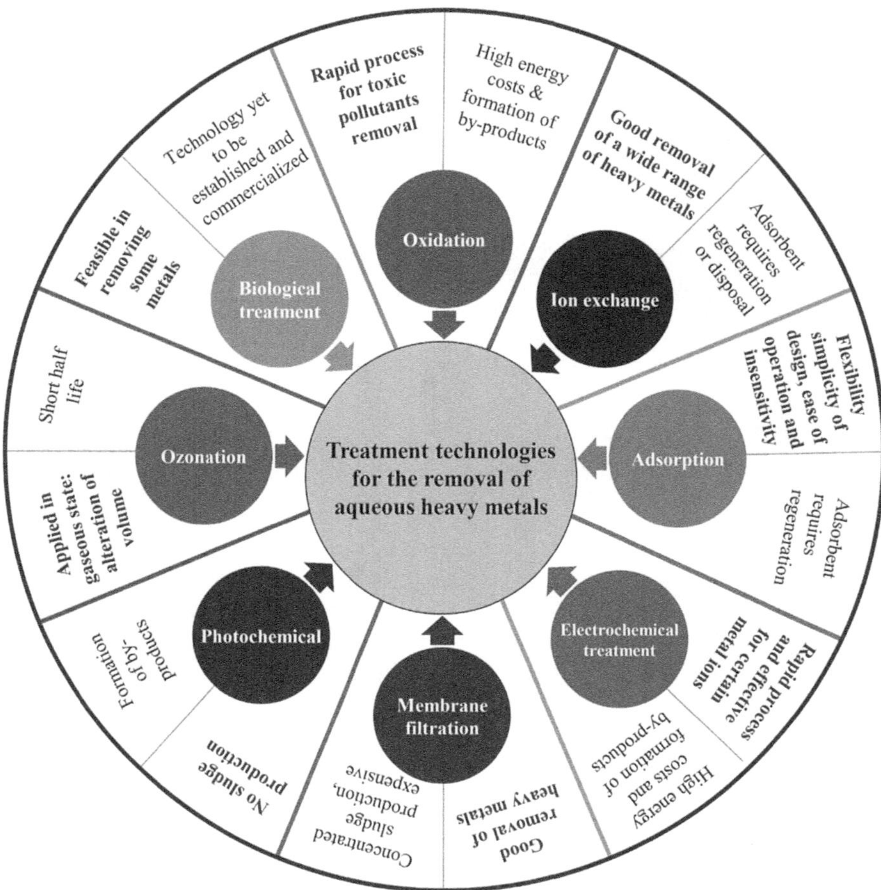

FIGURE 1.4 Main characteristics of technologies for heavy metal removal. (Reproduced with permission from Hoang et al.[22] Copyright 2022, Elsevier Science Ltd.)

traces of these emerging chemicals. Although found in small quantities in water, the human toxicity, ecological disruptions and bioaccumulation of pharmaceuticals, pesticides, plasticizers, antibiotics, etc., have been underestimated due to their complexity in the environment, which adds to the high costs of analysis and the time and methods required of monitoring them.[2] Antibiotic resistance is a concerning occurrence, and antibiotic-resistant strains have been facilitated by the overuse of antibiotics.[9–12,14,35] The majority of emerging contaminants are recalcitrant compounds that are often present in trace concentrations, making it difficult and challenging to thoroughly remove them from wastewater. Due to the inability of conventional technologies to completely eradicate contaminants available at low concentrations and other limitations demonstrated, efforts are made to compensate for the inadequacies of these technologies. A recent study by Zhu et al.[29] reported on the preparation of non-metallic hollow and porous spheres loaded with catalytic nitride nanotube to be used in the removal of specific emerging contaminants such as endocrine disrupter p-nitrophenol and antibiotic tetracycline hydrochloride. The synthesis of the material was eco-friendly and green in the selection of raw materials and the selected method of preparation. The authors claimed that the prepared material could effectively reach a complete photocatalytic reaction and enhance the efficiency of ozone utilization and has great results on emerging contaminants. They reported 90.1% degradation of tetracycline and 84.3% of p-nitrophenol in 20 minutes, with a system that followed a three-stage cyclic reaction mechanism. Furthermore, they alluded that their system could be employed not only for cutting-edge treatment with a good removal efficiency of secondary effluent of a sewage treatment plant but also as a secondary pre-treatment of quality produced water because of its good treatment effect.[29] The wide range of contaminants in water, each with a different chemical structure and properties, adds another layer of complexity to remove.[28,29] Therefore, this diversity of contaminants in water necessitates a wide range of treatment techniques that are both economical and environmentally friendly in addition to being effective.[12,14] The development in determining the most effective technique for wastewater treatment is encouraging. It has led to an expansion of research into emerging contaminants.

1.2.1.3 Dye Contaminants

Various industries, including textiles, paint, cosmetics, and chemicals, frequently use synthetic dyes as coloring agents. Due to the diverse chemical structure of dyes, synthetic dyes are identified as cationic, anionic, or non-ionic dyes. In general, ionic dyes are readily soluble in water, while non-ionic dyes, like vat and disperse dyes, are soluble in organic solvents but are insoluble in water. Anionic and cationic dyes are subcategories of ionic dyes. Cationic dyes, which include all basic dyes, are divided into three categories: acid, direct, and reactive dyes.[30,31,36] Even if their concentration was minimal, dye contamination would still change the color of the receiving water source and is easy to identify. This type of contamination slows the rate at which marine plants can photosynthesize, as the sunlight that enters the water is reflected and absorbed by the dye molecules. Long-term exposure to synthetic dyes can cause cancer, respiratory issues, and skin discomfort. Because synthetic dye molecules are intended to withstand washing water, chemicals, heat, and ultraviolet light, it is particularly challenging for traditional wastewater treatment techniques to handle them. Ikram et al.[26] reported on the findings of the World Bank Report, where it was found that about 17–20% of contaminants in water are channeled from the textile sector. Various textile techniques produce assorted dyes, including methylene blue, which accounts for 10–15% of being discharged directly into the effluent and can potentially cause harmful health risks, including allergies, cancer, etc. To address this problem, the authors used Hummer's method to successfully synthesize graphene oxide and its reduced graphene oxide derivative by using a thermal treatment on the graphene oxide. This carbon-based nanomaterial was incorporated with silver using the hydrothermal method to form a nanocatalyst to degrade methylene blue. Their results showed enhanced photocatalytic activity of the nanocatalyst with an increased silver loading ratio due to the reduced simple recombination in semiconductors. They concluded their study suggested that the prepared nanocatalysts demonstrated no harmful behaviour in water treatment and could be used as an efficient nanocatalyst for degrading dyes.[26] The adsorption process depends largely on

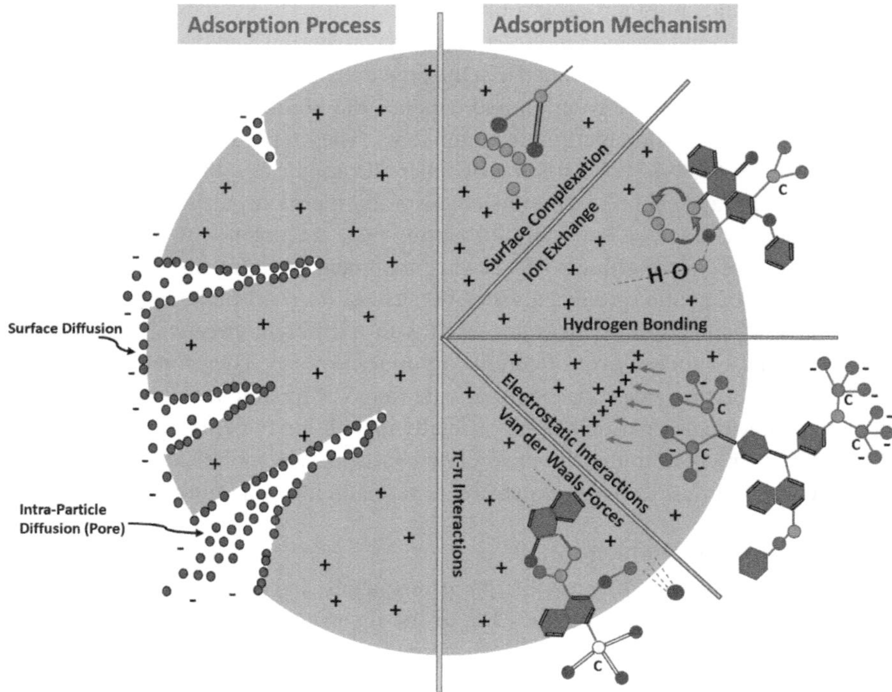

FIGURE 1.5 Adsorption process and adsorption mechanism for the removal of methylene blue. (Reproduced from permission from Hamad and Idrus.[20] Copyright 2022, the Authors.)

the nature of the adsorbent and the adsorbate. No process is similar to the other, as various factors are involved. Figure 1.5 above shows the mechanism and adsorption process of methylene blue on the carbon surfaces.

1.2.1.4 Phenols

Phenols are regarded as a priority contaminant among the several organic contaminants in wastewater because of their toxicity towards animals, humans, and plants, even when they are present in low quantities. Steel mills, oil refineries, pharmaceutical and petrochemical companies, paint manufacturers, coal gas producers, synthetic resin producers, plywood manufacturers, and mine discharge are the main producers of phenols. The coke processing often produces the wastewater with the greatest phenol concentration (>1000 mg/L). Resin producers release phenolic chemicals, which have a concentration between 12–300 mg/L. The Environmental Protection Agency (EPA) has regulated phenol contamination in wastewater to be less than 0.1 mg/L. The World Health Organization (WHO) is very stringent regarding phenol regulation. The allowable concentration in drinkable water is set at 0.001 mg/L for phenols.[15,30,37] For phenol removal in pure water and seawater, Xu et al.[37] prepared a titanium dioxide-based catalyst. First, holes and free radical traps were added to fully comprehend the photodegradation mechanism of various organic contaminants in seawater, using TiO_2-based catalysts exposed to visible light. Their results showed that the photogenerated holes created by the catalyst had a huge role in removing the organic contaminants, irrespective of where the photodegradation took place in pure water or seawater. The authors further mentioned that because the prepared catalyst was an excellent adsorbent for organic contaminants, the salt ions in the seawater did not interfere with the adsorption of the contaminants, which was attributed to the reduced graphene oxide in the catalyst. Although there was a slight hindrance in the adsorption capacity because of the abundance of salt ions available in seawater, the results of the two water systems were comparable.[37]

1.2.1.5 Pesticides

Pesticides intentionally released into the aquatic system are frequently found at low concentrations and frequently exist as complex mixtures. One of the most significant contributors to organic contamination in various water sources is the leaching of chemical fertilizers and pesticides used in agricultural and forest land. Because of their toxicity, carcinogenic and mutagenic, pesticides are hazardous to life. As a result of contaminating the environment, the toxicity of pesticides and their decomposition products render these chemical substances potentially hazardous. The cumulative long-term toxicological consequences of their individual and/or combined effects have caused major aquatic and human life concerns.[35] Considering the rising use of pesticides in the home and agricultural operations, there is a serious concern about contaminating other water bodies, including surface water, groundwater, and soil, using herbicides and pesticides. Mondol and Jhung[35] conducted a study that focused on the use of metal-organic framework-based materials and their composites to remove pesticides from aqueous solutions effectively. This was corroborated by previous reports that the gradual increase in living standards is also massively contributing to the contamination of water sources as the requirement for food production is increasing. The authors reported that millions of tons of pesticides were used globally in 2016. The use of metal-organic framework-based materials was because the materials have emerged as competitive adsorbents to activated carbon, zeolites, etc., as they are easy to functionalize and have a high specific surface area with tunable pore size. Adsorption was selected as the operation method for removing the contaminants. This is due to its cost-effectiveness, ease and straightforward operating system. Although the materials showed satisfactory results in the removal of pesticides from water, the comparison of these materials against other adsorbents, such as zeolites, porous organic polymers, mesoporous materials, etc., is very high in terms of the production of the materials, selectivity, and adsorption capacity, and their regeneration also being considered. The authors concluded that using metal-organic framework materials is relatively costly compared to natural adsorbents or agricultural waste by-products.[35]

1.2.1.6 Microorganisms

When open water sources like rivers and dams serve as an alternative for drinking water, the greatest threat comes from bacteria contamination. Unfortunately, not all open water sources are safe for humans to consume. Common issues with these water sources include waterborne illnesses that could be fatal and deteriorate the water quality. Millions of children die each year from diarrheal infections brought on by lack of potable water and inadequate sanitation protocols, making them the third biggest cause of mortality for children under 5 in developing countries. Contaminated water often contains bacteria contaminants, including *Escherichia coli*, *Salmonella*, *Vibrio cholerae*, *Shigella*, *Bacillus subtilis*, etc. As a result, effective water disinfection is a requirement everywhere.[38–40] Treatment procedures include various methods, including straightforward techniques (boiling of water), highly advanced (filtration using membrane systems), widely utilized methods (chlorination) and experimental size procedures (sonication). Drinking water is used for washing hands and preparing food; using contaminated water could easily cause infection without directly consuming the water. Zhang et al.[38] reported on a multifunctional composite prepared from reduced graphene oxide and montmorillonite aerogel to treat various wastewater contaminants. The composite was highly efficient in the degradation of methylene blue at 97.31% and hexavalent chromium at 94.8%. The adsorption mechanism of the dye and the metal ions by the composite was described through adsorption studies using Langmuir isotherm and the pseudo-second-order kinetics model. Furthermore, the composite was also used to inactivate bacteria like *Escherichia coli* and *Staphylococcus aureus*. The inactivation was recorded at 91.5% for *Escherichia coli* and 95.53% for *Staphylococcus aureus*. They concluded that the prepared composite was promising for water treatment as it could serve as a robust multifunctional material for inactivating pathogens and for degrading dyes and heavy metal ions in water. Moreover, the composite also exhibited excellent recyclable properties.[38]

1.3   WASTEWATER PURIFICATION

Multiple treatment techniques are available for the removal of contaminants from wastewater, including but not limited to the advanced oxidation process, adsorption, biological treatment, disinfection, coagulation, Fenton process, flocculation, photocatalytic degradation, membrane filtration, chemical oxidation, electrochemical degradation, etc. Table 1.1 summarizes some of the wastewater treatment processes and their advantages and disadvantages. Some of these treatment techniques incorporate primary, secondary, and tertiary treatment. In primary treatment, the initial steps involve screening and sedimentation of both larger and particulate particles from the influent sources. Secondary treatment involves the use of aerobic and anaerobic processes. All wastewater treatment technologies currently in use have various environmental effects, many heavily influenced by the technique and the contaminant being treated. Although the techniques are efficient in treating wastewater, some are not cost-effective or are ineffective and insufficient to remove or degrade emerging contaminants, including pharmaceuticals and personal care products (PCPPs) and endocrine-disrupting compounds (EDCs), prompting the need for new and advanced alternatives. Moreover, reusable materials have the prospects of reducing the waste generated and the energy consumption while increasing the treatment efficiency; hence, any material that could be efficiently regenerated after a great removal of contaminants is encouraged.

TABLE 1.1

Summary of the Advantages and Disadvantages of Various Wastewater Treatment Techniques

Treatment Technology	Advantages	Disadvantages
Adsorption	Efficient and simple operation.	Regeneration is expensive, high generation of concentrated sludge.
Ion exchange	Simple and effective, ease regeneration of resins.	Not economically feasible, adsorbent requires regeneration or safe disposal.
Biological treatment	Efficient and eco-friendly.	Slow process, difficult to scale-up, nutrient requirements, technology not yet commercialized.
Coagulation and flocculation	Simple and effective.	High sludge production, handling and disposal problems. High chemical reagents costs.
Filtration	Simple and cost effective.	Not feasible for most inorganic and organic contaminants, including emerging contaminants
Oxidation	Rapid and efficient process.	High operating costs, energy consumption, use of toxic chemical reagents, high maintenance.
Reverse osmosis	Simple and effective.	High operating costs.
Chemical precipitation	Simple and effective.	High sludge generation.
Membrane separations	Efficient and produce quality water.	Not economically feasible, short lifespan.
Photocatalysis	Efficient for pre-treated water.	High energy and maintenance cost.

1.3.1 OXIDATION

Various organic contaminants can be treated effectively by chemical oxidation, while inorganic contaminants such as dissolved minerals, salts, and heavy metals are typically more difficult to remove. Although present in low concentrations, oxidation could also remove emerging contaminants. However, efficiency is typically low. Chlorine, ozone, and hydrogen peroxide are common oxidizing agents (Figure 1.6). Both electrochemical oxidation and wet oxidation, in which an electrical current is employed to facilitate the redox reaction, are some of the frequently utilized treatment techniques.[16,17,41] For persistent contaminants present in wastewater in high concentrations, the use of electrochemical oxidation could be insufficient and ineffective for the degradation of the contaminants.[17] In addition to requiring large energy inputs, electrochemical oxidation frequently requires the replacement of the electrodes due to corrosion. The drawback of chlorine for oxidation is that it is a corrosive and extremely poisonous gas even in low quantities and produces hazardous by-products that need further treatment. Ozone, a strong oxidizing agent, can be used directly or indirectly to degrade contaminants. The exclusive use of ozone without any additional catalyst or photoactivation is called ozonation, whereas, with a catalyst addition or photoactivation, it is called an advanced oxidation process. Ozone and chlorine are widely utilized oxidation treatments for removing pharmaceuticals

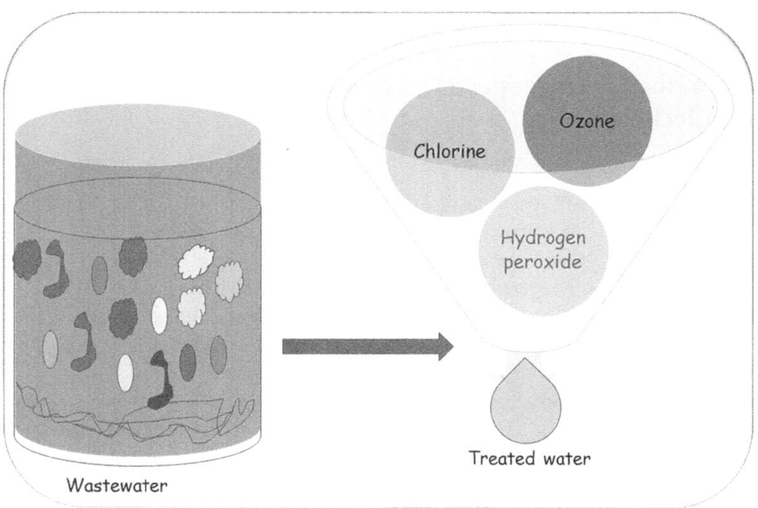

FIGURE 1.6   Typical oxidation process using chlorine, ozone, or hydrogen peroxide.

from wastewater, although ozone has been shown to be more effective in a wide range of pharmaceuticals than chlorine. Some factors that may influence both oxidation and advanced oxidation include oxidant dosage, level of contamination, operational mode and the quality of water, which includes salinity, pH, total suspended solids, and total dissolved solids.[17] To improve the effectiveness and reactivity of ozone, catalysts, hydrogen peroxide and ultraviolet irradiation are often applied simultaneously. Although most of the compounds can exclusively be degraded by ultraviolet irradiation, the technique is not sufficiently equipped for the degradation of xenobiotics. And also, hydrogen peroxide generates a powerful and nonspecific oxidant of hydroxyl radical when exposed to ultraviolet irradiation, which can be used for pharmaceutical contaminants. The generated hydroxyl radicals facilitate the conversion of contaminants to less harmful and biodegradable compounds.[17] However, ozone also has a variety of drawbacks, such as the high electrical requirements and costs involved in producing it on site, which are partly brought on by the high toxicity even at low quantities and harmful by-products. The production and delivery of concentrated hydrogen peroxide come at a comparatively high energy cost. Chemical oxidation involves handling reactive chemicals by nature, which raises some safety issues. It also generates sludge, due to interactions between the chemical reagents and other components of the water.

1.3.2 PHOTOCATALYSIS

Many halogenated organic compounds, certain non-halogenated organic compounds, heavy metals, dyes, and some pharmaceuticals and personal care products can be degraded in certain conditions by using different types of photocatalytic degradation (e.g., TiO_2 catalyzed UV photolysis, UV photolysis)[17,42,43] Figure 1.7 shows a typical photodegradation of dyes using TiO_2. Photolysis is frequently combined with other

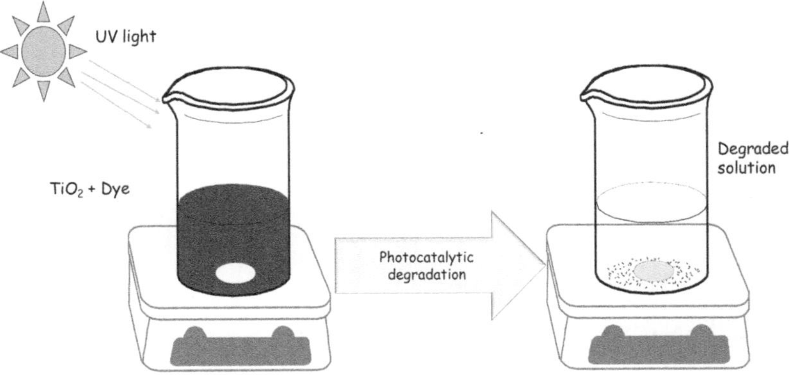

FIGURE 1.7 Photocatalytic degradation of dyes using TiO_2.

treatment methods since low quantities of volatile organic compounds (VOCs) might be challenging to decompose using photolysis alone. The quality and purity of the water being treated often pose a challenge to the UV light as it is difficult to penetrate through the murky water. Due to the high energy consumption, UV photolysis can significantly impact life cycles. Moreover, UV lamps require regular cleaning, which increases labor costs.[16,26,37] Water chemistry and the presence of other contaminants limit photocatalysis, which hinders its long-term performance. Direct and indirect photodegradation is the main removal mechanism for pharmaceuticals and their metabolites in water. The degradation time also heavily depends on the compounds being degraded. Antoniadou et al.[42] reported on applying carbon nanotubes and their composites to the photodegradation of emerging organic contaminants and pharmaceutical compounds in effluents and wastewater facilities. They used carbon nanotubes because of their high and accessible surface area, great adsorption capacity, and tunable surface areas, which allowed introduction of the photocatalytic properties through functionalization. The activity of the material was evaluated for photodegradation of the targeted contaminants, using solar light and UV light. The results demonstrated that the adsorption capacity of the carbon nanotubes with titania was enhanced, enabling the removal of diclofenac, a widely used nonsteroidal anti-inflammatory drug. Although the degradation method has not been confirmed for emerging contaminants of trace concentrations, the report indicated that various active species showed the dominance of hydroxyl radicals during the photodecomposition reaction process.

1.3.3 FENTON/PHOTO-FENTON

Iron and hydrogen peroxide are common catalysts used in the Fenton method to oxidize contaminants in wastewater. Due to their effectiveness and rapid speed, Fenton reactions are recommended alternatives for wastewater treatment, especially emerging contaminants. The electron-Fenton method has been developed to overcome the conventional Fenton process's limitations. Pesticides, herbicides, the majority of non-halogenated organics, dyes, and some emerging contaminants can

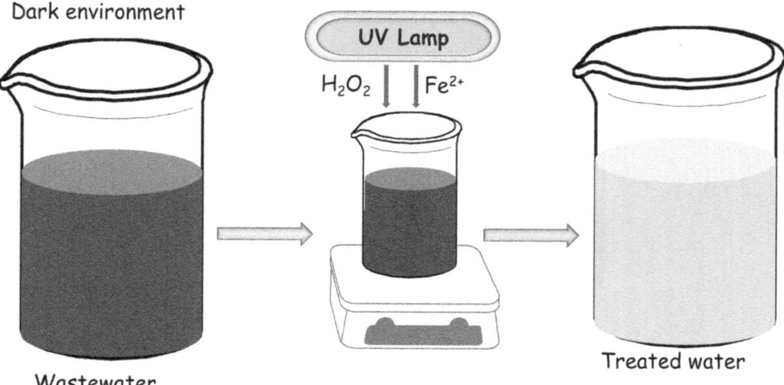

FIGURE 1.8   UV Fenton process.

all be removed using the Fenton and photo-Fenton reactions.[16,17,41,43] During this process; ultraviolet light is used to generate radicals due to the interactions with the iron and hydrogen peroxide catalyst. The rate of oxidant formation and, thus, the efficiency of the treatment process can be enhanced by adding UV irradiation to the process (Figure 1.8). Most photo-Fenton processes utilize lower UV energy, and some processes are powered by solar. Many photo-Fenton processes have been done in laboratory scales using synthetic water and targeting certain contaminants. Fenton and photo-Fenton reactions, however, suffer similar disadvantages to UV photolysis and hydrogen peroxide oxidation, including the high cost of the energy needed to produce hydrogen peroxide and the operation of the UV lights. Handling of highly reactive reagents is required when using the Fenton method.

1.3.4 Adsorption

Most contaminants, including organic and inorganic compounds, are effectively removed from wastewater using adsorption. Adsorption techniques trap contaminant molecules inside the pore structures of various porous materials. Activated carbon is one of the most porous, highly efficient, and widely used materials for treating various contaminants in wastewater when using the adsorption method.[36,38,44] However, it has been demonstrated that the amount of energy required for the production of high-quality activated carbon has a considerable effect on the environment because non-renewable energy sources are being used, and this is even though activated carbon is produced from cost-effective raw materials.

Furthermore, removing the adsorbed contaminant molecules on the activated carbon surfaces after treatment has proven to be tedious. As a result, the alternative materials that are sought after for adsorption should take into consideration the cost-effectiveness of producing and using the material and operation of the process.[9,11,12,21] Regeneration of the materials should also be prioritized to create a low-cost process because the adsorption method only removes the contaminant from the water and does not transform or degrade it. The adsorbent will either be regenerated onsite, treated as a hazardous waste that requires special disposal at an additional cost, or repurposed.[19] Studies on adsorption kinetics are crucial for understanding the mechanisms that govern adsorption processes and identifying the appropriate process parameters and the shortcomings of the overall process. Variables, including solution pH, adsorbate concentration, temperature, adsorbent particle size, system temperature, contact time, etc, directly influence the adsorption kinetics. A typical adsorption process is presented in Figure 1.9, with some of the process variables. The mechanisms that govern the adsorption process, as well as the response of the adsorbent, could be explained by applying the main kinetic models of pseudo-first-order and pseudo-second-order models. A study by Taamneh and Sharadqah[33] investigated the adsorption capacity and removal efficiency of copper and cadmium from water, using natural zeolite. Their results showed that zeolite was an excellent material for removing the specified metals from water. They assessed the efficiency of the adsorbent by adjusting the factors that influence the adsorption process, such as adsorbent mass, concentration of influent, and contact time.[33] It was observed from the results that with an adsorbent dosage of 30 mg, the adsorption equilibrium for both contaminants was reached in 20 minutes, with a maximum concentration value of 100 mg/L. Another study by Belova[32] also investigated natural zeolites for the removal of copper, nickel, cobalt, and iron ions from contaminated water with varying concentrations according to the composition of heavy metals in the mining wastewater. The author reported that zeolite was an appealing adsorbent for heavy metal ions removal, and the adsorption capacity was observed to increase with the metal ions concentration, and its effectiveness could be tested with industrial water from other areas.[32]

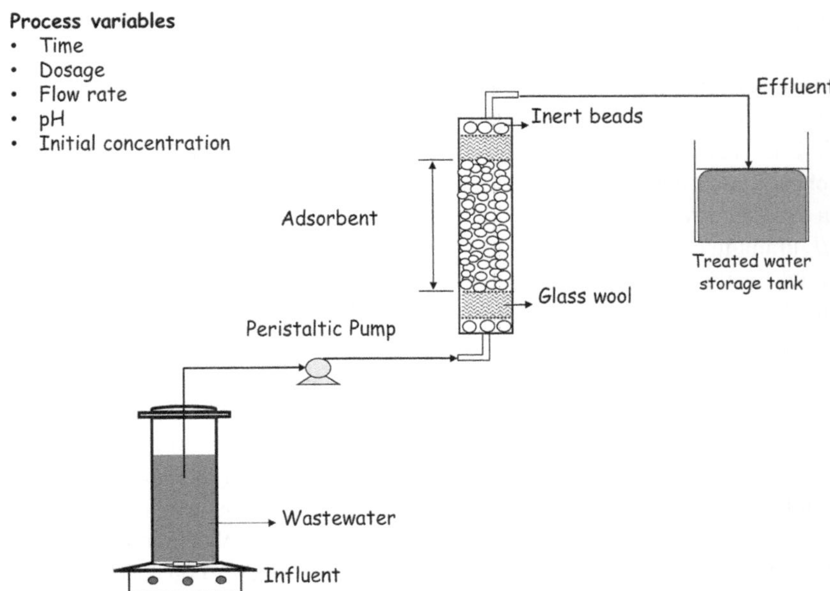

FIGURE 1.9 Typical adsorption process through a column experiment.

Clays are considered alternative adsorbents for treating various contaminants from wastewater, including dyes and heavy metals. This is due to the high adsorption capacity, properties of reactivity, and specific surface area that is characterized by great possibilities to modify and improve the targeted contaminants, using various methods. Common methods of clay modifications include thermal treatment, organic modifications, pillaring, acid treatment, etc. These methods have their benefits on the surface of the clay and their improved properties. Hence, the challenges encountered with clays during adsorption treatment are progressively being overcome by the different modification processes. A study reported on the successful preparations of halloysite incorporated iron oxide (Fe_3O_4) nanoparticles by in-situ preparation method in the internal hollow spaces of the halloysite and silane coupling agents on the external surfaces using condensation reaction for the removal of hexavalent chromium and antimony pentoxide. The composite showed high removal efficiency of both hexavalent chromium and antimony pentoxide. The authors further showed that the co-existence of hexavalent chromium enhanced the adsorption of antimony pentoxide with a removal efficiency of 67% in a single-solute system to over 98% in the presence of hexavalent chromium. It was concluded from the results that because of the composites' effectiveness in removing the specific contaminants, the low-cost materials and easy preparation process of the composites are promising to be used as efficient adsorbents. The evaluation method was through batch studies.[36] Another type of sorption is ion exchange, which is frequently used to remove hazardous ions, such as heavy metals and non-metallic ions, from aqueous solutions and substitute the ions with harmful ions. The cost of the ion exchange technology can differ greatly based on the type of ion exchange resin being used. Common cations such as Ca^{2+} and Na^+ are usually present in water in high concentrations. They will compete with other cations for exchange sites, making water's hardness an important factor in ion exchange. The ion exchange process is reversible by removing excessive undesirable ions. Moreover, the ion exchange resin can easily be regenerated.[16,17] However, the process generates significant waste concentrated with harmful contaminants. Due to the simplicity, effectiveness, and ease of operation of the adsorption method, Verma and Kim[27] conducted a study where they analyzed the removal of uranium using graphene-based adsorbents. The results of magnetic graphene oxide showed more than 90% removal of uranium within 30 minutes with synthetic wastewater of 10 mg/L uranium concentration at pH 6. The suitability of the magnetic graphene oxide was also evaluated with real water using radioactive wastewater from a mine with more than 100.8 µg/L of uranium concentration mixed with other common metal ions. The authors reported that the composite showed great selectivity of uranium adsorption from the mine water, with more than 93.68% removal efficiency. The remaining uranium concentration in the water was found to be 6.37 µg/L, which was very low compared to the allowable uranium concentration of 30 µg/L.[27] The selection of graphene-based adsorbents over other carbon-based adsorbents like carbon nanotubes or activated carbon was easy based on the previous studies, where it was shown that graphene oxide had a greater removal efficiency than copper at 75% with a pH of 5.5. In comparison, carbon nanotubes and activated carbon had 15% and 36% removal efficiency, respectively. The high performance of graphene oxide compared to carbon nanotubes or activated carbon is attributed to the higher BET surface area and its abundant oxygen moieties. The various functional groups introduced on the graphene surfaces contribute to different contaminants' high adsorption capacity.[27]

1.3.5 FILTRATION

One of the oldest and simplest ways of treating water is passing it through a bed of small particles. Most contaminants can be removed from wastewater using several filtering processes, including membrane filtration technologies (microfiltration (MF), nanofiltration (NF), ultrafiltration (UF), and reverse osmosis (RO)) and conventional sand filtration. Figure 1.10 shows a schematic diagram of the whole water treatment process in the wastewater treatment plant, with various sampling points. Filtration removes contaminants by retaining them between the pore spaces of the filter; hence, it heavily depends on the pore sizes of the filter material.[45,46] It is one of the primary steps of wastewater treatment in most groundwater and surface water treatment facilities. The water often contains particles of both organic and inorganic sediments and algae that can negatively hinder the subsequent steps. Removing emerging contaminants and other dissolved ions from wastewater is often impossible with standard filtering techniques alone; however, membrane filtration, ultrafiltration, and reverse osmosis are all comparatively efficient. The use of sand in filtration is called granular filtration, which could be rapid or slow sand filtration. As the name suggests, slow sand filtration is very slow compared to rapid filtration. Rapid filtration has a few characteristics that make it perform much faster than slow sand filtration, such as a filter bed with uniform size and processed granular particles not commonly found in nature, the conditioning of water by various coagulants, and mechanical systems that are effective in removing clogged solids from the filtering bed. Typically, the filtration process has two stages: the filtering stage, where contaminants accumulate on the filtering media, and the backwash stage, where the accumulated contaminants are flushed from the system. Backwashing is usually shorter than the filtration system, but it is a critical part of the filtration process. Granular activated carbon is commonly used as the preferred media for filtration in conventional water treatment

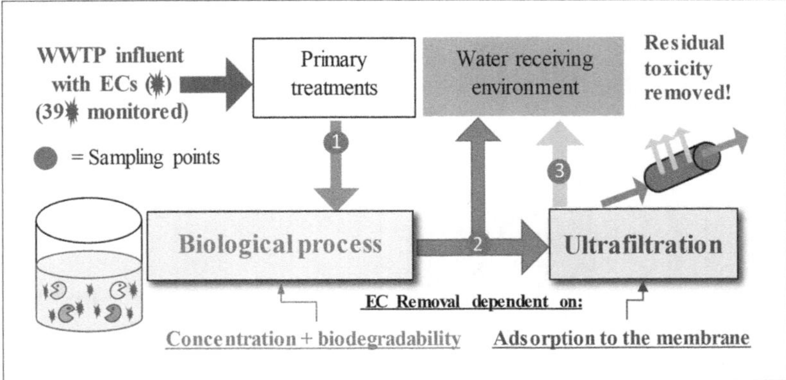

FIGURE 1.10 Schematical assembly of the different treatment processes. (Reproduced from Ferreiro et al.[46] Copyright 2020, the Authors.)

facilities because it offers great mechanical filtration of particulate matter and eliminates organic compounds that might lead to taste and odor issues.

1.3.6 BIOLOGICAL TREATMENT

Organic compounds are commonly removed from wastewater using biological treatment, although this technique is very slow but efficient. Additionally, the technique cannot remove halogenated organics and certain non-halogenated organics.[47] Aerobic and anaerobic processes are the two main processes of biological treatment. Membrane bioreactors, sequencing batch reactors, and active sludge are common for aerobic processes. For anaerobic processes, anaerobic film reactors and anaerobic sludge are also common and environmentally friendly. In Figure 1.11, we provide a conceptual flow diagram of the process. Bio-trickling filters have been used for decades in wastewater treatment plants to disinfect pathogens, remove biochemical and chemical oxygen demand, and control air pollution and smell. Although the technique can effectively remove the organic micropollutants, active sludge is more advantageous. However, for the removal of emerging contaminants, the technique has not produced good results so far, with less than 70% removal of emerging contaminants or no removal at all in some instances because of the resistance of the emerging contaminants to the microbial development of biodegradation.[17,41] Additionally, the method is typically ineffective in removing endocrine-disrupting compounds, pharmaceuticals, and personal care products because these contaminants are present in water at very low concentrations compared to most of the contaminants, limiting

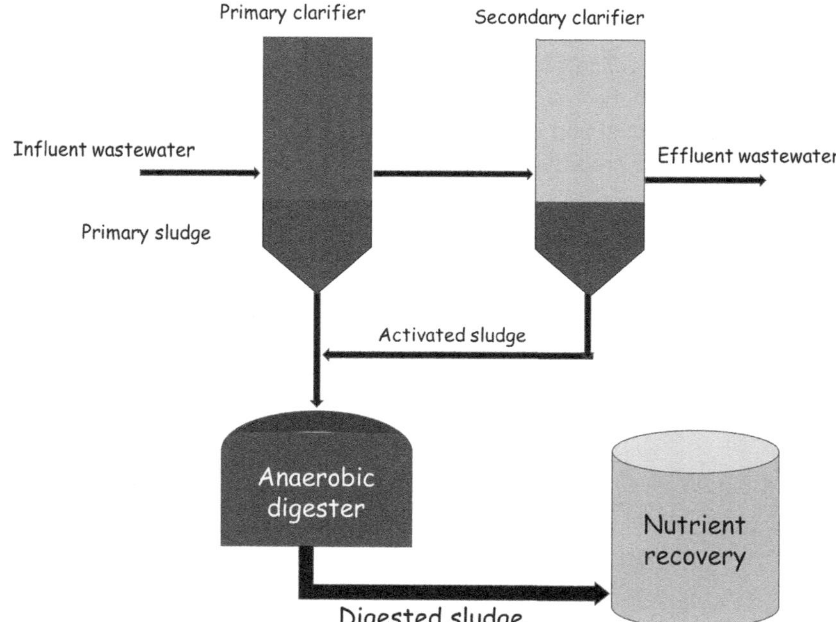

FIGURE 1.11   Conceptual flow diagram for anaerobic digestion.

the amount that may be degraded. While biological treatment alone is sometimes ineffective at removing organic contaminants from wastewater, when paired with filtration, it can be quite efficient, especially at getting rid of heavy metals.[17,41,47] Heavy metals can also be removed from water through methylation. The activity of microorganisms is imperative for biological treatment, as the technology depends on the microorganisms for efficient wastewater treatment. The effectiveness of the treatment process can be affected by several variables, including the loading rate, type of media, composition of water, temperature and degree of aeration. As a result, seasonal variations and eutrophication might affect biological treatment. Membrane fouling and clogging of filters are frequent problems encountered with biofilters or bioreactors, and this gradually reduces the performance and the durability of the system.[45]

1.3.7 CHEMICAL PRECIPITATION

In chemical precipitation, a chemical precipitant will react with heavy metal ions, creating an insoluble precipitate that is subsequently removed from the water by filtration or sedimentation (Figure 1.12). The filtered and treated water could either be repurposed or drained. The method is simple and effective and is mostly used for the degradation of heavy metals. However, the quantity of chemicals required makes the method very costly.[17,41,47] The method uses hydroxide precipitation and sulphide precipitation. The hydroxide precipitation technique is widely used because it is cost-effective, easy to adjust the pH and the hydroxide metals are easily removed by sedimentation or flocculation. Various hydroxides are used to precipitate metals, with calcium hydroxide being preferred, even in industry settings. For hydroxide precipitation processes, the removal efficiency is often improved by adding coagulants such as iron salts, alum, and some organic polymers. Although hydroxide is widely used, the technique has some limitations, such as producing large amounts of sludge, which is often highly toxic due to the contaminants; hence, it is expensive to manage and treat, which presents disposal challenges. Also, complex agents in the water hinder the metal hydroxide precipitation.[47] Furthermore, due to the amphoteric

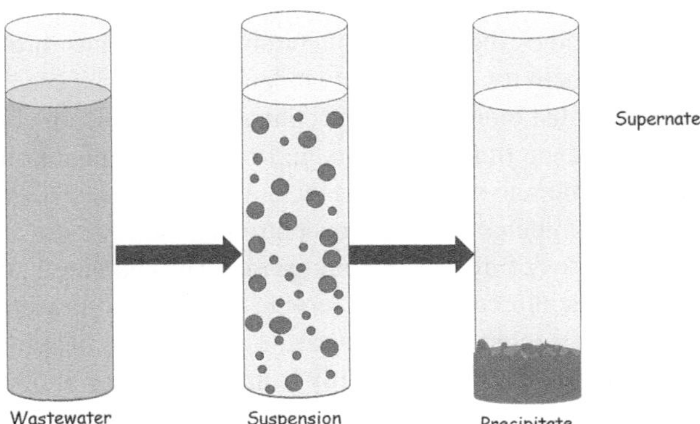

FIGURE 1.12 Typical chemical precipitation reaction.

nature of the various compounds, the existence of numerous metal species might make the treatment process challenging because optimizing the degradation of one species might make the degradation of the other species difficult, due to competing ions. Adjustment of pH in the effluent will also be required after treatment, increasing the cost of wastewater treatment. The sulphide precipitation technique is equally effective for the removal of toxic metal ions. In contrast to hydroxide precipitation, the solubilities of the metal sulphide precipitates are significantly lower, and sulphide precipitates are not amphoteric; hence, the sulphide precipitation technique can obtain high removal efficiencies of metal ions over a wide range of pH as compared to the hydroxide precipitation technique. The generated sludges in sulphide precipitation also indicate thicker and more dewatered characteristics than hydroxide sludges. However, the sulphide precipitation technique is without some limitations. Sulphide precipitants occurring in acidic conditions often generate poisonous and flammable hydrogen sulphide fumes when reacting with heavy metal ions in acidic conditions. It is suggested that the precipitation process takes place in neutral or basic conditions. Another limitation is that sulphide precipitation often produces colloidal precipitates that are either difficult to remove by filtration or settle by sedimentation.[47,48]

1.3.8 Membrane Technology

Membrane technology is a promising technology for effectively removing micro-contaminants from wastewater. Membrane technology is classified into biological processes, which involve membrane bioreactors, and nonbiological processes, including nanofiltration, ultrafiltration and reverse osmosis.[17,41] Membrane bioreactors (MBRs) are hybrid biological reactors with suspended growth and membrane-based filtration methods, such as microfiltration and ultrafiltration. Membrane bioreactors are widely used and well-established methods for achieving reasonably treated wastewater water by combining biological and membrane treatment techniques.[46,48] Compared to the conventional activated sludge alone, membrane bioreactors are more robust and effective in producing higher quality water with a small amount of sludge production and a smaller physical footprint. The effectiveness of a membrane bioreactor is determined by the residence time in which the sludge remains in the biological reactors. In addition to the ability of microfiltration or ultrafiltration techniques to block the pathway of certain contaminants through a filtration process, the microbial activity of activated sludge promotes the breakdown of contaminants in the membrane bioreactors. The contaminants' degradation rate is facilitated by the long residence time of the sludge in the membrane bioreactors. It has been noted that the membrane bioreactor is the only membrane separation technique that can degrade contaminants without concentrating them on the membrane surface.[2] There are two types of membrane bioreactors: aerobic membrane bioreactors and anaerobic membrane bioreactors. The aerobic membrane bioreactor is expensive to operate, as it is energy-intensive. In contrast, the anaerobic membrane bioreactor consumes less energy and differs significantly from the aerobic membrane bioreactor by efficiently removing the contaminant, throughput volume, and energy consumption.[48] Liu et al.[48] reported on the efficient degradation of organic contaminants, using aerobic and anaerobic membrane bioreactors. Their results demonstrated that using an aerobic membrane bioreactor was more effective in removing the contaminants than an anaerobic membrane bioreactor. The removal effectiveness when aerobic was employed was 98% ibuprofen, 98% naproxen, 92% triclosan, and 95% ketoprofen, and the effectiveness declined when anaerobic was employed with 28% ibuprofen, 52% naproxen, 68% triclosan, and 25% ketoprofen. The results also revealed that whether the aerobic membrane bioreactor or the anaerobic membrane reactor is being used, the membrane bioreactor showed consistency in the ability to degrade diclofenac in the study. The main mechanism for effectively removing organic contaminants in their low concentrations by aerobic or anaerobic membrane bioreactor was attributed to biodegradation and/or biotransformation. Furthermore, the two systems showed comparable results in the efficiency of removing bulk organic materials, which was observed through the total organic and nitrogen content.[48]

Reverse osmosis, nanofiltration, membrane pressure filtration, or ultrafiltration are nonbiological procedures or pressure-driven membrane technologies that remove contaminants from generated water by applying high pressures across the membranes (Figure 1.13). These are the most widely used membrane water filtration technologies. The membranes are also regularly upgraded or modified to enhance usage and performance.[17,41] As a result, turbidity and microbiological pollutants are best removed via membrane techniques. Although nanofiltration membranes operate at low pressures, the technique can effectively remove very small organic compounds and highly complex inorganic compounds. As a result, the nanofiltration technique is extremely competitive with the operational costs and selectivity compared to conventional filtration techniques. Licona et al.[49] reported on removing active pharmaceutical contaminants that are usually present in natural waters and are not easy to remove or degrade by conventional water treatment processes because they are present at trace concentrations. The authors evaluated the removal of five nonsteroidal anti-inflammatory medicines, analgesics, and anti-pyretic: ibuprofen, dipyrone, acetaminophen, diclofenac,

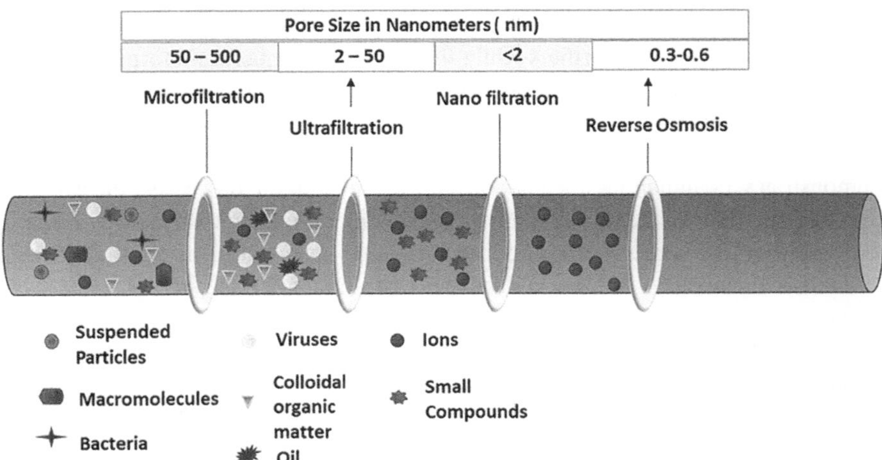

FIGURE 1.13 Conceptual diagram of pressure membrane filtration. (Reproduced from Musa and Idrus.[50] Copyright 2021, the Authors.)

and caffeine using nanofiltration and reverse osmosis. The authors used NF90 and BW30 membranes, which both showed a removal efficiency of over 80% for all selected contaminants. For non-ionic compounds such as caffeine and acetaminophen, the particle sizes were the main mechanism of removal by the NF90 membrane that determined the removal efficiency. In contrast, the primary mechanism was electrostatic interaction in the removal efficiency for the anionic compounds of ibuprofen, dipyrone, and diclofenac. Furthermore, the removal results by the NF90 membrane showed hydrophobicity to have played a critical role because of the adsorption process that took place on the surfaces of the membrane. However, the high operating costs still prevent the widespread use of both nanofiltration and reverse osmosis, although reverse osmosis has higher operating costs than nanofiltration because of the requirement of high operating pressure. Membrane fouling is another important factor that influences the efficiency of both nanofiltration and reverse osmosis processes. Membrane fouling is a natural occurrence caused by organic matter in water. It has been reported that membrane fouling plays an important role in the removal efficiencies of emerging contaminants such as pharmaceuticals and endocrine-disrupting contaminants; in other words, it influences the improvement removal efficiencies of nanofiltration and reverse osmosis, although the effect is not significant.[2]

1.3.9 Coagulation and Flocculation

Chemical coagulants are classified as organic polymers, synthetic polyelectrolytes, or inorganic electrolytes, which are used to mitigate the electrostatic repulsion between the particles. Although it might be used to remove dyes from wastewater, using coagulation to remove organic contaminants is uncommon. Coagulation and flocculation techniques are mostly used to remove heavy metals and inorganic compounds, followed by sedimentation or filtration.[16,41,47] In a rapid mixing unit, the coagulant is often added to the water for quick and thorough mixing. The mixing speed is then decreased to let the developing floc come into contact with the water and grow in size. The subsequent mixing must be gentle enough to keep the floc suspended until it is ready to settle in the clarifiers while still allowed to expand and form larger flocs. Some widely used coagulants in conventional wastewater treatment are ferrous sulphate, aluminium, and ferric chloride, which have indicated effectiveness in removing contaminants by forming amorphous metal hydroxide precipitate. Figure 1.14 demonstrates coagulation–flocculation and sedimentation processes. Because the required chemical reagents as expensive, coagulation treatment technology is highly expensive.[17,47] The process must be monitored and managed to ensure that the right amount of chemicals is added to induce the right amount of removal, but not excess.[47] The amphoteric nature of the various substances available in water and the presence of multiple metal compounds could make treatment challenging, as maximizing the treatment of one species could prevent efficient treatment of the other. Moreover, the effluent requires pH adjustment after each, which increases the overall treatment cost. Another disadvantage of coagulation is the production of large proportions of sludge, which is costly to manage and could typically be hazardous due to the treatment of heavy metals. The water can be filtered to eliminate flocs before or after sedimentation. The technique is controlled to ensure that the coagulant chemicals

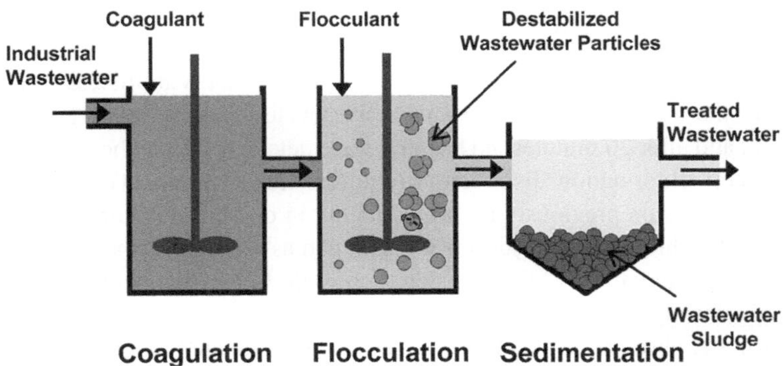

FIGURE 1.14 Schematic representation of coagulation–flocculation processes. (Reproduced from Musa and Idrus.[50] Copyright 2021, the Authors.)

and the contaminants are removed. The removal efficiencies of medicines (warfarin, hydrochlorothiazide, and betaxolol) by coagulation were reported to be more than 80% with aluminium sulphate, while for personal care products from hospital wastewater, the removal efficiencies in high quantities were reported with an average of more than 78%. Coagulation showed ineffectiveness in removing endocrine disruptive compounds with contaminants such as sulfadimethoxine, carbadox, and trimethoprim, which were not effectively removed by the coagulants of ferric sulphate and aluminium sulphate. Electrocoagulation is another technique where an electric current is passed through the water, and the coagulant precursor is discharged by electrolytic oxidation of the anode material, which is either aluminium or iron.[41]

1.3.10 DISINFECTION

Waterborne illnesses and even death are commonly caused by bacteria contamination, which greatly threatens human health. Microorganisms could be pathogenic or non-pathogenic and are highly susceptible to disinfection. The disinfection process will either remove the microbial contaminants completely or minimize them.[17,41] Some disinfection techniques include ozonation, UV light, chlorination, ultrasonication, membrane separation, etc. These techniques can successfully control and remove bacteria contaminants in water to the desired levels. Still, some of them have a tendency to produce disinfection by-products (DBPs). For some, the bacteria have grown resistant to disinfectants, necessitating higher dosages and more by-product production.[39,40] Alternative materials with less processing time that produce fewer by-products due to bacteria resistance are searched for. Waterborne infections become an increasing threat if they are not treated. Hussain et al.[39] used multifunctional graphite flakes to disinfect *Escherichia coli* using the electrochemical treatment. *Escherichia coli* is a common and widely studied gram-negative bacteria that is generally not harmful. Some of these bacteria strains are used as indicators to monitor the presence and quality of other microbial contaminants in water. In the study, *Escherichia coli* was adsorbed on the graphite flakes and in a sequential batch reactor with various experimental conditions, followed by the electrochemical treatment.

The results showed that within 5 minutes, more than 99.98% of the bacteria were removed, indicating that the process was efficient, fast, and capable of removing bacterial cells from the solution. The authors claimed that with the electrochemical treatment, the adsorbent was reused over five cycles and still gave good bacteria reduction, and after 20 minutes, no bacteria were detected. The authors further investigated electrochlorination disinfection (chloride-free experiment) using the same bacterial suspension previously used inoculating in deionized water. This was done to remove the chloride in the suspension as much as possible, although not guaranteed for complete removal. After this step, a chloride-free experiment for *Escherichia coli* removal using the adsorption process showed that the adsorption process was not efficient in the absence of the chloride. The results indicated that about 65% of the bacteria were inactivated during the first cycle of adsorption. However, adsorption capacity increased in the second or subsequent cycles of adsorption by more than 98%. Disinfection continued with each electrochemical treatment cycle until no bacteria was observed in the supernatant following the subsequent electrochemical treatment cycles. The authors also claimed that the substantial decrease of bacteria in the supernatant during the electrochemical process implied disinfection was indirectly continuing even when there was no chloride present, which could be linked to other disinfecting agents such as peroxide or persulphate that, after several cycles, are likely to accumulate on the adsorbent. The authors recommended the material after effectively removing the bacteria with high initial concentrations and the potential for producing fewer disinfection by-products. This technique was, however, found to consume a lot of energy that could not be sustainable.[39] Additionally, the lack of sufficient light and rapid electron-hole recombination hinders the extensive use of photocatalysts, and the authors have suggested that photocatalytic disinfection could be an alternative water treatment.

1.4 CONCLUSION

This chapter has highlighted the plight water sources worldwide face because of population growth, global climate, and the expansion of industries. Treatment of wastewater before distribution is necessary, as the challenge of developing treatment systems for the end-user might be too expensive. Various treatment methods have been outlined, showing how cost-effective they may be to operate, their effectiveness in removing various contaminants in water, and the quality of the water. As most of these methods require a lot of chemicals or expensive machinery to operate, often requiring a lot of energy, adsorption has been considered the most effective method for most contaminants in water. Adsorption is a well-established method that has been proven universally due to its simplicity of operation, cost-effective process, the production of high-quality water, and the removal of contaminants, including emerging contaminants in their low concentration. The subsequent chapters will explore adsorption technology further, including how certain contaminants interact with their adsorbents, the interaction mechanism, and detailed factors that influence adsorption and adsorption kinetics. Carbon-based adsorbents, especially graphene and its derivatives, will be explored further as the adsorbent of interest for various contaminants.

REFERENCES

(1) Wang, B.; Su, H.; Zhang, B. Hydrodynamic cavitation as a promising route for wastewater treatment–A review. *Chemical Engineering Journal* 2021, *412*, 128685.

(2) Olasupo, A.; Suah, F. B. M. Recent advances in the removal of pharmaceuticals and endocrine-disrupting compounds in the aquatic system: A case of polymer inclusion membranes. *Journal of Hazardous Materials* 2021, *406*, 124317.

(3) Rivas, F. J. Monopersulfate in water treatment: Kinetics. *Journal of Hazardous Materials* 2022, *430*, 128383.

(4) Han, B.; Weatherley, A. J.; Mumford, K.; Bolan, N.; He, J.-Z.; Stevens, G. W.; Chen, D. Modification of naturally abundant resources for remediation of potentially toxic elements: A review. *Journal of Hazardous Materials* 2022, *421*, 126755.

(5) Zhao, X.; Yu, X.; Wang, X.; Lai, S.; Sun, Y.; Yang, D. Recent advances in metal-organic frameworks for the removal of heavy metal oxoanions from water. *Chemical Engineering Journal* 2021, *407*, 127221.

(6) Yoo, D. K.; Bhadra, B. N.; Jhung, S. H. Adsorptive removal of hazardous organics from water and fuel with functionalized metal-organic frameworks: Contribution of functional groups. *Journal of Hazardous Materials* 2021, *403*, 123655.

(7) Feng, Z.; Xu, Y.; Yue, W.; Adolfsson, K. H.; Wu, M. Recent progress in the use of graphene/polymer composites to remove oil contaminants from water. *New Carbon Materials* 2021, *36*, 235–252.

(8) Yadav, V. B.; Gadi, R.; Kalra, S. Clay based nanocomposites for removal of heavy metals from water: A review. *Journal of Environmental Management* 2019, *232*, 803–817.

(9) Cheng, N.; Wang, B.; Wu, P.; Lee, X.; Xing, Y.; Chen, M.; Gao, B. Adsorption of emerging contaminants from water and wastewater by modified biochar: A review. *Environmental Pollution* 2021, *273*, 116448.

(10) Zhu, F.; Zheng, Y.-M.; Zhang, B.-G.; Dai, Y.-R. A critical review on the electrospun nanofibrous membranes for the adsorption of heavy metals in water treatment. *Journal of Hazardous Materials* 2021, *401*, 123608.

(11) Rangabhashiyam, S.; dos Santos Lins, P. V.; de Magalhães Oliveira, L. M.; Sepulveda, P.; Ighalo, J. O.; Rajapaksha, A. U.; Meili, L. Sewage sludge-derived biochar for the adsorptive removal of wastewater pollutants: A critical review. *Environmental Pollution* 2022, *293*, 118581.

(12) Krasucka, P.; Pan, B.; Ok, Y. S.; Mohan, D.; Sarkar, B.; Oleszczuk, P. Engineered biochar–A sustainable solution for the removal of antibiotics from water. *Chemical Engineering Journal* 2021, *405*, 126926.

(13) Naghdi, M.; Taheran, M.; Brar, S. K.; Kermanshahi-Pour, A.; Verma, M.; Surampalli, R. Y. Removal of pharmaceutical compounds in water and wastewater using fungal oxidoreductase enzymes. *Environmental Pollution* 2018, *234*, 190–213.

(14) Wilkinson, J.; Hooda, P. S.; Barker, J.; Barton, S.; Swinden, J. Occurrence, fate and transformation of emerging contaminants in water: An overarching review of the field. *Environmental Pollution* 2017, *231*, 954–970.

(15) Qiu, B.; Shao, Q.; Shi, J.; Yang, C.; Chu, H. Application of biochar for the adsorption of organic pollutants from wastewater: Modification strategies, mechanisms and challenges. *Separation and Purification Technology* 2022, *300*, 121925.

(16) Boulkhessaim, S.; Gacem, A.; Khan, S. H.; Amari, A.; Yadav, V. K.; Harharah, H. N.; Elkhaleefa, A. M.; Yadav, K. K.; Rather, S.-U.; Ahn, H.-J. Emerging trends in the remediation of persistent organic pollutants using nanomaterials and related processes: A review. *Nanomaterials* 2022, *12*, 2148.

(17) Patel, M.; Kumar, R.; Kishor, K.; Mlsna, T.; Pittman Jr, C. U.; Mohan, D. Pharmaceuticals of emerging concern in aquatic systems: Chemistry, occurrence, effects, and removal methods. *Chemical Reviews* 2019, *119*, 3510–3673.

(18) Solangi, N. H.; Kumar, J.; Mazari, S. A.; Ahmed, S.; Fatima, N.; Mubarak, N. M. Development of fruit waste derived bio-adsorbents for wastewater treatment: A review. *Journal of Hazardous Materials* 2021, *416*, 125848.

(19) Mo, Z.; Tai, D.; Zhang, H.; Shahab, A. A comprehensive review on the adsorption of heavy metals by zeolite imidazole framework (ZIF-8) based nanocomposite in water. *Chemical Engineering Journal* 2022, *443*, 136320.

(20) Hamad, H. N.; Idrus, S. Recent developments in the application of bio-waste-derived adsorbents for the removal of methylene blue from wastewater: A review. *Polymers* 2022, *14* (4), 783.

(21) Singh, R.; Dutta, R. K.; Naik, D. V.; Ray, A.; Kanaujia, P. K. High surface area Eucalyptus wood biochar for the removal of phenol from petroleum refinery wastewater. *Environmental Challenges* 2021, *5*, 100353.

(22) Hoang, A. T.; Nižetić, S.; Cheng, C. K.; Luque, R.; Thomas, S.; Banh, T. L.; Nguyen, X. P. Heavy metal removal by biomass-derived carbon nanotubes as a greener environmental remediation: A comprehensive review. *Chemosphere* 2022, *287*, 131959.

(23) Benis, K. Z.; McPhedran, K. N.; Soltan, J. Selenium removal from water using adsorbents: A critical review. *Journal of Hazardous Materials* 2022, *424*, 127603.

(24) Jiang, Z.; Ho, S.-H.; Wang, X.; Li, Y.; Wang, C. Application of biodegradable cellulose-based biomass materials in wastewater treatment. *Environmental Pollution* 2021, *290*, 118087.

(25) Saheed, I. O.; Da Oh, W.; Suah, F. B. M. Chitosan modifications for adsorption of pollutants—A review. *Journal of Hazardous Materials* 2021, *408*, 124889.

(26) Ikram, M.; Raza, A.; Imran, M.; Ul-Hamid, A.; Shahbaz, A.; Ali, S. Hydrothermal synthesis of silver decorated reduced graphene oxide (rGO) nanoflakes with effective photocatalytic activity for wastewater treatment. *Nanoscale Research Letters* 2020, *15*, 1–11.

(27) Verma, S.; Kim, K.-H. Graphene-based materials for the adsorptive removal of uranium in aqueous solutions. *Environment International* 2022, *158*, 106944.

(28) Li, N.; Yang, H. Construction of natural polymeric imprinted materials and their applications in water treatment: A review. *Journal of Hazardous Materials* 2021, *403*, 123643.

(29) Zhu, B.; Chen, S.; Li, C.; Jiang, G.; Liu, F.; Zhao, R.; Liu, C. Non-metallic hollow porous sphere loaded CN/catalytic ozonation synergistic photocatalytic system: Enhanced treatment of emerging pollutants by three-stage cyclic reaction mechanism. *Applied Catalysis B: Environmental* 2022, *318*, 121881.

(30) Liu, Y.; Chen, Y.; Li, Y.; Chen, L.; Jiang, H.; Li, H.; Luo, X.; Tang, P.; Yan, H.; Zhao, M. Fabrication, application, and mechanism of metal and heteroatom co-doped biochar composites (MHBCs) for the removal of contaminants in water: A review. *Journal of Hazardous Materials* 2022, *431*, 128584.

(31) Qin, H.; Hu, T.; Zhai, Y.; Lu, N.; Aliyeva, J. The improved methods of heavy metals removal by biosorbents: A review. *Environmental Pollution* 2020, *258*, 113777.

(32) Belova, T. Adsorption of heavy metal ions (Cu^{2+}, Ni^{2+}, Co^{2+} and Fe^{2+}) from aqueous solutions by natural zeolite. *Heliyon* 2019, *5*, e02320.

(33) Taamneh, Y.; Sharadqah, S. The removal of heavy metals from aqueous solution using natural Jordanian zeolite. *Applied Water Science* 2017, *7*, 2021–2028.

(34) Jiang, Q.; He, Y.; Wu, Y.; Dian, B.; Zhang, J.; Li, T.; Jiang, M. Solidification/stabilization of soil heavy metals by alkaline industrial wastes: A critical review. *Environmental Pollution* 2022, *312*, 120094.

(35) Mondol, M. M. H.; Jhung, S. H. Adsorptive removal of pesticides from water with metal–organic framework-based materials. *Chemical Engineering Journal* 2021, *421*, 129688.

(36) Zhang, T.; Wang, W.; Zhao, Y.; Bai, H.; Wen, T.; Kang, S.; Song, G.; Song, S.; Komarneni, S. Removal of heavy metals and dyes by clay-based adsorbents: From natural clays to 1D and 2D nano-composites. *Chemical Engineering Journal* 2021, *420*, 127574.

(37) Xu, H.; Hao, Z.; Feng, W.; Wang, T.; Li, Y. Mechanism of photodegradation of organic pollutants in seawater by TiO_2-based photocatalysts and improvement in their performance. *ACS Omega* 2021, *6*, 30698–30707.

(38) Zhang, Y.; Yan, X.; Yan, Y.; Chen, D.; Huang, L.; Zhang, J.; Ke, Y.; Tan, S. The utilization of a three-dimensional reduced graphene oxide and montmorillonite composite aerogel as a multifunctional agent for wastewater treatment. *RSC Advances* 2018, *8*, 4239–4248.

(39) Hussain, S.; de Las Heras, N.; Asghar, H.; Brown, N.; Roberts, E. Disinfection of water by adsorption combined with electrochemical treatment. *Water Research* 2014, *54*, 170–178.

(40) Omran, B.; Baek, K.-H. Graphene-derived antibacterial nanocomposites for water disinfection: Current and future perspectives. *Environmental Pollution* 2022, *261*, 118836.

(41) Kumar, R.; Qureshi, M.; Vishwakarma, D. K.; Al-Ansari, N.; Kuriqi, A.; Elbeltagi, A.; Saraswat, A. A review on emerging water contaminants and the application of sustainable removal technologies. *Case Studies in Chemical and Environmental Engineering* 2022, *6*, 100219.

(42) Antoniadou, M.; Falara, P. P.; Likodimos, V. Photocatalytic degradation of pharmaceuticals and organic contaminants of emerging concern using nanotubular structures. *Current Opinion in Green and Sustainable Chemistry* 2021, *29*, 100470.

(43) Lee, B. C. Y.; Lim, F. Y.; Loh, W. H.; Ong, S. L.; Hu, J. Emerging contaminants: An overview of recent trends for their treatment and management using light-driven processes. *Water* 2021, *13*, 2340.

(44) Gong, Y.; Wang, Y.; Lin, N.; Wang, R.; Wang, M.; Zhang, X. Iron-based materials for simultaneous removal of heavy metal (loid) s and emerging organic contaminants from the aquatic environment: Recent advances and perspectives. *Environmental Pollution* 2022, *299*, 118871.

(45) Chadha, U.; Selvaraj, S. K.; Thanu, S. V.; Cholapadath, V.; Abraham, A. M.; Manoharan, M.; Paramsivam, V. A review of the function of using carbon nanomaterials in membrane filtration for contaminant removal from wastewater. *Materials Research Express* 2022, *9*, 012003.

(46) Ferreiro, C.; Gómez-Motos, I.; Lombraña, J. I.; de Luis, A.; Villota, N.; Ros, O.; Etxebarria, N. Contaminants of emerging concern removal in an effluent of wastewater treatment plant under biological and continuous mode ultrafiltration treatment. *Sustainability* 2020, *12*, 725.

(47) Adeleye, A. S.; Conway, J. R.; Garner, K.; Huang, Y.; Su, Y.; Keller, A. A. Engineered nanomaterials for water treatment and remediation: Costs, benefits, and applicability. *Chemical Engineering Journal* 2016, *286*, 640–662.

(48) Liu, W.; Song, X.; Huda, N.; Xie, M.; Li, G.; Luo, W. Comparison between aerobic and anaerobic membrane bioreactors for trace organic contaminant removal in wastewater treatment. *Environmental Technology & Innovation* 2020, *17*, 100564.

(49) Licona, K.; Geaquinto, L. d. O; Nicolini, J.; Figueiredo, N.; Chiapetta, S.; Habert, A.; Yokoyama, L. Assessing potential of nanofiltration and reverse osmosis for removal of toxic pharmaceuticals from water. *Journal of Water Process Engineering* 2018, *25*, 195–204.

(50) Musa, M. A.; Idrus, S. Physical and biological treatment technologies of slaughterhouse wastewater: A review. *Sustainability* 2021, *13*, 4656.

2 Adsorbents for Wastewater Decontamination

2.1 INTRODUCTION

Wastewater treatment is a pressing issue since it harms humans, animals and the environment. Various treatment methods are available for treating contaminants in wastewater, including chemical precipitation, advanced oxidation processes, adsorption, coagulation, reverse osmosis, electrolysis, ion exchange, membrane, biological processes, etc.[1–3] However, most of these techniques do not always offer the most economical ways of removing common contaminants. In most cases, they are also not economical when eliminating contaminants present in trace amounts. Some of these methods require many chemicals, are energy intensive, and produce high quantities of toxic sludge and other hazardous wastes. Moreover, many of these methods are already at their breaking points and might be unable to accommodate the increasing requirements for water quality standards.[1–3] Because contaminants are highly complex mixtures of organic and inorganic compounds, the treatment methods are based on combining multiple techniques that allow for cost-effectively achieving the desired water quality.

Due to its simplicity in operation, the chemical precipitation method is a typical traditional technique used for contaminant removal. However, the process generates enormous quantities of sludge that should be discarded, and its capacity to remove contaminants at high concentrations limits its industrial application.[4] The ion exchange technique, however, does not treat vast amounts of wastewater as the method is not economically feasible, and also produces resins that, if they are not recycled, could lead to secondary contamination. Therefore, among these treatment methods, the adsorption technique has been regarded as an alternative for removing contaminants in wastewater, whether in low or high concentrations.[1–3] The technology is well-established and has contributed effectively to the remediation of wastewater sources. The process has gained considerable attention because of its simplicity, cost-effective approach, and production of high-quality water. The adsorption process is economical regarding maintaining the operating equipment and regenerating the spent adsorbents.[5–7] Furthermore, the treatment process is fast, universal, and applicable to most water contaminants, even at low concentrations. Significant work has been done in recent years to develop eco-friendly and highly effective adsorbents for treating various pollutants in water. Several materials are being explored to be used as adsorbents for contaminants from wastewater, especially the treatment of emerging contaminants in low concentrations.[8]

DOI: 10.1201/9781032621302-2

The current chapter addresses several adsorbents utilized for adsorptive wastewater decontamination and their benefits over G-based materials in recent decades, according to the currently available literature.

2.2 CONVENTIONAL APPROACHES FOR WASTEWATER DECONTAMINATION

Population growth, industrialization and urbanization have placed a significant demand on the accessibility of safe and potable water globally. The global challenge is to keep up with the demand and supply of clean water, given that commercial and industrial activities directly or indirectly discharge contaminants into water sources. Human activities are not exempt from the constant contamination of water sources either. Given the high demand for clean water, the various conventional methods for wastewater decontamination have become very costly or ineffective in removing the contaminants to allowable limits. These methods were developed in ancient times to remove the suspended and floating materials in water, remove pathogenic microorganisms and remove biodegradable organics (BOD).[1-3] Today, there is much more to remove from wastewater, with the earlier specific objectives still being addressed at high levels. The removal and recovery of value-added compounds (nitrogen and phosphorus are the essential compounds in nutrient recovery) are also being addressed, together with the removal of emerging contaminants available in trace amounts.

Conventional methods, including but not limited to chemical precipitation, biological treatment, oxidation, ion exchange, reverse osmosis, etc., for wastewater treatment are either high in operational costs or need to be more effective to meet the stringent regulatory effluent limits in place. Some of these methods could be more efficient in treating a wide range of contaminants, including emerging contaminants such as pharmaceuticals and personal care products and persistent organic pollutants.[2,6] As a result, there is an urgent need to improve these methods or develop new technologies that can address the growing water dema, which are cost-effective, effective in improving water quality, and sustainable.[1-3] The availability of more scientific knowledge and skills in recent years has expanded the information base on wastewater decontamination, which now focuses on health-related issues, due to potentially toxic chemicals discharged into the environment. For instance, adsorption has been significantly boosted in how it operates from the laboratory to the pilot scale.[4] And more emphasis is placed on exploring materials that could be sustainable adsorbents for specific water contaminants. Figure 2.1 gives some of the carbon-based materials used for adsorption. Using nanomaterials in adsorption is an excellent development as the nanomaterials enhance the removal efficiency of contaminants. This also shows that the water quality improvement objectives of the early years have been maintained. Still, the emphasis has shifted to developing new materials for various treatment techniques for removing toxic and trace compounds that could potentially cause adverse long-term health effects and environmental impacts. Consequently, stringent water quality laws are in place, with additional objectives and goals for the treatment process. There is also a vast knowledge of the various

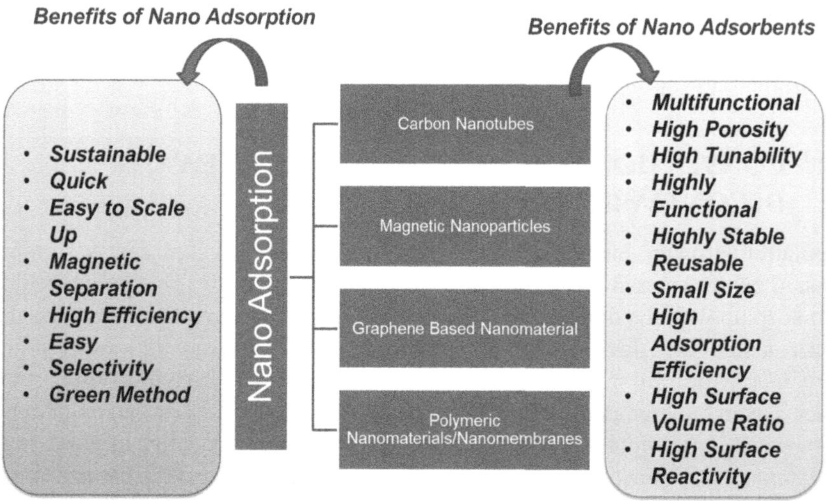

FIGURE 2.1 Carbon-based nanomaterials and their benefits in adsorption. (Reproduced from Boulkessaim et al.[2] Copyright 2022, the Authors.)

materials that could be used as adsorbents because of their inherent physical and chemical properties, availability, and the prospect of fine-tuning them to the desired properties.

2.3 WHY ADSORPTION FOR WASTEWATER MANAGEMENT

The adsorption process entails the adsorbate accumulating on the surfaces of the adsorbent material. An adsorbent is a solid surface; an adsorbate is a substance (contaminant) that attaches to or adheres to it. Adsorption is classified into two main types: chemical and physical. In physical adsorption, an adsorbate binds to an adsorbent surface due to van der Waals force attraction or dispersion. Physical adsorption is distinguished by having an enthalpy of adsorption that is relatively low and an adsorbed layer that can range in thickness from monolayer to multilayer.[9] Adsorbed layers can be removed more quickly than with chemisorption at an average temperature. Conversely, chemisorption only creates a monolayer and requires the chemical synthesis of ions or covalent bonds. The difficulty in removing the adsorbed layer is what distinguishes the two methods. Physical adsorption frequently occurs in the second layer, as the chemisorbed substance often occupies the first layer. The characteristics of the adsorbent and the composition of the adsorbate, other additional contaminants, temperature, ambient conditions, and experimental settings all have an impact on adsorption (contact time between the adsorbent and adsorbate, pH of the solution, the concentration of the contaminated solution and particle size of the adsorbent).[10] Figure 2.2 gives an overview of the inherent parameters of the adsorption process. The incorporated functional groups also greatly contribute to the adsorbents' performance and how the contaminants are removed. Pre-filtration is usually necessary to remove the substances that could hinder and decrease the

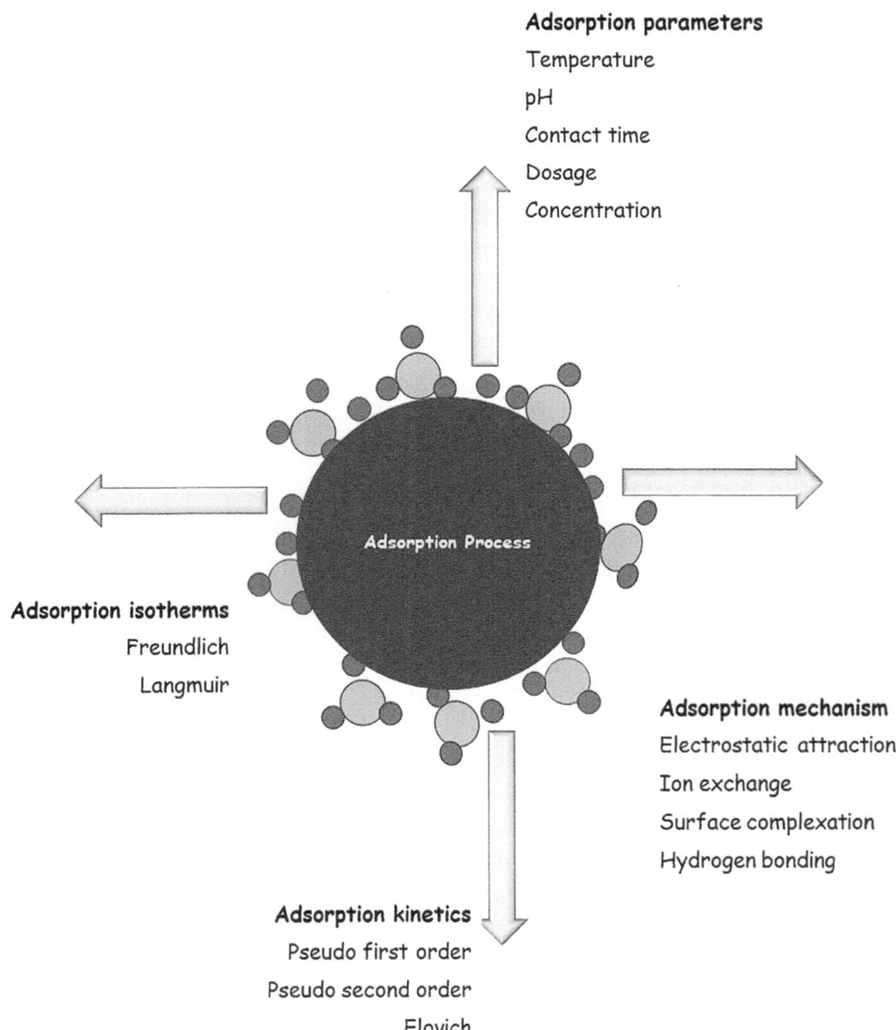

FIGURE 2.2   Overall adsorption process.

efficiency of the adsorption process; this includes greases and oils and other suspended materials. When the contaminated water is passed through a finely powdered
adsorbent, the contaminants attach to the surfaces of the solids and an equilibrium
state is created.[1–3,11] At this point, the concentration of the adsorbed contaminant and
the water solution are constant. An adsorption process can either take place through
batch operation or continuous flow operations. Batch adsorption studies are commonly used to treat small quantities of water on a laboratory scale, where various
parameters that could affect the adsorption process are evaluated to assess when the
equilibrium condition is reached. These parameters are the effect of adsorbent mass,
adsorbate concentration, solution pH, contact time, temperature, and agitation speed.
The continuous adsorption process is applied on pilot scales in the laboratory and on

large industrial scales. The fixed bed method is one of the methods employed in the adsorption process for establishing contact between the adsorbent and the adsorbate. Similar to batch studies, influencing parameters such as column bed height, initial concentration, solution pH, adsorbent particle sizes, the flow rate of the solution, temperature, exhaustion points, and breakthrough points are all evaluated in a continuous column adsorption process. An adsorption isotherm is a relationship between the equilibrium levels of the contaminant adsorbed in the water at a specific temperature. Well-known models like Langmuir and Freundlich can easily describe how well a contaminant is adsorbed.[12] In the laboratory, batch and column studies are used to develop and optimize the adsorption parameters. Large columns are then used to implement the adsorption technology at pilot and industrial scales. Furthermore, because the adsorption process is often reversible, adsorbents are regenerated using various methods and chemicals by the desorption process.

2.3.1   FACTORS INFLUENCING ADSORPTION

For experimental adsorption analyses, researchers typically vary different reaction parameters, including contact time, temperature, adsorbent dosage, solution pH, ionic strength, and initial concentration of the solution, to determine the optimum operating conditions for the adsorption process, which may hinder the process from operating smoothly. In addition, the generated experimental data will be evaluated for the microscopic interaction mechanisms employing isotherms, kinetics, and thermodynamics models.

2.3.1.1   Effect of Solution pH

The solution pH plays a crucial role during the adsorption process because even a slight change in the surface characteristics of the adsorbent or the adsorbate will influence the solution pH to change. For instance, an increase or a decrease in the concentration of the hydroxyl ions in a solution will alter the solution's pH, affecting the adsorption capacity. Also, water sources have various pH levels, which often influence the metal ions, as well as their characteristics. Moreover, pH influences how heavy metal ions associate in aqueous solution, from low to neutral pH. For metal ions adsorption, an increase in pH contributes to the creation of negative charges on the surfaces of the adsorbent, which enhances the adsorption of positively charged metal ions.[13] The characteristics of the functional groups and surface charges of the adsorbent, along with the contaminants in the solution, would be impacted by the solution pH, which would subsequently directly influence the adsorption capacity. Cationic impurities may be drawn to and concentrated on the surfaces of electronegative materials by the electrostatic interaction. The deprotonation of the hydroxyl ions at high pH values is primarily responsible for the observed increases in the electronegativity of the adsorbents with rising solution pH.[14] By protonating hydroxyls, high concentrations of H^+ at low pH levels, however, would lower the negative charge of the adsorbent's electronegative surface and compete with cationic contaminants for the active adsorption sites. With such effects, the surface charge may be impacted by the corresponding pH of heavy metals, the quantity of the accessible, functional groups on the adsorbent and the ionization state. When the pH of the solution increases, the adsorption capacity also increases because of the improved electrostatic contact between the negatively charged surfaces of the adsorbent and the positively charged adsorbate.

2.3.1.2 Effect of Ionic Strength

The electrical double layers and structural characteristics of hydrated particles can be considerably changed by foreign ions in water, which may influence the binding of adsorbing species. In addition to contaminants, wastewater contains other ions, including Na^+, K^+, Cl^-, Ca^{2+}, NO_3, etc., which may influence the removal of contaminants. It was established that Cl^- can combine with some metal ions such as Ni^{2+} and Cu^{2+} to form a soluble; also, it is challenging to remove heavy metal-chloride complexes that are negatively or neutrally charged. Foreign cations tend to compete for the accessible and active sites of the adsorbent with other cationic contaminants, including dyes and heavy metals, which would lower the adsorption performance. Furthermore, contaminants would partially aggregate, and the ionic strength might lower the double electric layer of the adsorbent, both of which can reduce the electrostatic interaction.[14] It was also established that the ionic strength of some water sources impacts contaminants and how they are removed. The following are the primary causes of how ionic strength affects metal ion adsorption: (i) the attraction of metal ions to the adsorbent is facilitated by the affinity between the adsorbent and the ionic species; the foreign ions usually compete for a limited number of active sites on the surfaces of the adsorbent with the contaminants, which in return reduces the adsorption capacity; (ii) the agglomeration of particles might be influenced by the electrostatic interactions which the concentration of foreign ions may impact. (iii) the concentration of the other metal ions in the solution affects the contaminants' activity coefficient, which limits how the contaminants can move freely from the solution onto the surfaces of the adsorbent.[14]

2.3.1.3 Effect of Temperature

Temperature is one of the crucial elements that have a substantial impact on the adsorption process. The system energy reaction, the rate and amount of sorption, and the sorption isotherms that describe the sorption behaviours can all be affected by temperature. The temperature will be established when the adsorption process is endothermic or exothermic. Usually, adsorption processes are endothermic. First, raising the temperature might cause the viscosity of the solution to decrease, intensifying the diffusion of contaminants such as heavy meals and dyes. Due to this, pollutants reach the surfaces of the adsorbent more quickly, increasing adsorption effectiveness. High temperatures, however, will also cause contaminants to desorb from adsorbent surfaces.[14] Moreover, some of the functional groups on the adsorbent are readily broken or deprotonated, increasing the number of accessible adsorption sites and enhancing adsorption effectiveness. In addition, energy fluctuation is generally present when contaminants are combined onto an adsorbent through surface complexation, ion exchange, or chemisorption. For most endothermic processes, an increase in temperature favors adsorption, whereas the adsorption capacity decreases for exothermic processes. Thermodynamic models, which primarily comprise factors like entropy change (ΔS°), enthalpy change (ΔH°), and Gibbs free energy change (ΔG°), can be used to evaluate the endothermic or exothermic adsorption process. While a negative ΔH° indicates an exothermic process, a positive ΔH° denotes an endothermic process. The positive ΔS° indicates the disordered adsorption. Negative ΔG° denotes spontaneous adsorption of contaminants onto an adsorbent surface, while positive ΔG° denotes non-spontaneous adsorption.[14] Two typical sorption isotherms are the Freundlich model and the Langmuir model.

2.3.1.4   Effect of Contact Time

To evaluate the reaction rate and the practicality of materials in removing contaminants in water, various models have been examined to explore the effect of contact time between the pollutants and the adsorbent during adsorption. The adsorption capacity changes with contact time and can give more details about how rapid the adsorption process is and the equilibrium time. The adsorption capacity will increase with an increase in contact time until the optimum value is reached. In general, the adsorption process involves a quick adsorption response within the first few minutes due to more active sites, then a slower reaction, which eventually leads to an equilibrium phase at a specific time and may be linked to the appropriate number of active adsorption sites. Initially, the contaminants interact with the active surface sites on the adsorbents when placed in wastewater, leading to rapid removal.[14] Then, the number of active surface sites becomes less and less, eventually reaching saturation. Due to saturation, the adsorption process is slowed down, causing the contaminants to permeate all over the adsorbent, thus further decreasing adsorption. It is crucial to optimize the right time for the adsorption process because it reduces operating costs and saves time. The optimum time required for adsorption equilibrium to be reached will vary from adsorbent to contaminant, as it is unusual for adsorbents to perform the same way with different pollutants. The adsorbent structure is a crucial factor affecting this process, i.e., macroporous adsorbents achieve the adsorption equilibrium more rapidly than materials with smaller pores. An intra-particle diffusion model is a multistep adsorption procedure that includes the diffusion of contaminants from waste solution to the surfaces of the adsorbent before they further penetrate into the interior pores and interact with the adsorption sites.[14]

2.3.1.5   Effect of Initial Concentration of Contaminants

Wastewater typically has different initial contaminant concentrations, which may require additional requirements of the adsorbent. Conversely, the higher the concentration of the contaminants in the wastewater, the more impurities are removed and vice versa. Therefore, the adsorption kinetics of various pollutants at different concentrations of the adsorbents have substantial practical significance. Because of a limited number of active surface sites in adsorbents, the removal of the contaminants decreases as the concentration rises, while adsorption capacity and rates increase. As a result of the initial low solution concentration and the moderate ratio of ions to accessible adsorptive sites, contaminants can be removed rapidly owing to the good adsorption sites.[15] However, low concentration could result in insufficient driving force, decreasing the adsorption rate. Contaminants move more rapidly as the concentration rises, increasing the likelihood that they may come into contact with the active adsorption sites. Therefore, each adsorbent unit mass is exposed to more ions, and the accessible sites are quickly occupied and saturated. The different contaminants will remain in the solution because of the fixed number of active adsorption sites, leading to reduced percentage removal. To fully utilize the adsorption sites quickly, an optimum concentration of the contaminants is necessary.[16] This maximizes the advantages of an adsorbent.

2.3.1.6 Adsorbent Dosage

To produce cost-optimal adsorption of contaminants from wastewater, the minimum amount of the adsorbent that could facilitate effective removal of the pollutants is determined. Optimization of the adsorbent dosage is significant in developing an effective adsorption system. Finding the equilibrium between the adsorbent mass and the adsorbent helps estimate the cost required for the adsorbent, particularly the treatment system. An increase in the adsorbent amount increases the available active sites, which improves the adsorption capacity through the efficient removal of the contaminants. Moreover, a high adsorbent dosage might accelerate the adsorption process since increasing active and available sites improves the likelihood of contaminant interaction and activity, leading to a quick removal.[16,17] However, some small-sized adsorbents, particularly nanomaterials, tend to form aggregates at high dose concentrations that block the active sites and lower adsorption capacity. To solve the agglomeration problem with small-sized adsorbents, other porous materials and biocompatible polymers are being investigated to form composites with small-sized materials. The porous structure of polymer composites effectively prevents aggregation and allows contaminants to access the active sites adsorbent.

2.3.2 ADSORPTION MECHANISM

Adsorption is one of the most widely used and effective methods for removing the majority of contaminants; its effectiveness can be attributed to sufficient functional groups with enough oxygen in the adsorbent. The specific properties of an adsorbent to effectively remove contaminants have been attributed to their great active sites (made possible by the high specific surface area), pore volume, and functional groups. Physical adsorption is more dominant with adsorbents with more micropores due to increased high surface area. At the same time, mesopores promote the diffusion of the contaminant so that the adsorption kinetics can be accelerated. However, porosity and high surface area are not the primary factors influencing great contaminant adsorption. A robust connection between the contaminant ions and the adsorbent material is also desired. Different surface functional groups on adsorbent materials could effectively improve the adsorption capacity and strongly co-ordinate with contaminants through various adsorption mechanisms. The possible means of interaction include ion exchange, electrostatic attraction, hydrogen bonding, and surface complexation.[17,18] Advanced spectroscopic analysis, surface complexation models and theoretical calculations are used to determine the adsorption behavior to identify the primary adsorption mechanism.

2.3.2.1 Electrostatic Attraction

An electrostatic attraction is created when the adsorbent and the contaminants have opposite surface charges. For some adsorbents, the positive surface charges are preferable for removing anionic pollutants, whereas the negative surface charges are preferable for removing cationic metal ions and other organic pollutants.[19] Electrostatic interaction is the primary mechanism for the adsorption of ionizable organic compounds and ionic organic compounds, which includes electrostatic attraction and electrostatic repulsion. In most instances, these organic compounds are attracted to the opposite charge of the surfaces of the adsorbent.[17] The electrostatic attraction, which is strongly influenced by a permanent negative surface charge, plays a crucial role when cationic contaminants are adsorbed onto adsorbents. Depending on the solution pH and particle sizes, adsorbents may also have a pH-dependent charge. Due to the negative charge of the adsorbent, the adsorbent often retains the negative charge across the entire pH range. The electronegativity of the adsorbent would increase with a pH increase because of the deprotonation of the available functional groups at high pH.[13,17,18] More contaminants would be drawn to adsorbents due to their increased electronegativity, which may also make it difficult to remove the pollutants from the adsorbents.[20]

2.3.2.2 Ion Exchange

Contaminants are often adsorbed on the surfaces of the adsorbents by altering the functional groups on the adsorbent or by exchanging the metal ions. The exchange of the metal ions between the adsorbents and the contaminants is directly associated with the decrease in concentration of the target contaminants in the solution.[19] Some adsorbents (mainly clays) contain a persistent negative charge on their surfaces, balanced by exchangeable cations. The cationic contaminants can take the place of the adsorbed exchangeable cations. Due to this characteristic, clays are one of the most common materials used as adsorbents, ion exchange plays a crucial role in removing contaminants.[20] Moreover, the ion exchange between the contaminants and pollutant protons in functional groups like -COOH or -OH are incorporated by integrating organic agents.[13,17,18] The ion exchange interaction between ions solid and liquid phases is highly reversible. This means that the electrical neutrality of the aqueous solution is always maintained because when the ion exchange solid phase absorbs the ions from the liquid phase, equivalent ions will be released back into the solution by the ion exchange of the solid.[21]

2.3.2.3 Surface Complexation

Surface complexation occurs mainly between the surface functional groups of the adsorbents and the contaminants in the solution. Selective metal-ligand associations create complexions. For instance, the surface functional groups of biochar may allow heavy metal ions to form metal-ligand complexes.[19] The degree to which different functional groups can bind various contaminants varies. Functional groups are incorporated on the surfaces of the adsorbent by first treating the adsorbent with the precursor of that particular active group. Adsorbents are commonly treated with the -COOH, -OH, and -NH$_2$ groups, with the former being the most popular.[13,17,18]

2.3.2.4 Hydrogen Bonding

Hydrogen bonding is a particular class of either intramolecular or intermolecular interaction that takes place between hydrogen atoms and other atoms, including nitrogen, fluorine, oxygen, etc. Organic contaminants are composed and contain a lot of functional groups, such as amino and hydroxyl, which, during the interaction with adsorbents with lots of functional groups, contribute to hydrogen bonding. The high presence of oxygen-containing functional groups facilitates adsorption. For adsorbent materials with less functional groups on their surfaces, these groups could be introduced through a modification that will foster the necessary hydrogen bonding.[17] For instance, the layered aluminosilicate structure of the clay is composed of the Al-OH group, which is found in large quantities on the edges of the clay, such as montmorillonite, and the surfaces of the clay, such as kaolinite, in the solution environment. In addition, more functional groups, including -OH, $-NH_2$, -COOH, -HF, etc., are introduced by incorporating precursors of these functional groups in clay-based adsorbents. Combining these groups, along with the H, O, N, and F-groups found in organic and inorganic contaminants, is relatively easy to establish a hydrogen bonding required for contaminants removal.[13,17,18,20]

2.3.3 Regeneration

The economic feasibility of the adsorption technology is influenced by the regeneration of the adsorbent, which allows the use of the adsorbent to its full capacity. Although adsorption and rate capacity are substantial factors in removing contaminants, the regeneration of the adsorbents is also a crucial factor to be considered. Adsorption and desorption typically co-occur and, to some extent, maintain equilibrium. Several variables can be adjusted to favor adsorption in the removal of contaminants. The dominance of desorption can also be realized by altering the solution environment, hence realizing the regeneration of adsorbents. For wastewater treatment, the adsorbent regeneration is crucial for the process's high efficiency as it reduces the waste generated. In particular, the regeneration of agricultural-based adsorbents not only prolongs the treatment cycle but also reduces the operational costs of the treatment process. A high adsorption capacity with an efficient adsorption-desorption process of the ideal adsorbent is crucial for increasing adsorption effectiveness and lowering operational costs.[22,23] The characteristics of the adsorbent, the nature of the adsorbate and the adsorption mechanism all have an imperative role in selecting an appropriate eluent for effective desorption. The most common eluents are alkalis, acids, and organic solutions. An acid solution produces abundant H^+, which, through ion exchanges, can replace the contaminants on the adsorption sites.[8] Since most adsorbents primarily absorb contaminants through the ion exchange process, this technique works well to regenerate most adsorbents. Additionally, salt solutions, including KCl or NaCl, could be utilized to regenerate adsorbents through the ion exchange of K^+ or Na^+ with the contaminants.[18] In contrast to using either salt or acid solutions for regeneration techniques, organic solutions can solvate organic contaminants from the adsorbent to obtain regeneration. It is important to remember that the eluent desorption technique suffers from poor efficiency, requiring multiple elution cycles to obtain maximum regeneration. The efficacy of the adsorbent will also decrease after several adsorption-desorption cycles.

2.3.4 ADSORPTION PROTOCOL

Adsorption technology includes in-depth water treatment experiments. This will involve establishing adsorption technologies using batch studies and fixed-bed column processes on a laboratory, pilot, and industrial scale, as well as synthesizing the adsorbents and their characterization, reagents, and other related activities. Developing and characterizing materials for adsorption are well-established and well-documented processes. For example, because of its robust adsorption performance and high specific surface area, activated carbon has been utilized to treat and recycle municipal and industrial water into potable water. However, because of high production and regeneration costs, activated carbon is rarely used for the treatment of wastewater on large scales. These shortcomings have prompted researchers to develop more inexpensive adsorbents that are effective at producing high-quality potable water and are simple to prepare. Adsorbents can either be natural or synthetic adsorbents. Although natural adsorbents are in abundance and cost-effective, they often provide an adsorption capacity that is comparable to that of modified adsorbents or even synthetic adsorbents. The primary advantage of preferring natural adsorbents over synthetic adsorbents is their potential to modify their physicochemical structure easily, and they are also biocompatible in nature. Hence, they are effective as adsorbents for the removal of emerging contaminants, heavy metals, dyes, and other contaminants from wastewater. Selecting a suitable adsorbent depends mainly on the characteristics of the adsorbent, the pore structure, and the nature of the adsorbate.[5–7,24] The adsorbent's cost-effectiveness and the adsorption process's efficiency are some of the factors that make the adsorption process economical. However, as a comparison of the adsorbents, the adsorption capacity differs from adsorbent to adsorbent, comprising different factors. Various physical and chemical properties uniquely characterize adsorbents, and modifying them optimizes their conditions for optimum performance. Some of the pre-defined characteristics that should define a material as a suitable adsorbent for contaminants removal from wastewater are: (i) they should exhibit high removal capacities and quick adsorption rates; (ii) they should be affordable; (iii) they should be environmentally friendly; (iv) they should have a robust mechanical structure and robustness; and (v) they should have a high surface area. These characteristics are well-defined for carbon-based materials, clay minerals, ion exchange resins, zeolites, and other inorganic materials or agricultural waste products. The materials are readily available and easy to use for wastewater treatment.

2.4 PREFERENCE FOR GRAPHENE MATERIALS FOR WASTEWATER MANAGEMENT

2.4.1 ADSORBENTS FOR WASTEWATER DECONTAMINATION

Adsorption shows more benefits than conventional methods in that the process is less expensive and more efficient, and the possibility of metal recovery is high, with high efficiency in adsorbent regeneration, reducing chemical usage and sludge generation. Various natural and synthetic adsorbents are available for wastewater decontamination. Most adsorbents have a high specific surface area and multiple active sites on their surfaces, which enables them to remove targeted contaminants in water effectively. The subsequent sections list some of the available adsorbents used for wastewater decontamination.

2.4.1.1 Carbon-Based Adsorbents

Carbon is one of the most crucial elements in the world and a readily available non-metal material, second to oxygen. Because of carbon's self-bonding capacity, the material has various forms of allotropes and compounds ranging from zero to three dimensions. Carbon materials' extraordinary properties have made them of interest and widely researched for various applications, such as biological and medical science, personal care products, electronics, catalysts, batteries, etc. Carbon materials' morphology, orientation, size, and shape determine their exceptional properties. Although there could be slight variations in the properties, based on the application, the materials allow modifications to alter the properties by optimizing the experimental conditions. Apart from modifying the various properties for potential applications, new and enhanced cost-effective methods for preparing the materials are being developed. Carbon-based materials have become more prominent in wastewater treatment of various contaminants. Factors including the particle sizes, porosity and surface area, surface functional groups, etc., influence the treatment process. The quality of carbon-based adsorbents determines the treatment's performance and the adsorption process. The chemical and physical properties of the adsorbent also influence the performance of the adsorption-desorption process. The study of novel carbon structures began in the middle of the 1980s with the lab-produced fullerene (C_{60}), a superstable structure with 60 carbon atoms with seemingly unrecognized properties, including interactions with various compounds like transition metals and oxygen. From this, numerous other studies were conducted, to the point where single-walled carbon nanotubes (SWCNTs) and multi-walled carbon nanotubes (MWCNTs) were produced in the early 1990s, with properties similar to those of fullerene but with additional ones that are very significant in the biological, electrical, and environmental fields. Activated carbon (AC), biochar (BC), CNTs, graphene, and its derivatives are some of the carbon-rich materials currently being used as adsorbents. Due to their restricted functional groups and pore shape, most of these materials are modified to improve their adsorption capacity. In addition to hydrophobicity, modification can change the surface functional groups, charge ability, and metallic content, which leads to increased interactions between contaminants and adsorbents. Adsorption on various porous materials like carbon-based materials can be described in a few steps: (i) movement of the adsorbate molecules from the liquid phase to the adsorbent; (ii) distribution of the adsorbate molecules to surround the adsorbent forming a layer; (iii) movement of the adsorbate molecule from the adsorbent surfaces to the accessible active sites; and (iv) occurrence of the adsorption process as a result of the adsorbent – adsorbate interaction. The adsorption mechanism and performance will depend on the interaction between the contaminants in the solution and the adsorbent, as well as many other factors that are influenced by the same interaction. Although carbon-based adsorbents are highly effective in removing their targeted contaminants in water through the adsorption method, the vast production costs of the materials on a large scale are adversely affecting their use industrially and commercially. Below is a summary of some materials used as adsorbents to remove various contaminants in wastewater. It is important to emphasize that this study will focus on graphene and its derivatives as adsorbents for various pollutants.

2.4.1.1.1 Activated carbons (ACs)

Activated carbon, also known as "activated charcoal", is produced by the pluralization of carbon-rich materials, mostly from natural sources (including wheat, peats, coal, or coconut shells) at very high temperatures between 600 and 900 °C in inert environment. By pyrolyzing carbon-rich natural sources like bamboo, coconut shells, wood, wheat, peat, or coal at high temperatures between 600 and 900 °C in an inert atmosphere, activated carbon, sometimes known as "activated charcoal," can be produced. Before the recent developments of producing activated carbon cost-effectively, the material was previously prepared by the expensive and unrenewable carbon materials such as natural coal, residues from petroleum industries, and wood. Today, there are many alternatives where activated carbon can easily be produced from agricultural by-products and industrial and municipal waste products. Activated carbon offers a high surface area, which is necessary for better adsorption of contaminants due to its reduced pore volume. The effectiveness of activated carbon as an adsorbent depends mainly on the material it was produced from, and the adsorption capacity will vary from contaminant to contaminant. Because of its long history and high exploration in wastewater treatment, activated carbon has always been the adsorbent in which other adsorbents are bench-marked. Besides the production cost of activated carbon, another limitation is that the material gives metal ions a very low adsorption capacity. To overcome this drawback, activated carbon is chemically modified with various substances to enhance and create active sites on the material for better interaction with the cations. Granular activated carbon (GAC), powdered activated carbon (PAC), extruded activated carbon, and impregnated carbon are types of activated carbon that are categorized by their physical appearance.[25,26] Although powdered activated carbon is mainly used due to the low initial operating costs, its usage becomes more expensive in large quantities and the lack of efficient regeneration, while on the other hand, granular activated carbon has excessive starting operational costs. Still, the material can be used for an extended period with minimum maintenance, and the granular activated carbon can be regenerated. The raw activated carbon can be activated, or the surfaces modified by adding different oxidizing chemicals such as strong alkalis, acids, or other salts, which are subsequently pyrolyzed at moderate temperatures in the 400–800 °C range. However, the primary disadvantage of activated carbon as an adsorbent is that a considerable amount of it is lost in the process since it is difficult to separate from the aqueous medium.[26,27] Except for activated carbon that has been magnetically modified, activated carbon is not entirely regenerated, which adds to its high production cost.

2.4.1.1.2 Biochar

Biochar is a carbon-rich material produced through slow pyrolysis at a temperature below 700 °C from various types of biomasses in an inert environment.[19,22,23,28] Even though activated carbon and biochar are produced from the same materials and prepared with the same methods; biochar is produced at lower temperatures than activated carbon, making it cost-effective. Furthermore, producing biochar does not involve the activation step like in activated carbon, although the production heavily depends on the starting material. Previously, biochar was used to fertilize the soil to boost the nutrient content and for the degradation of both organic and inorganic contaminants in wastewater. The material has shown to be a competitor to other adsorbents, primarily activated carbon because it is inexpensive and consumes less production energy. Biochar is produced from many different feedstocks, such as livestock waste, plant biomass, municipal solid waste, and marine microalgae.[29] The processing of various biomasses for biochar production can recover waste biomass and regenerate precious resources, reducing the amount of waste generated from these facilities. Different techniques for biochar production include fast pyrolysis, slow pyrolysis, hydrothermal carbonization, etc. Using biochar for wastewater treatment is both affordable and environmentally friendly. The operating technology is also quite comparable to that of commercial activated carbon. Some of biochar's few chemical and physical properties include being readily abundant in nature, a high micropore density, high specific surface area, good adsorption capacity, and stable carbon structure. To overcome the limitations of biochar as an adsorbent, various modification techniques are available to acquire biochar with enhanced properties and adsorption efficiency, with improved financial benefits. Figure 2.3 below shows the adsorption mechanism between emerging contaminants and modified biochar. Some modification methods commonly used include acid treatment, magnetic modification, alkali modification, polymer modification, and nanoparticle modification. The modification ensures that biochar is well tailored for removing the contaminant

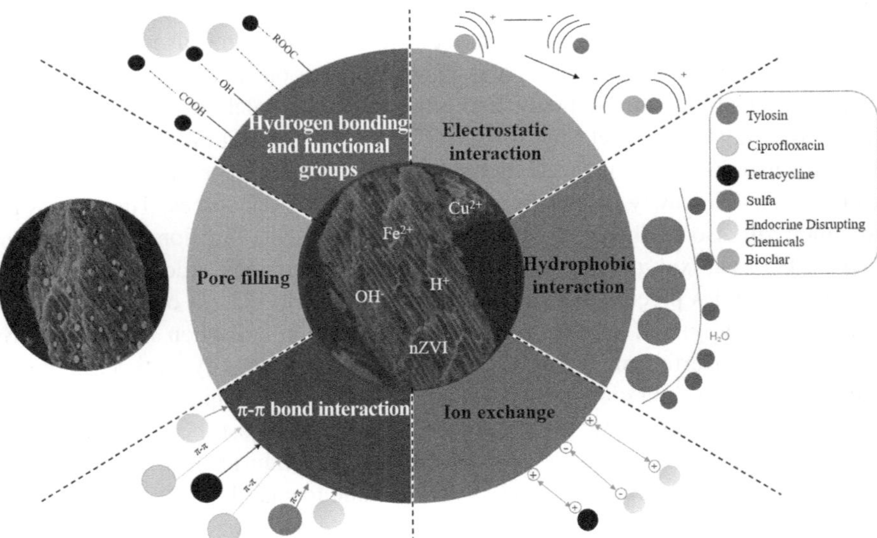

FIGURE 2.3 Adsorption mechanism on the removal of emerging contaminants using modified biochar. (Reproduced with permission from Cheng et al.[31] Copyright 2021, Elsevier Science Ltd.)

of interest with more improved surface functional groups, high specific area, and porosity. Biochar is typically alkaline, and it is primarily characterized by its pH value, which influences the quality of the material. For instance, a high pH value indicates a high degree of carbonization.[13,17] Additionally, biochar is a renewable resource with positive economic and environmental outcomes, making it a suitable and eco-friendly material for the degradation of contaminants from wastewater.[10,30] A recent growth trend of using biochar indicates that the material is being noticed.

2.4.1.1.3 *Carbon Nanotubes*

Carbon nanotubes are sp^2-hybridized carbon sheets rolled into hollow tubes of unlimited length. Nanotubes have a high elastic modulus and tensile strength, and because of their hydrophobic properties, they usually aggregate when dispersed in water. Therefore, depending on the application, they can be used directly as they are or with a surface modification.[32] Although activated carbon has a significant surface area that gives it similar and comparable properties to carbon nanotubes, activated carbon contains many micro-pores inaccessible to large organic molecules such as pharmaceuticals and antibiotics. Therefore, carbon nanotubes offer a better adsorption potential for several organic materials due to their broader active sites and more open sorption locations. Different types of nanotubes (i.e., single-walled carbon nanotubes and multi-walled carbon nanotubes) are used to treat water.[33,34] Single-walled carbon nanotubes are made of a single-layer graphene sheet of a cylindrical shape. In contrast, multi-walled carbon nanotubes are made of numerous single-walled carbon nanotubes stacked and rolled up into cylindrical shapes. Like many other adsorbents, carbon nanotubes can be modified or functionalized by chemical or physical methods to enhance their adsorption capacity. One of the most common and studied methods of carbon nanotube functionalization is acid treatment that introduces the carboxylic groups on the carbon nanotubes. Acid treatment improves the dispersion capability of carbon nanotubes in polar organic solvents due to the hydrophilic nature of the carboxylic group. Sulphuric and nitric acids are widely used and refluxed at high temperatures for a set time for surface functionalization of carbon nanotubes. Functionalization by the physical method includes wrapping the modifier within the carbon nanotube cavity or absorbing it into the carbon nanotubes. This means the modifier will also be chemically bonded with the available functional groups of the carboxylic, hydroxyl, amine, and other inorganic compounds. Additional functionalization includes precipitation, hydrothermal treatment, plasma oxidation, in situ, chemical grafting, and surface-initiated polymerization.[35] Carbon nanotubes have exceptional properties, including excellent thermal conductivity, optical, high specific surface area, vibrational, mechanical properties, etc., which distinguish them from other materials. However, on a broad spectrum of adsorbents, carbon nanotubes are not a suitable replacement for activated carbon. Instead, since their surface chemistry can be modified to target specific contaminants, they are possibly suited for analytical functions in polishing steps to remove resistant compounds or in the pre-concentration of trace organic contaminants. Carbon nanotubes are not attractive for large-scale wastewater treatment operations because of their high production and other operational costs, making them difficult to develop. The purification of carbon nanotubes is also a critical issue, as the materials are synthesized with many metallic or carbonaceous contaminants from the metallic catalysts and carbon sources used. Furthermore, using large quantities of carbon nanotubes for treating contaminants in wastewater can lead to environmental toxicity.[35]

2.4.1.1.4 Graphene

Graphene is a carbon allotrope with the monoatomic layers of sp^2 carbon atoms assembled in a honeycomb framework. It is one of the most stable allotropes of carbon. The material has unique electrical, thermal, mechanical, and surface properties that can be beneficially applied in various ways. Some of the graphene derivatives are reduced graphene and graphene oxide, which all have a high surface area that is excellent for the removal of dyes, metal ions, oils, and other chemical compounds from wastewater.[36] Graphene oxide consists of various functional groups, including hydroxylic, carboxylic, and epoxy groups. It is synthesized by the well-known Hummers method (Figure 2.4), which involves oxidizing graphite to create its oxide form.[37,38] These functional groups' availability makes graphene oxide extremely hydrophilic, enabling water molecules to move between the graphene oxide sheets freely. Unlike other carbon-rich materials like carbon nanotubes, the aspect ratio of graphene oxide is not disturbed as the graphene oxide is already functionalized during the synthesis process. Although the Hummer's method is well known, the technique uses a lot of toxic chemical reagents which are harmful to the environment. Hence, studies are also exploring the use of green methods for synthesizing these nanomaterials. Green methods for producing adsorbent materials are geared towards developing adsorbents that could be produced from plant extracts or materials that could be used at room temperature or solar light. This method aims to reduce the toxic effects of chemical reagents during the preparation processes. The emphasis is on synthesizing nanomaterial adsorbents with enhanced properties that are desirable for various applications. Natural agricultural residues, microorganisms, and biodegradable materials are used to produce green nanomaterials. Like any other synthesis method, the synthesis of

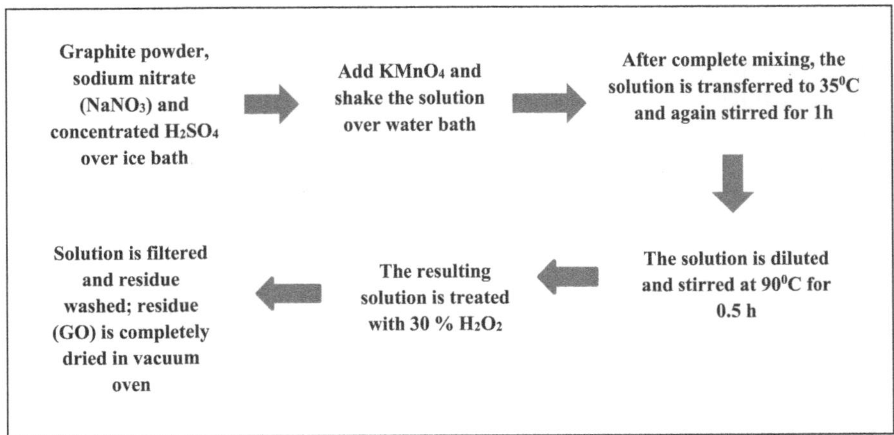

FIGURE 2.4 Synthesis of graphene oxide using the modified Hummer's method. (Reproduced with permission from Joshi and Gururani.[42] Copyright 2022, Elsevier Science Ltd.)

green materials is influenced by several factors that directly impact their properties and the final intended application. Three components are required when synthesizing graphene materials, whether using the green methods or the conventional methods: the reducing agent, the capping agent, and the solvent used. In green synthesis, the reducing and capping agents are often not harmful and produce fewer secondary contaminants or nothing, whereas the solvent used is also naturally sourced. Capping agents are utilized in the synthesis to stabilize the graphene and control the morphology to avoid any agglomeration between the nanomaterials. Because of their low toxicity and biocompatibility, biomolecules and polysaccharides are employed as capping agents in green synthesis. And green reducing agents of polysaccharides nature, such as β-D-glucose, peptides, and proteins, are used for chemical reduction. The dissolution of precursors, the dispersion of nanomaterials and assisting in the heat transfer are accomplished by the use of solvents; water, a natural source, is one of the most used solvents because of its inherent characteristics of non-toxicity, eco-friendliness, and non-flammability. The individual graphene sheets produced can be either single-layer or several-layer sheets, easily dispersed in water to create a stable colloidal graphene oxide solution. This aqueous colloidal graphene oxide suspension creates an appropriate environment for reducing graphene oxide to electrochemically reduced graphene oxide using electrochemical methods.[36] Reduced graphene oxide is prepared by chemical reduction or thermal annealing at very high temperatures (Figure 2.5). One effective thermal annealing approach for reduced graphene oxide is the thermal deoxygenation of graphene oxide at a gradual increase in temperatures to remove the oxygen-based moieties like the hydroxyl group. However, the process requires high operational energy, and the degree of oxidation is difficult to maintain. Preparing reduced graphene oxide by the chemical reduction method requires low-temperature ranges and uses reducing agents such as hydrazine, hydroiodic acid, and metal hydrides.[36] Due to the existence of electroactive sites and a substantial specific surface area, reduced graphene oxide exhibits strong surface reactivity and is biocompatible. Furthermore, the material shares several properties similar to those of pristine graphene. Although graphene and its derivatives have a high surface area and high adsorption properties, strong inter-functional bonds between graphene sheets cause inactive surface chemical characteristics, reduced surface area, poor dispersion, and aggregation in aqueous solutions, decreased performance of its adsorption capacity, additional applications in water. To address this drawback, graphene and its derivatives are systematically synthesized as highly effective adsorbents to remove the various contaminants from wastewater.[39–41]

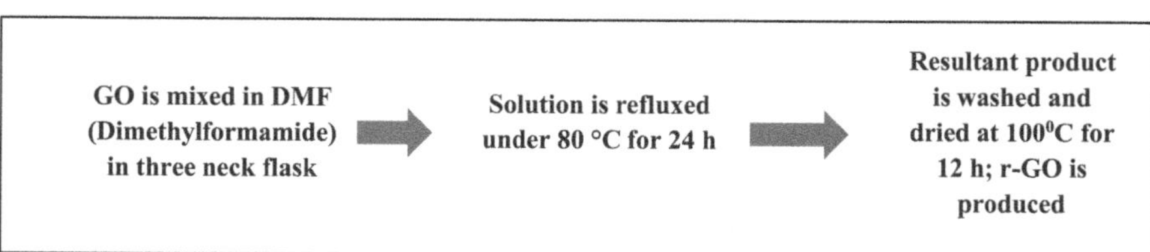

FIGURE 2.5 Typical synthesis process for reduced graphene oxide. (Reproduced with permission from Joshi and Gururani.[42] Copyright 2022, Elsevier Science Ltd.)

2.4.2 BIOPOLYMERS

Biopolymers are natural polymers that are sourced from animal or plant products; for instance, polynucleotides (DNA), polypeptides (proteins), and polysaccharides (cellulose, starch) are all biopolymers. They are ideal for use as they are non-toxic, biodegradable, and eco-friendly materials capable of removing contaminants from wastewater according to how simple it is to modify them and their biocompatibility.[12,21,34] Biopolymers have already been established as packaging materials in food industries and biomedical applications. For water treatment, there is still much that needs to be developed with the biopolymers. For instance, the second largest naturally occurring polysaccharide in the world is chitin, which is readily available and extracted from shrimp or crab shells. Chitin can be converted into chitosan (in an essential medium at a temperature of more than 60 °C), a polymer made of deacetylated b-glucosamine, through a straightforward deacetylation process. Chitosan is cost-effective, has a wide range of sources, and is easy to biodegrade. Its structure contains a considerable amount of amino and hydroxyl groups, which are attractive for forming strong and stable bonds with metal ions. It is also an alternative adsorbent for removing colloidal compounds using either the adsorption and coagulation or flocculation method.[43,44] Several contaminants, including organic dyes, micropollutants, toxic heavy metals, and hydrocarbons, were tested against a wide range of chitosan-based adsorbents. The adsorbents were used to facilitate the movement of the ionic species between the liquid and the solid interfaces due to the abundant 1 and 2 hydroxyl and amine functional groups on their surfaces.[45,46] Due to their biocompatibility, ease of production, low toxicity, and high abundance, chitosan-based compounds are widely used to remove various contaminants. Despite these advantages, poor mechanical properties have limited their use to larger scales. Furthermore, the specific surface area and porosity of chitosan also play a crucial part during the adsorption process that could significantly affect the adsorption capacity. Hence, chitosan composites are being explored, and chitosan nanofibers are being developed since the nanofibers have an increased surface area compared to pristine chitosan, which is favorable for the adsorption capacity. Cellulose is another natural polysaccharide that is readily available and is made of long linear chains of β-D-Glucose. Cellulose is among the critical materials in plant cells and is responsible for the structure and durability of plants. It also contains numerous hydroxyl groups that serve as the basis for possible material modification. It has also been shown to be an excellent adsorbent in combination with other materials for the removal of various contaminants in wastewater.[12] The main components of nanocellulose consist of cellulose nanofibrils, cellulose nanocrystals, and bacterial cellulose, each possessing its unique properties, morphology, and production process. Nanocellulose materials have found applications as absorbents, membrane filters, and composites for wastewater treatment and air purification. To achieve high efficiency and improve the adsorption capacity of nanocellulose for various contaminants, various functional groups are introduced on the surface of nanocellulose.[46] Starch is another plant-based polysaccharide that contains numerous glucose chains. Starch is a complex non-ionic type of polymer that must be modified first to be used as either a flocculant agent or an adsorbent. Cationic starch is a non-toxic, sustainable flocculant agent that works in a wide range of pH.[45] As an adsorbent, starch has been tested by removing dyes, antibiotics, and heavy metal ions. Even though the full potential of starch has not been realized in removing most contaminants, it has been reported that the material is non-hazardous as an adsorbent and suitable for wastewater treatment.[12] Most biopolymers have the potential to be used as adsorbents for the removal of various contaminants in water. Due to their given characteristics, the adsorption rate of contaminants on biopolymers is not economically effective. This is due to the practical application, where it is a challenge to recover biopolymer adsorbents and their poor mechanical strength after adsorption. Furthermore, the water molecules decrease the contact between the dyes in aqueous solutions and the biopolymer adsorbent, which interferes with adsorption.[46] Consequently, combining biopolymers with other materials (e.g., carbon-based material, clays, nanoparticles, etc.) is imperative to enhance their chemical stability and mechanical properties.

2.4.3 NANOCLAYS

Nanoclays are environmentally friendly, low-cost, and non-toxic natural phyllosilicate materials with particle sizes in the microns range. The structure of clays is almost similar and consists of layered units best described with one or two tetrahedral silica sheets attached to an octahedral aluminium sheet. The tetrahedral sheets comprise $Si_2O_6(OH)_4$ units, and the octahedral sheets comprise $Al_2(OH)_6$ units. Because of their distinct layered structures, exchangeable ions and tuneable lamellar spacing, clays have an excellent capacity for adsorbing various water contaminants. The complex porous structure and substantial surface area of clays make it easy for dissolved species to interact with the surfaces of the clay material.[47,48] Because of the isomorphic substitution that occurs in the octahedral sheets (Mg^{2+} for Al^{3+}) and the tetrahedral sheets (Al^{3+} for Si^{4+}), the majority of clay minerals contain negative surface charges, which can attract cationic contaminants. Clays can exchange ions with contaminants because naturally existing cations like Ca^{2+}, Na^+ and Mg^{2+} occur in the surfaces or between layers to maintain the balance of charges. Moreover, precipitation and substitution also appear to impact the removal behavior of contaminants. Due to a growing understanding of clay structure, clay is being evaluated for potential applications such as food additives, antibacterial agents, pharmaceuticals, personal care products, etc. Clays such as bentonite, montmorillonite, kaolinite, and halloysite have been widely used for wastewater treatment containing organic contaminants. The main difference between these clays is their percentage compositions of SiO_2 and Al_2O between their layers. Due to the nature of the structure and the surface charge, the common modification methods for clays include acid treatment, thermal treatment, organic modification, pillaring, etc., Figure 2.6.[49,50] These methods have certain benefits. As a result, research advances are gradually overcoming the adsorption limitations of clay minerals. Some studies have combined two or more modification methods to produce multifunctional materials, such as inorganic-organic-clay, that can simultaneously adsorb organic and inorganic contaminants. Like many other adsorbents, the regeneration of spent clay composites is imperative as it lowers the operational costs of the adsorption process. Figure 2.7 gives some techniques used to regenerate clay composites.[48]

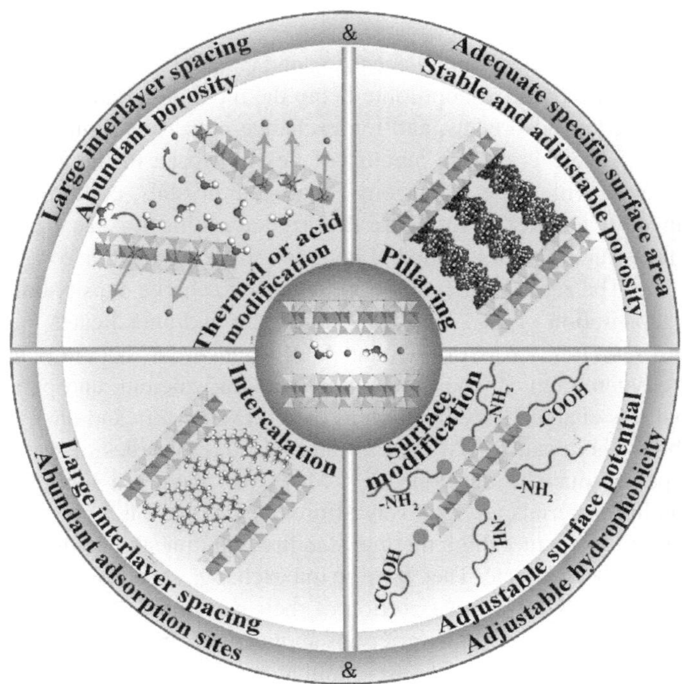

FIGURE 2.6 Common modification methods used for clays. (Reproduced with permission from Zhang et al.[51] Copyright 2021, Elsevier Science Ltd.)

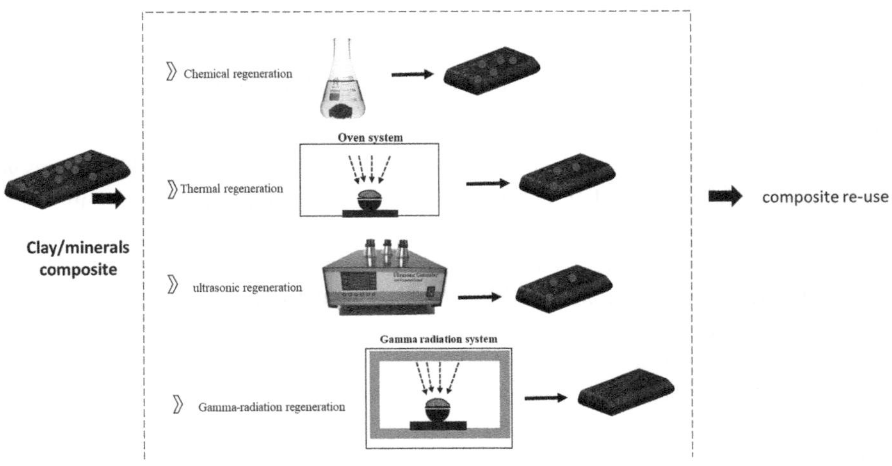

FIGURE 2.7 Typical regeneration techniques are used for spent adsorbents. (Reproduced with permission from Ewis et al.[48] Copyright 2022, Elsevier Science Ltd.)

2.4.4 ZEOLITES

Zeolites are microporous, naturally occurring and crystalline aluminosilicates. There are different zeolites, with clinoptilolite being the most investigated. They are typical minerals used as adsorbents, and ion exchange power determines their adsorption properties. There is a large amount of flexible space in the zeolite structure for ion exchange with other metal ions. Aluminate crystals, composed of oxygen atoms arranged in tetrahedral form, are the original zeolite crystals. The main component of the structure is a tetrahedral complex made up of four oxygen ions and the cation Si^{4+}. The zeolite lattice develops a negative charge density due to the isomorphous substitution of Al^{3+} for Si^{4+}. Zeolites are aluminosilicates made up of a three-dimensional framework of SiO_4 and AlO_4 tetrahedrons that are joined together by mutual oxygen atoms.[14] The exchange of sodium, calcium, and potassium ions balances out this charge. They function as cation exchangers and have cavities and tiny chambers where cations (Mg^{2+}, Ca^{2+} and NH^{4+}), molecules of water and other available species may be enveloped.[52,53] Moreover, having a high ion-exchange capacity, zeolite materials are relatively affordable and have a high specific surface area, regular pore structure, high thermal stability with tuneable hydrophobicity and hydrophilicity properties, etc. They have an ion exchange capacity that is comparable to clays and accounts for a large portion of their capacity for adsorption. Zeolites are widely used in industry for a variety of applications, including water treatment, biomass conversion, separation of gases, animal husbandry, agriculture, etc. Natural zeolites contain several crucial physical and chemical characteristics, but they also have certain drawbacks. The adsorption efficacy of zeolites on organics and anions is unsatisfactory. Therefore, surface modification of the material is required to change the hydrophobic or hydrophilic nature and to increase the adsorption efficacy.[14,18] To improve the pristine surface structure of zeolites, the material is treated with organic or inorganic compounds, which improve their surface characteristics and enhance their adsorption capacity.

2.4.5 PREFERENCE OF GRAPHENE FOR WASTEWATER DECONTAMINATION

As a carbon-based material, graphene (and its derivatives) has received attention in various applications, including wastewater treatment, because of their effective and adaptable properties when it comes to the degradation of various contaminants in water. These properties include chemical stability, substantial surface area, and improved active sites, which are necessary for a highly effective adsorptive material. The properties also include an abundance of functional groups on the surfaces, which provide a significant contaminant-binding capacity, a carbon structure with sp^2 conjugation that has a strong attraction for aromatic organic contaminants and a simple surface structure that is easy to modify and functionalize to add and enhance the surfaces with the desired functionalities.[32,36,54,55] However, the infamous aggregation and agglomeration of graphene materials severely degrades their inherent distinctive structure and properties (particularly a significant reduction on the surface area, which is directly related to the adsorption efficacy, because the aggregation hinders the accessibility of the active adsorption sites). The preparation of graphene-based adsorbents requires confinement and immobilization of graphene constituents to avoid aggregation and maintain the accessibility of the adsorption sites as much as possible.[6–8,55,56] Chemical exfoliation of graphite results in the production of graphene oxide, which makes it an extremely oxidative form of graphene. There are sufficient oxygen-containing surface functions on graphite oxide surfaces, including carboxyl and carbonyl groups at the end of the layers and hydroxyl and epoxy groups on the base plane. Because of its high specific surface area, inexpensive production, high water solubility, excellent chemical stability, and oxygen-containing surface functions, graphene oxide is used as an adsorbent, just like graphene. However, graphene oxide presents challenges to recycling and reusing the colloidal particles following the adsorption process because it typically requires rapid centrifugation and a prolonged filtering process.[21,57] Although much work has been done in advancing graphene and its derivatives as adsorbents, more work is still required to develop effective graphene-based adsorbents. The adsorbents are anticipated to have simple processing characteristics for in-house treatment, excellent selectivity to certain contaminants, high efficiency in treating a wide range of contaminants, and robust materials reduce the operational costs of the treatment process. Table 2.1 summarises some carbon-based adsorbents employed to remove various wastewater contaminants. Furthermore, the superior properties of graphene and its derivatives have suggested that they could be used in other applications, such as sensors, solar cells, fuel cells, supercapacitors, etc. A study by Abdelkhalek et al.[58] reported on the preparation of graphene oxide with silica hybrid nanocomposites for the degradation of trypan blue (TB) and bisphenol A (BPA) from water. Both trypan blue and bisphenol A are regarded as emerging contaminants that are being discharged into the aquatic environment in large quantities, although their concentration is in trace amounts. Due to the large production scale and use of bisphenol A, contamination is found all over the environment. Exposure to this contaminant in high volumes has been found to cause carcinogenic and mutagenic changes, while trypan blue is a diazo dye that has been found to have the same harmful effects on humans and the environment as azo dyes. The graphene oxide/silica hybrid nanocomposites were prepared from the low-cost and abundant agricultural by-product of rice husk. According to their experimental results, the authors reported a significant adsorption capacity for both contaminants, which was suggested to be attributed to the various oxygen-containing surface groups, large surface area, the electrostatic interaction in an acidic environment, electron donor-acceptor, and hydrogen bonding. The surface area of the prepared nanocomposites was reported as 1768 m^2/g. The adsorption thermodynamics showed that the adsorption of trypan blue and bisphenol A using the prepared nanocomposites is a spontaneous exothermic process facilitated by reducing the temperature. The authors concluded that the findings of their study suggest that the functionalized graphene oxide/silica hybrid nanocomposites might be a sustainable adsorbent that is economical and effective in removing aromatic compounds from wastewater. Sabzevari et al.[7] developed a composite of graphene oxide crosslinked with chitosan for the adsorption of methylene blue. The study revealed that crosslinking graphene oxide with other materials could enhance the sorption properties of the composite. Hence, the removal efficiency of methylene blue was evaluated using the composite at kinetic and equilibrium conditions. This was indicated by their results, where the composite

TABLE 2.1

Summary of Selected Carbon-Based Adsorbents for Treatment of Various Contaminants

Material	Contaminant	Adsorption Capacity/ Removal Rate	Ref.
Biochar composite	Ciprofloxacin	47–71%	10
Graphene oxide	Hexavalent chromium	92.8%	42
Graphene oxide composite	Methylene blue	84%	42
Graphene catalyst	Methylene blue and Rhodamine B	98–99%	42
Graphene oxide nanocomposite	Trypan blue (TB) and Bisphenol A (BPA)	455 and 500 mg/g	58
Multi-walled carbon nanotubes	Alizarin red S and alizarin yellow R	257.73 and 45.39 mg/g	27
Reduced graphene oxide	Malachite green dye	476.2 mg/g	27
Aerogels (cellulose, carbon nanotubes and graphene)	Methylene blue and Congo red	1178.5 and 585.3 mg/g	27
Graphene oxide-chitosan composite	Methylene blue	402.6 mg/g	7
Reduced graphene oxide composite	Methylene blue	100%	40
Biochar	Phosphate	620 mg/g	29
Graphene oxide, magnetic graphene oxide and dendrimer-graphene oxide	Selenium	15, 25 and 61 mg/g	60
Biochar	Phenol	95%	13
Graphene oxide			
Single-walled carbon nanotubes photocatalyst	Total organic carbon and chemical oxygen demand	89 and 93%	61
Biochar	Methylene blue	100%	17
Nitrogen-doped biochar	Methylene blue and methylene orange	326.90 and 906.52 mg/g	17
Graphene oxide	Lead and zinc	987.33 and 313.43 mg/g	24
Activated carbon	Tetracycline hydrochloride	239.6 mg/g	27
Biochar	p-nitrophenol	622.73 mg/g	62
Biochar	Cobalt	72.27 mg/g	63
Magnetic multi-walled carbon nanotubes	Lead	215.05 mg/g	33
Biochar	Lead	137.1 mg/g	64
Activated carbon	Phenol	434 mg/g	65
Activated carbon	Carbon dioxide	217 mg/g	26
Activated carbon and magnetic activated carbon	Crystal violet dye	23.6 and 35.3 mg/g	25

material resulted in a higher adsorption capacity for the methylene blue compared to pristine graphene oxide due to the available active sites and enhanced stability and mechanical properties. The adsorption capacity of the composites was reported to be 402.6 mg/g, which far exceeds the 286.9 mg/g of pure graphene oxide. The authors further indicated that the improved adsorption capacity of the crosslinked graphene oxide composites could be because of the large surface functional groups introduced to the composite material providing additional active sites that favor adsorption, and the crosslinking effects contributed to the unrestricted interaction of the adsorbent and the methylene blue due to the surface structural changes which serve to disperse the graphene oxide and prevent agglomeration in aqueous media. Both chitosan and graphene oxide are renowned for removing various dyes through electrostatic interactions. Hence, the combination of these two precursor materials contributes to the excellent performance of methylene blue removal by enhancing the overall adsorption capacity of the composites. Moreover, the crosslinked composites were more stable and reusable in multiple adsorption-desorption processes.[7]

Khalil et al.[6] evaluated the effectiveness of graphene as a filtering medium for water and wastewater in removing emerging contaminants. Several pharmaceutical contaminants, such as ciprofloxacin, diclofenac, ibuprofen, gemfibrozil, atenolol, and carbamazepine, were used to assess the efficacy of the porous graphene as a filtering medium through column studies. Because of the interactions between hydrophobic molecules and heterogeneous adsorption, the kinetic studies indicated that the adsorption process followed the pseudo-second-order model and the adsorption technique by porous graphene for the emerging contaminants was well described by the adsorption isotherms of Toth and Sips. It was reported that a sand column with a double layer of the porous graphene dose proved to be more effective in the removal of the contaminants from water. The removal efficiency of the pollutants showed improvement with an increase in the porous graphene dosages and the increase of the filter size and bed height where the contact time was prolonged. The removal effectiveness remained above 90% for over 100 min in distilled and grey water. Furthermore, it was demonstrated from the results that the removal efficiency of the contaminants in grey water was not severely affected, as it was assumed that treatment of mixed emerging contaminants from various water sources could be impacted by the different competing compounds in water, causing negative interference. The porous graphene was a powerful filtering medium that could find more application in the widespread emerging contaminants from wastewater.[6]

Khan et al.[59] used date syrup, a readily available and eco-friendly carbon source, through the pyrolysis method, to synthesize a graphene adsorbent for the removal of cationic (methyl violet) and anionic (Congo red) dye and heavy metals (Pb^{2+} and Cd^{2+}) from water. The study used batch techniques to evaluate the efficiency of the adsorbent against the contaminants, where it was observed that the adsorbent had the exceptional potential for removing the targeted contaminants with fast adsorption that was well described by the pseudo-second-order kinetics. The adsorption capacities were reported as 2564, 781, and 793 mg/g at 25 C° for methyl violet, Pb^{2+}, and Cd^{2+}, respectively. The authors stated that these were the highest adsorption capacities for graphene-based adsorbents. For Congo red, the adsorption capacity was observed to be 333 mg/g, and the authors also alluded that this was the most superior compared to what has been confirmed in the literature on graphene-based adsorbents. Although the specific area of the graphene adsorbent was found to be low at 22.4 m²/g, the adsorption results obtained make this graphene more interesting, considering that the date syrup is also available in abundance as the carbon source. The authors attributed the high removal efficiency of the contaminants and the excellent regeneration of the graphene to its carbon source, which played a crucial part in the improved properties of graphene due to the difference in structure, morphology, composition, and surface properties of the date syrup.

Furthermore, the graphene adsorbent exhibited good recyclability, with three cycles run for all the tested contaminants. Thermodynamic investigations also demonstrated that the adsorption process for the contaminants was spontaneous, thermodynamically viable and endothermic. It was concluded that utilizing the simple one-step process for preparing a graphene-based adsorbent with date syrup as the carbon source has proved to be convenient, economical, and practical for removing the selected contaminants.[59]

2.5 CONCLUSION

The limitations of conventional adsorbents in removing contaminants, especially emerging contaminants from contaminated wastewaters, have prompted research in developing new and novel adsorbents. Hence, research and development of new alternative and cost-effective adsorbents is continuing. The use of nanomaterials as alternative adsorbents is also being explored extensively. However, the path to using these materials on a large scale is very long. It requires adequate knowledge of the potential impact on the environment and the feasibility of being scaled up to industrial scales, regenerated and recycled. Despite these limitations, using nanomaterials as potential adsorbents for contaminants in wastewater is promising, as the materials can remove various pollutants at low concentrations, with high selectivity and adsorption capacity. Although adsorption selectivity and capacity remain the primary concerns of most materials in the adsorption process, in-depth exploration of various materials has become more comprehensive.

REFERENCES

(1) Kumar, R.; Qureshi, M.; Vishwakarma, D. K.; Al-Ansari, N.; Kuriqi, A.; Elbeltagi, A.; Saraswat, A. A review on emerging water contaminants and the application of sustainable removal technologies. *Case Studies in Chemical and Environmental Engineering* 2022, *6*, 100219.

(2) Boulkhessaim, S.; Gacem, A.; Khan, S. H.; Amari, A.; Yadav, V. K.; Harharah, H. N.; Elkhaleefa, A. M.; Yadav, K. K.; Rather, S.-U.; Ahn, H.-J. Emerging trends in the remediation of persistent organic pollutants using nanomaterials and related processes: A review. *Nanomaterials* 2022, *12* (13), 2148.

(3) Patel, M.; Kumar, R.; Kishor, K.; Mlsna, T.; Pittman Jr, C. U.; Mohan, D. Pharmaceuticals of emerging concern in aquatic systems: Chemistry, occurrence, effects, and removal methods. *Chemical reviews* 2019, *119* (6), 3510–3673.

(4) Adeleye, A. S.; Conway, J. R.; Garner, K.; Huang, Y.; Su, Y.; Keller, A. A. Engineered nanomaterials for water treatment and remediation: Costs, benefits, and applicability. *Chemical Engineering Journal* 2016, *286*, 640–662.

(5) Mundkur, N.; Khan, A. S.; Khamis, M. I.; Ibrahim, T. H.; Nancarrow, P. Synthesis and characterization of clay-based adsorbents modified with alginate, surfactants, and nanoparticles for methylene blue removal. *Environmental Nanotechnology, Monitoring & Management* 2022, *17*, 100644.

(6) Khalil, A. M.; Memon, F. A.; Tabish, T. A.; Fenton, B.; Salmon, D.; Zhang, S.; Butler, D. Performance evaluation of porous graphene as filter media for the removal of pharmaceutical/emerging contaminants from water and wastewater. *Nanomaterials* 2021, *11* (1), 79.

(7) Sabzevari, M.; Cree, D. E.; Wilson, L. D. Graphene oxide–chitosan composite material for treatment of a model dye effluent. *ACS Omega* 2018, *3* (10), 13045–13054.

(8) Hossain, M. F.; Akther, N.; Zhou, Y. Recent advancements in graphene adsorbents for wastewater treatment: Current status and challenges. *Chinese Chemical Letters* 2020, *31* (10), 2525–2538.

(9) Zhu, B.; Chen, S.; Li, C.; Jiang, G.; Liu, F.; Zhao, R.; Liu, C. Non-metallic hollow porous sphere loaded CN/catalytic ozonation synergistic photocatalytic system: Enhanced treatment of emerging pollutants by three-stage cyclic reaction mechanism. *Applied Catalysis B: Environmental* 2022, *318*, 121881.

(10) Zheng, D.; Wu, M.; Zheng, E.; Wang, Y.; Feng, C.; Zou, J.; Juan, M.; Bai, X.; Wang, T.; Shi, Y. Adsorption and oxidation of ciprofloxacin by a novel layered double hydroxides modified sludge biochar. *Journal of Colloid and Interface Science* 2022, *625*, 596–605.

(11) Feng, Z.; Xu, Y.; Yue, W.; Adolfsson, K. H.; Wu, M. Recent progress in the use of graphene/polymer composites to remove oil contaminants from water. *New Carbon Materials* 2021, *36* (2), 235–252.

(12) Mangla, D.; Sharma, A.; Ikram, S. Critical review on adsorptive removal of antibiotics: Present situation, challenges and future perspective. *Journal of Hazardous Materials* 2022, *425*, 127946.

(13) Singh, R.; Dutta, R. K.; Naik, D. V.; Ray, A.; Kanaujia, P. K. High surface area Eucalyptus wood biochar for the removal of phenol from petroleum refinery wastewater. *Environmental Challenges* 2021, *5*, 100353.

(14) Belova, T. Adsorption of heavy metal ions (Cu^{2+}, Ni^{2+}, Co^{2+} and Fe^{2+}) from aqueous solutions by natural zeolite. *Heliyon* 2019, *5* (9), e02320.

(15) Hamad, H. N.; Idrus, S. Recent developments in the application of bio-waste-derived adsorbents for the removal of methylene blue from wastewater: A review. *Polymers* 2022, *14* (4), 783.

(16) Solangi, N. H.; Kumar, J.; Mazari, S. A.; Ahmed, S.; Fatima, N.; Mubarak, N. M. Development of fruit waste derived bio-adsorbents for wastewater treatment: A review. *Journal of Hazardous Materials* 2021, *416*, 125848.

(17) Qiu, B.; Shao, Q.; Shi, J.; Yang, C.; Chu, H. Application of biochar for the adsorption of organic pollutants from wastewater: Modification strategies, mechanisms and challenges. *Separation and Purification Technology* 2022, *300*, 121925.

(18) Mo, Z.; Tai, D.; Zhang, H.; Shahab, A. A comprehensive review on the adsorption of heavy metals by zeolite imidazole framework (ZIF-8) based nanocomposite in water. *Chemical Engineering Journal* 2022, *443*, 136320.

(19) Liu, Y.; Chen, Y.; Li, Y.; Chen, L.; Jiang, H.; Li, H.; Luo, X.; Tang, P.; Yan, H.; Zhao, M. Fabrication, application, and mechanism of metal and heteroatom co-doped biochar composites (MHBCs) for the removal of contaminants in water: A review. *Journal of Hazardous Materials* 2022, *431*, 128584.

(20) Gong, Y.; Wang, Y.; Lin, N.; Wang, R.; Wang, M.; Zhang, X. Iron-based materials for simultaneous removal of heavy metal (loid) s and emerging organic contaminants from the aquatic environment: Recent advances and perspectives. *Environmental Pollution* 2022, *299*, 118871.

(21) Alves, D. C. d. S.; Healy, B.; Yu, T.; Breslin, C. B. Graphene-based materials immobilized within chitosan: Applications as adsorbents for the removal of aquatic pollutants. *Materials* 2021, *14* (13), 3655.

(22) Rangabhashiyam, S.; dos Santos Lins, P. V.; de Magalhães Oliveira, L. M.; Sepulveda, P.; Ighalo, J. O.; Rajapaksha, A. U.; Meili, L. Sewage sludge-derived biochar for the adsorptive removal of wastewater pollutants: A critical review. *Environmental Pollution* 2022, *293*, 118581.

(23) Zhu, F.; Zheng, Y.-M.; Zhang, B.-G.; Dai, Y.-R. A critical review on the electrospun nanofibrous membranes for the adsorption of heavy metals in water treatment. *Journal of Hazardous Materials* 2021, *401*, 123608.

(24) Guerrero-Fajardo, C. A.; Giraldo, L.; Moreno-Piraján, J. C. Preparation and characterization of graphene oxide for Pb (II) and Zn (II) ions adsorption from aqueous solution: Experimental, thermodynamic and kinetic study. *Nanomaterials* 2020, *10* (6), 1022.

(25) Foroutan, R.; Peighambardoust, S. J.; Peighambardoust, S. H.; Pateiro, M.; Lorenzo, J. M. Adsorption of crystal violet dye using activated carbon of lemon wood and activated carbon/Fe_3O_4 magnetic nanocomposite from aqueous solutions: A kinetic, equilibrium and thermodynamic study. *Molecules* 2021, *26* (8), 2241.

(26) Acevedo, S.; Giraldo, L.; Moreno-Piraján, J. C. Adsorption of CO_2 on activated carbons prepared by chemical activation with cupric nitrate. *ACS Omega* 2020, *5* (18), 10423–10432.

(27) Cai, Y.; Liu, L.; Tian, H.; Yang, Z.; Luo, X. Adsorption and desorption performance and mechanism of tetracycline hydrochloride by activated carbon-based adsorbents derived from sugar cane bagasse activated with $ZnCl_2$. *Molecules* 2019, *24* (24), 4534.

(28) Krasucka, P.; Pan, B.; Ok, Y. S.; Mohan, D.; Sarkar, B.; Oleszczuk, P. Engineered biochar–A sustainable solution for the removal of antibiotics from water. *Chemical Engineering Journal* 2021, *405*, 126926.

(29) Almanassra, I. W.; Mckay, G.; Kochkodan, V.; Atieh, M. A.; Al-Ansari, T. A state of the art review on phosphate removal from water by biochars. *Chemical Engineering Journal* 2021, *409*, 128211.

(30) Saravanan, A.; Kumar, P. S. Biochar derived carbonaceous material for various environmental applications: Systematic review. *Environmental Research* 2022, *214*, 113857.

(31) Cheng, N.; Wang, B.; Wu, P.; Lee, X.; Xing, Y.; Chen, M.; Gao, B. Adsorption of emerging contaminants from water and wastewater by modified biochar: A review. *Environmental Pollution* 2021, *273*, 116448.

(32) Khan, F. S. A.; Mubarak, N. M.; Tan, Y. H.; Khalid, M.; Karri, R. R.; Walvekar, R.; Abdullah, E. C.; Nizamuddin, S.; Mazari, S. A. A comprehensive review on magnetic carbon nanotubes and carbon nanotube-based buckypaper for removal of heavy metals and dyes. *Journal of Hazardous Materials* 2021, *413*, 125375.

(33) Wang, Z.; Xu, W.; Jie, F.; Zhao, Z.; Zhou, K.; Liu, H. The selective adsorption performance and mechanism of multiwall magnetic carbon nanotubes for heavy metals in wastewater. *Scientific Reports* 2021, *11* (1), 16878.

(34) Fiyadh, S. S.; AlSaadi, M. A.; Jaafar, W. Z.; AlOmar, M. K.; Fayaed, S. S.; Mohd, N. S.; Hin, L. S.; El-Shafie, A. Review on heavy metal adsorption processes by carbon nanotubes. *Journal of Cleaner Production* 2019, *230*, 783–793.

(35) Chadha, U.; Selvaraj, S. K.; Thanu, S. V.; Cholapadath, V.; Abraham, A. M.; Manoharan, M.; Paramsivam, V. A review of the function of using carbon nanomaterials in membrane filtration for contaminant removal from wastewater. *Materials Research Express* 2022, *9* (1), 012003.

(36) Asghar, F.; Shakoor, B.; Fatima, S.; Munir, S.; Razzaq, H.; Naheed, S.; Butler, I. S. Fabrication and prospective applications of graphene oxide-modified nanocomposites for wastewater remediation. *RSC Advances* 2022, *12* (19), 11750–11768.

(37) Ali, I.; Mbianda, X.; Burakov, A.; Galunin, E.; Burakova, I.; Mkrtchyan, E.; Tkachev, A.; Grachev, V. Graphene based adsorbents for remediation of noxious pollutants from wastewater. *Environment International* 2019, *127*, 160–180.

(38) Sharif Nasirian, V.; Shahidi, S. A.; Tahermansouri, H.; Chekin, F. Application of graphene oxide in the adsorption and extraction of bioactive compounds from lemon peel. *Food Science & Nutrition* 2021, *9* (7), 3852–3862.

(39) Baig, N.; Sajid, M.; Saleh, T. A. Graphene-based adsorbents for the removal of toxic organic pollutants: A review. *Journal of Environmental Management* 2019, *244*, 370–382.

(40) Ikram, M.; Raza, A.; Imran, M.; Ul-Hamid, A.; Shahbaz, A.; Ali, S. Hydrothermal synthesis of silver decorated reduced graphene oxide (rGO) nanoflakes with effective photocatalytic activity for wastewater treatment. *Nanoscale Research Letters* 2020, *15*, 1–11.

(41) Farooq, S.; Aziz, H.; Ali, S.; Murtaza, G.; Rizwan, M.; Saleem, M. H.; Mahboob, S.; Al-Ghanim, K. A.; Riaz, M. N.; Murtaza, B. Synthesis of functionalized carboxylated graphene oxide for the remediation of Pb and Cr contaminated water. *International Journal of Environmental Research and Public Health* 2022, *19* (17), 10610.

(42) Joshi, N. C.; Gururani, P. Advances of graphene oxide based nanocomposite materials in the treatment of wastewater containing heavy metal ions and dyes. *Current Research in Green and Sustainable Chemistry* 2022, *5*, 100306.

(43) Matei, E.; Predescu, A. M.; Râpă, M.; Țurcanu, A. A.; Mateș, I.; Constantin, N.; Predescu, C. Natural polymers and their nanocomposites used for environmental applications. *Nanomaterials* 2022, *12* (10), 1707.

(44) Russo, T.; Fucile, P.; Giacometti, R.; Sannino, F. Sustainable removal of contaminants by biopolymers: A novel approach for wastewater treatment. Current state and future perspectives. *Processes* 2021, *9* (4), 719.

(45) Biswas, S.; Pal, A. Application of biopolymers as a new age sustainable material for surfactant adsorption: A brief review. *Carbohydrate Polymer Technologies and Applications* 2021, *2*, 100145.

(46) Choi, W. S.; Lee, H.-J. Nanostructured materials for water purification: Adsorption of heavy metal ions and organic dyes. *Polymers* 2022, *14* (11), 2183.

(47) Adeyemo, A. A.; Adeoye, I. O.; Bello, O. S. Adsorption of dyes using different types of clay: A review. *Applied Water Science* 2017, *7*, 543–568.

(48) Ewis, D.; Ba-Abbad, M. M.; Benamor, A.; El-Naas, M. H. Adsorption of organic water pollutants by clays and clay minerals composites: A comprehensive review. *Applied Clay Science* 2022, *229*, 106686.

(49) Han, B.; Weatherley, A. J.; Mumford, K.; Bolan, N.; He, J.-Z.; Stevens, G. W.; Chen, D. Modification of naturally abundant resources for remediation of potentially toxic elements: A review. *Journal of Hazardous Materials* 2022, *421*, 126755.

(50) Yadav, V. B.; Gadi, R.; Kalra, S. Clay based nanocomposites for removal of heavy metals from water: A review. *Journal of Environmental Management* 2019, *232*, 803–817.

(51) Zhang, T.; Wang, W.; Zhao, Y.; Bai, H.; Wen, T.; Kang, S.; Song, G.; Song, S.; Komarneni, S. Removal of heavy metals and dyes by clay-based adsorbents: From natural clays to 1D and 2D nanocomposites. *Chemical Engineering Journal* 2021, *420*, 127574.

(52) Solińska, A.; Bajda, T. Modified zeolite as a sorbent for removal of contaminants from wet flue gas desulphurization wastewater. *Chemosphere* 2022, *286*, 131772.

(53) Hong, M.; Yu, L.; Wang, Y.; Zhang, J.; Chen, Z.; Dong, L.; Zan, Q.; Li, R. Heavy metal adsorption with zeolites: The role of hierarchical pore architecture. *Chemical Engineering Journal* 2019, *359*, 363–372.

(54) Queiroz, R. N.; Prediger, P.; Vieira, M. G. A. Adsorption of polycyclic aromatic hydrocarbons from wastewater using graphene-based nanomaterials synthesized by conventional chemistry and green synthesis: A critical review. *Journal of Hazardous Materials* 2022, *422*, 126904.

(55) Jiang, J.; Pachter, R.; Selhorst, R. C.; Susner, M. A.; Maruyama, B.; Rao, R. Patterned graphene: Analysis of the electronic structure and electron transport by first principles computational modeling. *Applied Surface Science* 2022, *589*, 152953.

(56) Zhang, Y.; Yan, X.; Yan, Y.; Chen, D.; Huang, L.; Zhang, J.; Ke, Y.; Tan, S. The utilization of a three-dimensional reduced graphene oxide and montmorillonite composite aerogel as a multifunctional agent for wastewater treatment. *RSC Advances* 2018, *8* (8), 4239–4248.

(57) Omran, B.; Baek, K.-H. Graphene-derived antibacterial nanocomposites for water disinfection: Current and future perspectives. *Environmental Pollution* 2022, *298*, 118836.

(58) Abdelkhalek, A.; El-Latif, M. A.; Ibrahim, H.; Hamad, H.; Showman, M. Controlled synthesis of graphene oxide/silica hybrid nanocomposites for removal of aromatic pollutants in water. *Scientific Reports* 2022, *12* (1), 7060.

(59) Khan, S.; Achazhiyath Edathil, A.; Banat, F. Sustainable synthesis of graphene-based adsorbent using date syrup. *Scientific Reports* 2019, *9* (1), 18106.

(60) Benis, K. Z.; McPhedran, K. N.; Soltan, J. Selenium removal from water using adsorbents: A critical review. *Journal of Hazardous Materials* 2022, *424*, 127603.

(61) Antoniadou, M.; Falara, P. P.; Likodimos, V. Photocatalytic degradation of pharmaceuticals and organic contaminants of emerging concern using nanotubular structures. *Current Opinion in Green and Sustainable Chemistry* 2021, *29*, 100470.

(62) Ma, H.; Xu, Z.; Wang, W.; Gao, X.; Ma, H. Adsorption and regeneration of leaf-based biochar for p-nitrophenol adsorption from aqueous solution. *RSC Advances* 2019, *9* (67), 39282–39293.

(63) Hu, C.; Zhang, W.; Chen, Y.; Ye, N.; YangJi, D.; Jia, H.; Shen, Y.; Song, M. Adsorption of Co (II) from aqueous solution using municipal sludge biochar modified by HNO_3. *Water Science and Technology* 2021, *84* (1), 251–261.

(64) Gao, J.; Liu, Y.; Li, X.; Yang, M.; Wang, J.; Chen, Y. A promising and cost-effective biochar adsorbent derived from jujube pit for the removal of Pb (II) from aqueous solution. *Scientific Reports* 2020, *10* (1), 7473.

(65) Mojoudi, N.; Mirghaffari, N.; Soleimani, M.; Shariatmadari, H.; Belver, C.; Bedia, J. Phenol adsorption on high microporous activated carbons prepared from oily sludge: Equilibrium, kinetic and thermodynamic studies. *Scientific Reports* 2019, *9* (1), 1–12.

3 Graphene, Its History, and Derivatives

3.1 HISTORY OF GRAPHENE AND ITS DERIVATIVES

Graphene is a sheet-like substance made of carbon atoms organized in a mono-layer to create a hexagonal honeycomb lattice that is securely linked. The carbon atoms in graphene are actually sp^2-bonded. Graphene monolayers are layered on top of one another to form graphite. It is the second-highest concentration of carbon, the fourth most prevalent element in the universe, and is found in the human body. Every known form of life on Earth has been connected to carbon.[1]

Carbon in graphene is bound to another carbon approximately 0.142 nanometers long. Only 0.33 nm of layer height is present. It is one of the strongest and thinnest materials currently known. Nearly all of graphene is transparent. Due to its tremendous density, not even the smallest helium atom can get through it.[2] The use of graphene can help with both the creation of new technologies and the improvement of already existing ones.[1]

In chronological order, Figure 3.1 depicts the history of specific graphene preparation, isolation, and characterization events. Early accounts of graphene oxide (GO) and graphene intercalated compounds (GICs) began in the 1840s, when the German researcher Schafhaeutl described the intercalation and exfoliation of graphite with sulfuric and nitric acids. When a tiny molecular species, such as acid or alkali metal, is inserted between the carbon lamellae, this process is known as intercalation,[3,4] and the use of nitric and sulfuric acids have been used to exfoliate graphite. Since then, a wide variety of intercalants and exfoliants have been employed, including potassium (together with other alkali metals), different forms of fluoride salts,[5] transition metals, and others (iron, nickel, and many others),[6–8] a wide range of organic organisms, etc.[9] Although GICs maintain the stacked structure of graphite, the interlayer gap is expanded, frequently by several angstroms or more. This causes the individual layers to become electronically decoupled from one another. In some situations, this electronic decoupling results in fascinating superconducting effects,[10] which are a prelude to the amazing electronic properties that freestanding graphene would eventually be shown to possess.

In fact, once a vocabulary to describe the dissociated layers became apparent, the chemistry of GICs gave rise to the term "graphene".[11,12] Boehm et al.[12] are credited with coining the term "graphene" in 1986, to the best of our knowledge. Later, it was hypothesized that pristine graphene could be produced, provided the interlayer spacing of GICs could be expanded all through the structure and the tiny molecule spacers were eliminated.[13]

Attempting to determine the molecular weight of graphite in 1859, the British scientist Brodie modified Schafhaeutl's reported procedures by employing strong acids (such

DOI: 10.1201/9781032621302-3

"

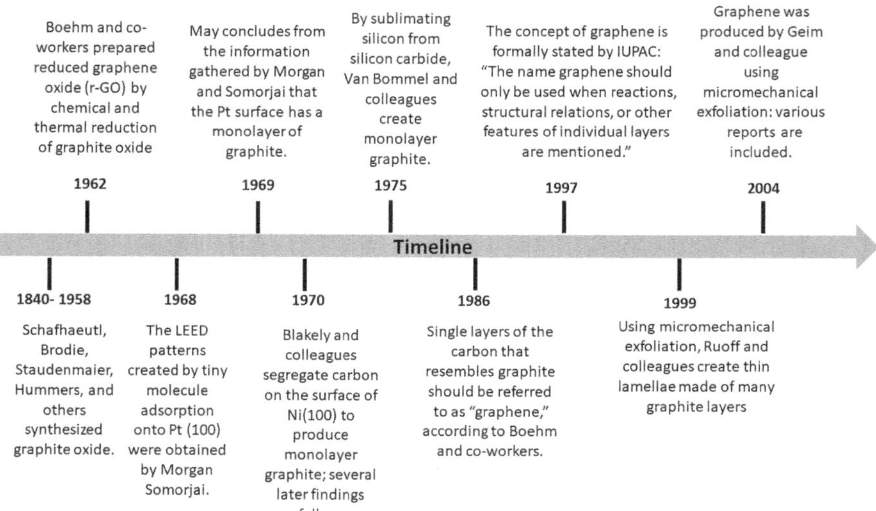

FIGURE 3.1   History of selected graphene preparation, isolation, and characterization events in chronological order.

as sulfuric and nitric) and oxidants, including $KClO_3$.[14] These circumstances brought about the intercalation of the graphite layers as well as its surface being oxidized chemically, which ultimately led to the creation of GO. This method of chemically modifying graphite's surface has proven to be useful for a number of processes, including the production of mono-layer reduced graphene oxide (r-GO), GICs, GO, and other related materials, as well as the application of GO as chemical oxidants in chemical synthesis processes.[15,16] This surface modification of the graphite surface reduces the interplanar forces that lead to lamellar stacking, allowing for easy exfoliation of the oxidized layers in extreme environments such as ultrasonic, thermal, or others. Approximately four decades later, Staudenmaier identified a marginally modified version of Brodie's oxidation procedure for producing GO by introducing the chlorate salt in a series of aliquots throughout the reaction as opposed to all at once.[17] The first instances of graphite delamination into its individual lamellae can be seen in these intercalation and oxidation tests. The preparation of r-GO and other chemically modified graphenes (CMGs) is still done using many of these techniques or variations of them. Kohlschutter and Haenni finished a preliminary investigation into the characteristics of this GO paper in 1919.[2]

In 1916, powder diffraction was utilized to reveal the surface structure of graphene, and in 1924, single-crystal X-ray diffraction methods were employed to discover the internal structure. In 1947, Wallace began researching the graphene theory in order to understand the electrical characteristics of three-dimensional (3D) graphite, while the revolutionary massless Dirac equation was initially revealed by Semenoff, DiVincenzo, and Mele.[1,18] Ruess and Vogt reported the earliest TEM images of few-layer graphite in 1948. The basal-plane conductivity of graphite intercalation compounds was surprisingly higher than that of the original graphite after that, according to Ubbelohde and Lewis (1960), who isolated one atom of graphite.[1,2]

In 1962, Boehm and collaborators conducted the initial in-depth research on the few-layer graphite with decreasing GO monolayer flakes. By chemically reducing GO dispersions in diluted alkaline media with hydrazine, hydrogen sulfide, or iron (II) salts, it was discovered that the resulting thin, lamellar carbon had very little hydrogen and oxygen.[19] In a different work, Morgan and Somorjai utilized low-energy electron diffraction (LEED) to examine how different gaseous organic molecules, such as CO, C_2H_4, and C_2H_2, adsorb on top of a platinum (100) surface at elevated temperatures.[20] Upon examining the LEED data, May hypothesized in 1969 that these adsorption processes resulted in the presence of both single and multiple layers of a material with a graphitic structure.[21] Furthermore, he concluded that "the first monolayer of graphite minimizes its energy of placement on each of the examined faces of platinum," which effectively satisfied the IUPAC classification of graphene, even though the classification was not yet created.

Several investigations on the surface separation of mono- and multilayers of carbon from numerous crystalline faces of transition-metal supports, including Ni (100) and (111), Pt (111), Pd (100), and Co (0001), were published by Blakely and coworkers.[22–25] It was found by employing scanning tunneling microscopy (STM) (Figure 3.2), LEED, and Auger electron spectroscopy that the carbon dissolved in these metal alloys phases split when treated to high temperatures, generating one or more layers of carbon on the metal surface.[26]

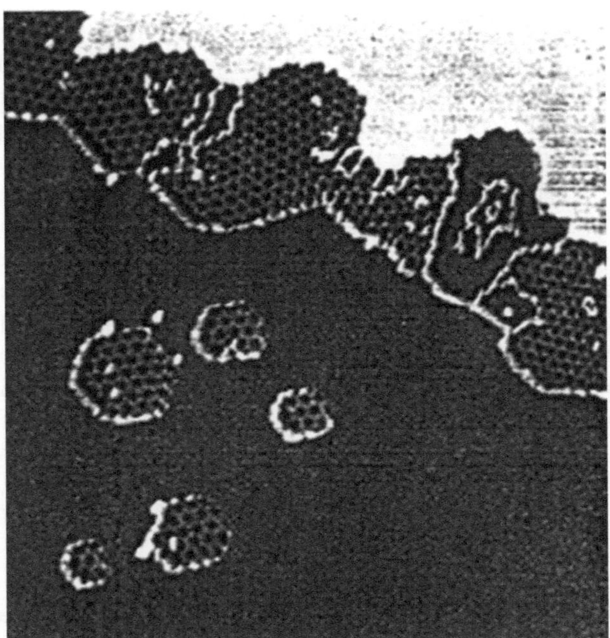

FIGURE 3.2 STM image (1000 × 1000 Å²) showing the formation of a graphitic structure on a metal surface; the image was obtained at room temperature after annealing ethylene over Pt (111) at 1230 K. (Reproduced with permission from Land et al.[26] Copyright 1992, Elsevier Science Ltd.)

The epitaxial sublimation of silicon from single crystals of silicon carbide (0001) was first documented by van Bommel et al. in 1975. By using LEED and Auger electron spectroscopy, as well as high temperatures and extremely-high vacuum (UHV; $<10^{-10}$ Torr), monolayered carbon flakes with the same structure as graphene were produced.[27] The number of layers created in multilayered carbon materials depended on the experimental conditions used; for example, at temperatures lower than 800°C, SiC generally maintained its original structure. However, at higher temperatures, "graphite rings" in the LEED pattern appeared.[27] Additionally, it was claimed that the appearance of a graphite peak coincided with the carbide peak's removal in the Auger spectrum. The authors emphasized a Badami-proposed method for graphitization in which, following sublimation of the silicon, three layers of retained carbon disintegrate over each other to produce graphitic sheets.[28] By using X-ray diffraction analysis, Badami found that in the investigations that supported this concept, the C–C distance was roughly 1.85 Å only in cases where a few layers of carbon disintegrated. The C–C distance, however, dropped to 1.42 Å with the dissolution of the third layer. The expected and empirically established bond measurements in graphene (which range from 1.41 to 1.43 Å) were both consistent with this procedure.[29]

Even while it had not yet been possible to successfully make layered graphite with fewer than 50–100 layers, mechanical exfoliation had already been used to build very thin graphite films as early as 1990. To begin with, exfoliation operations resembling drawing techniques were used to create nanoscale graphitic films. Later, it was possible to create multilayer samples with a 10 nm thickness.[1]

It is significant to highlight that Dr. Bor Jang, who holds the original graphene patent, has extensively studied the substance. He was awarded the first patent for mono-layer graphene in 2002, as well as the first patent for mono-layer graphene strengthened metal, glass, carbon, and ceramic-matrix composites and polymer composites. He currently holds more than forty patents relating to the production and application of graphene. However, because he hardly ever published any academic papers, Dr. Jang is practically unknown in the academic community.[2]

Applications for patents covering the manufacturing of graphene were first announced in October 2002 and were first granted in 2006. Geim and Novoselov ultimately isolated single-atom-thick crystallites from bulk graphite following two years of attempts. These crystallites were then put onto a silicon wafer on which the graphene is electrically insulated, or onto a thin layer of silicon dioxide (SiO_2). Graphene's peculiar quantum Hall effect was discovered for the first time using the cleavage technique. Berry's phase of the massless Dirac fermion, which directly demonstrated the existence of graphene, was theoretically predicted, theoretically stated, and published in Nature in 2005. This result was described by Geim's group, Kim, and Zhang.[1]

Micromechanical exfoliation is a novel, recently published technique for the solution of graphene, r-GO, or CMGs, in addition to epitaxial growth and chemical/thermal reduction of GO. Naturally occurring graphite, kish graphite (precipitated from molten iron),[30] and highly ordered pyrolytic graphite (HOPG) are some of the carbon sources that are available for this process. Because of its high atomic purity and smooth surface, HOPG is frequently used because the weak van derWaals forces that hold carbon layers together allow for easy delamination of the layers.[31]

Multiple layers of graphene were obtained in 1999 via a micromechanical method, although these lamellae weren't completely exfoliated into their individual monolayers.[32,33] In this technique, pillars were produced by lithographic patterning of HOPG mixed with oxygen-plasma etching, which were then transformed into thin lamellae by rubbing. By demonstrating in 2004 that thin flakes of graphene could be found by optical microscopy, and their electric-field effects could be studied when a HOPG surface was pressed against a silicon wafer surface (i.e., silicon dioxide on silicon) and then removed, Geim, Novoselov, and coworkers realized the potential of this mechanical approach.[34]

Geim and Novoselov received the Nobel Prize in Physics in 2010 for their innovative research on graphene. In 2013, the European Commission launched a massive research program with a €1 billion budget, a graphene Flagship, and 150 partner organizations. After industrial sized production of the material was shown, graphene's introduction into commerce advanced swiftly. A very recent development is the commercial development of an integrated graphene electronics device in 2017, 13 years after graphene was first produced in a lab. Pharmaceutical developers can get this product from San Diego's Nanomedical Diagnostics.[1]

3.2 VARIOUS METHODS OF FORMULATION OF GRAPHENE

As shown in Figure 3.3, there are two main methods for creating graphene: top-down (destruction) and bottom-up (construction). Graphite layers are typically separated and delaminated into single-, bi-, and few-layer graphene using top-down techniques like mechanical exfoliation,[35] arc discharge,[36] oxidative exfoliation-reduction,[37] liquid-phase exfoliation[38] and unzipping of CNT.[39] These methods essentially degrade larger predecessors like graphite and other carbon-based materials in order to create nano graphene. Although many top-down techniques are well recognized to be very scalable and to make items of the highest quality, they struggle to produce goods of a consistently high quality, have a limited output, and significantly depend on the limited source of graphite.[40]

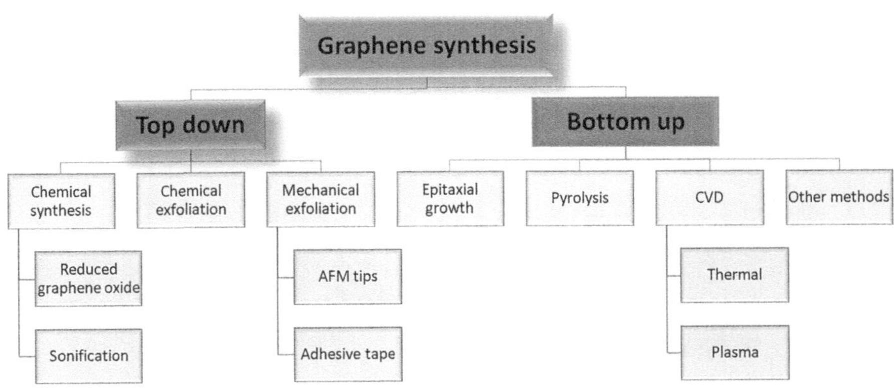

FIGURE 3.3 Schematic diagram for the different method of graphene synthesis.

In contrast, bottom-up strategies utilize carbon sources other than graphite to produce graphene and its byproducts. Graphene compounds are created using these techniques from atomically tiny precursors. Some bottom-up techniques involve chemical vapor deposition (CVD),[41] epitaxial growth,[42] substrate-free gas-phase synthesis (SFGP),[43] the template route, [44] and complete organic synthesis.[45] Despite producing graphene materials with a large surface area and almost no flaws, bottom-up approaches usually demand high manufacturing costs.

3.2.1 MECHANICAL EXFOLIATION

Mechanical exfoliation can be split into various categories based on directed approaches, such as normal force and shear force vectors.[40]

3.2.1.1 Micromechanical Cleavage

The Scotch-tape method (Figure 3.4), commonly referred to as micro-mechanical exfoliation, was developed by Novoselov and Geim in 2004 and uses sticky tape to exfoliate graphene from graphite crystals. It is one of the normal force procedures. During the initial peel, layers of graphene detach from graphite. To produce mono, bi, or a few layers of graphene, the graphene is continuously removed from the sticky tape. Finally, after the tape has been placed on a substrate, the adhesive is removed by dissolving it in a solvent like acetone.[2]

3.2.1.2 Sonification

Graphite may be successfully exfoliated under liquid circumstances by using ultrasonic methods to separate individual layers. In order to prepare monolayer or few-layer graphene sheets by exfoliating graphene, two different types of sonication methods, bath sonication and tip sonication, have been used.[38] The liquid-phase exfoliation procedure is shown in Figure 3.5.

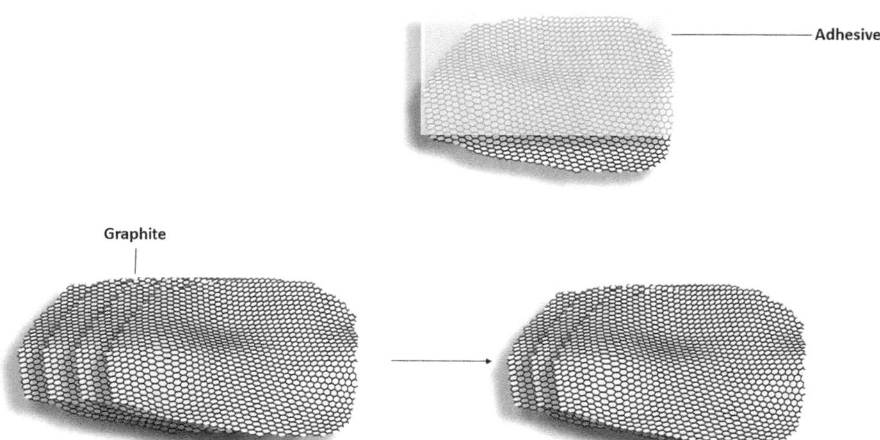

FIGURE 3.4 An illustration of the micromechanical cleavage process for producing graphene using Scotch tape.

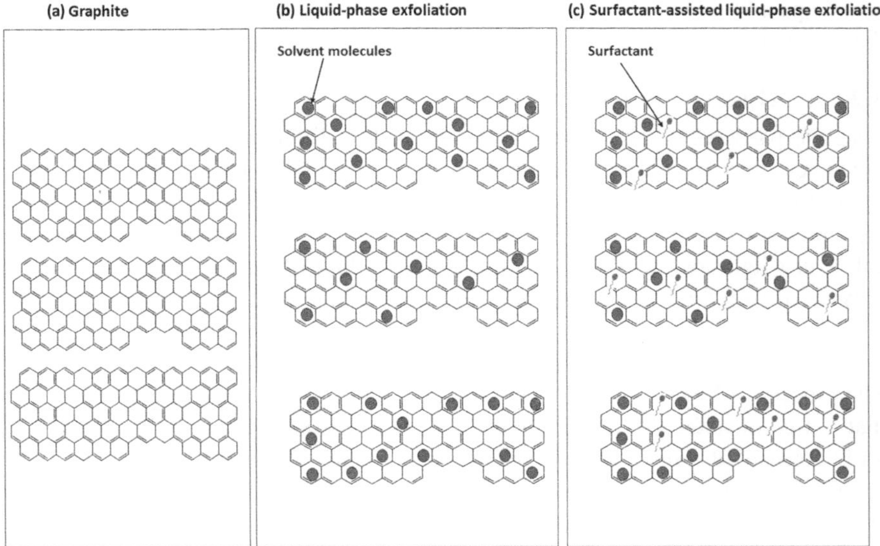

FIGURE 3.5 Schematic representation of the liquid-phase exfoliation process of graphite (a) in the absence (b) and presence (c) of surfactant molecules.

Typically, ultrasonic exfoliation consists of dispersing graphite in a solvent in step 1, exfoliating in step 2, and purifying in step 3. Flakes of graphene can be created without surfactants by chemically wetting down graphite, treating it ultrasonically in organic solvents, and then drying it. Forces of shear and cavitation, which results in the creation and burst of micrometer-sized bubbles or spaces in liquids due to pressure fluctuations, are what cause the exfoliation process during ultrasonication. Following exfoliation by the solvent-graphene interaction, the attractive forces between the graphene sheets need to be balanced. The best solvents for dispersing graphene are those that lower the interfacial tension [mN/m] between the liquid and the graphene flakes or the force that separates the two.[46]

3.2.1.3 Ball Milling

Shear force is another method for lateral exfoliation of graphite into graphene particles. Ball milling is an efficient way to generate shear force. It is a mechanical technique for dividing graphene layers from unprocessed graphite. The graphene sheets' crystalline structure is however vulnerable to dissolution in the event of normal-force-induced exfoliation caused by ball contact with the sheets' surface during milling. In this instance, fragmentation instead of exfoliation occurs.[47] The exfoliation and fragmentation effects are produced by the majority of ball milling equipment through two possible mechanisms. The most important is shear force, which is regarded as the best mechanical mechanism of exfoliation. This method is ideal for making huge amounts of graphene flakes. The second one is brought on by the balls rolling and clashing or making vertical collisions. In this way, large flakes can be split into smaller ones and crystalline structures can occasionally be destroyed and changed into amorphous or non-equilibrium phases. Consequently, it is anticipated that acquiring high-quality and huge quantities of graphene will reduce the secondary effect.[48]

3.2.1.4 Fluid Dynamics

Graphite particles can be repeatedly exfoliated at different sites because, in fluid dynamics, they may travel together with the liquid. Because of this characteristic, it differs fundamentally from sonication and ball milling and could be a useful technique for generating graphene at a large scale. Fluid dynamics has two different levels: light and heavy.[48] The three types of fluid dynamics are vortex fluidics, pressure-driven fluid dynamics, and mixer-driven fluid dynamics.

3.2.1.4.1 Vortex Fluidics Devices (VFD)

The process of VFD exfoliation uses the least amount of energy to create graphene that is free of defects.[49] In Figure 3.6, a VFD's schematic diagram is displayed. A VFD consists of a tube that has one end that is closed and the other open. A thin layer forms in the liquid as a result of the tube's rapid rotational shear stress. Small amounts of milliliters make up the thin film. The tube's velocity, alignment, and other operational factors can be changed to alter the shear fluidic film.[50]

Fast-spinning fluids, according to fluid dynamics, is the creation of boundary and shear layers known as Stewartson/Ekman layers depending on the direction of the fluid's spin. In comparison to the tube surface, the liquid surface has a downward flow direction. Graphene that has been exfoliated by Chen et al.[51] shearing vortex fluidic sheets using N-methylpyrrolidone as the solvent. First, a coating made of scattered graphite particles will form on the tube wall due to centrifugal force. The tube wall will start to experience partial lifting and sliding of the graphite layers. By now, the centrifugal force of the moving graphite particles against the wall will start a shear-triggered motion along the tube. Wahid et al.,[52] in contrast, have effectively prepared a graphene hybrid in water by employing the fluidics of vortex in continuous flow. The continuous flow model has the benefit that additional liquid can be supplied to the VFD, increasing the shear force in the film as a result of the frictional force produced as the liquid rotates the tube. As seen in Figure 3.6, the restricted

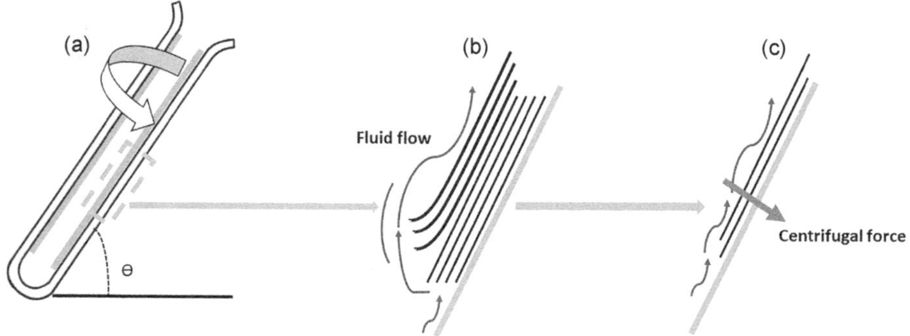

FIGURE 3.6   This diagram shows (a) the vortex fluidic device, (b) peeling off and partial lifting of the graphene layers, and (c) sliding on the inner surface of the graphene layers.

mode is first employed to transform graphite into multilayer graphene in water. The multi-layer graphene produced in the restricted mode is then injected into the liquid in the continuous flow mode of the VFD. More recirculation later, Tran and colleagues' combined prod VFD utilizing Taylor-Couette flow reactor is used to produce bulk few-layer graphene with a high yield. Pumped into the reactor, the mixture of graphite and solvent is shred between an outer cylinder that is stationary and an inner cylinder that rotates. As a result, the enormous shear and torque forces generated by the vortex movement will peal graphite flakes. This technique can successfully produce few-layer graphene with fewer defects. The capacity is limited since only 2.5 mm separates the moving and fixed cylinders.[53]

3.2.1.4.2 Pressure-Driven Fluid Dynamics

Pressure-driven fluid dynamics (PFD) is a method utilized to produce graphene on a large scale. Graphene is exfoliated by PFD using several flow channels. The pressure difference among the inflow and outflow might result in intricate fluid dynamics in the stream channels. Because the graphite particles in the PFD device can move together with the fluid in the flow channel[54] and the flow channel's size might range from millimeter to nanometer. PFD devices' compression and expansion tubes' volume can be modified; while cavitation, pressure release, viscous shear stress, turbulence, and collision are among the key fluid dynamics characteristics. These characteristics of fluid dynamics result in normal and shear forces that function as the exfoliation's driving factor. For instance, cavitation and pressure-induced normal force may cause exfoliation. Furthermore, due to viscous and Reynolds shear stresses brought on by velocity gradient and turbulence, respectively, graphite self-exfoliates to generate single- or few-layer graphene.

3.2.1.4.3 Mixer-Driven Fluid Dynamics

Mixer-driven fluid dynamics (MFD) could be used as an alternative to the sonication process for exfoliating graphene. The MFD method is effective and simple for separating poorly connected nanoparticles. To set up an MFD system, use a rotor/stator blender that is available for purchase. The mixer's rotor-based head[54,55] is the primary exfoliating component, and the diameter of the rotor can be changed as needed. The MFD technique has been employed by researchers to exfoliate graphite in the N-methylpyrrolidone dispersion media.

3.2.2 CHEMICAL EXFOLIATION

Similar to mechanical exfoliation, chemical exfoliation is a well-known method of creating graphene. It is a method that isolates few-layer graphene that is dispersed in solution and intercalates alkali metals with the graphite structure. To easily build intercalated structures with graphite at various stoichiometric ratios, alkali metals are the periodic table elements that can easily mix with it. Because they have smaller ionic radii than the graphite interlayer gap, alkali metals are able to fit inside the interlayer spacing with ease.[56]

In 2003, Kaner and colleagues disclosed the first chemical exfoliation of few-layer graphite, which eventually became known as "graphene," using potassium (K) as the intercalating agent to produce alkali metal.[57] At 200 °C and in the presence of inert helium, potassium (K) reacts with graphite to produce the chemical KC_8 (less than 1 ppm H_2O and O_2).

3.2.2.1 Thermal Chemical Vapor Deposition Process

The chemical vapor deposition (CVD) method makes it simple to produce graphene on a large scale. A high-temperature, high-vacuum, and chemically inert atmosphere are required for the CVD method (Figure 3.7) to produce graphene from extremely volatile carbon sources. Methane (CH_4), methanol, ethane, ethylene, ethanol, acetylene, polymers, and waste plastic are typical carbon precursors used in the production of graphene. As shown in Figure 3.8, there are numerous CVD subtypes, depending on the following factors:

- Utilizing activated techniques, such as heat, plasma, light, or ions, results in thermal CVD, plasma-assisted CVD, microwave (MW)-assisted CVD, photo-/laser-assisted CVD, and ion-assisted CVD, respectively. Both heat and plasma modes can be used for atomic layer deposition (ALD).
- Typically, metal halides and hydrides are utilized as precursors in traditional CVD. Metalorganic CVD was a result of using metalorganic precursors (MOCVD).
- Deposition pressure: Deposition can occur at atmospheric or normal pressure, low pressure, or ultra-high vacuum (UHV), resulting in normal pressure CVD or low-pressure CVD, respectively. Both molecular-beam epitaxy (MBE) and chemical-beam epitaxy are UHV CVD techniques (CBE).
- Deposition mechanism: In traditional CVD, the precursors react simultaneously on the substrate. In contrast, the precursors react independently in ALD.

The initial CVD approach, known as thermal CVD, was introduced in 1966 to create very crystalline graphite coatings on Ni surfaces.[54] Despite their exceptional quality and low flaw density, the preparation costs of this approach are rather significant (cost: 1–2 USD cm^2; sale price: 10 USD cm^2).[58,59]

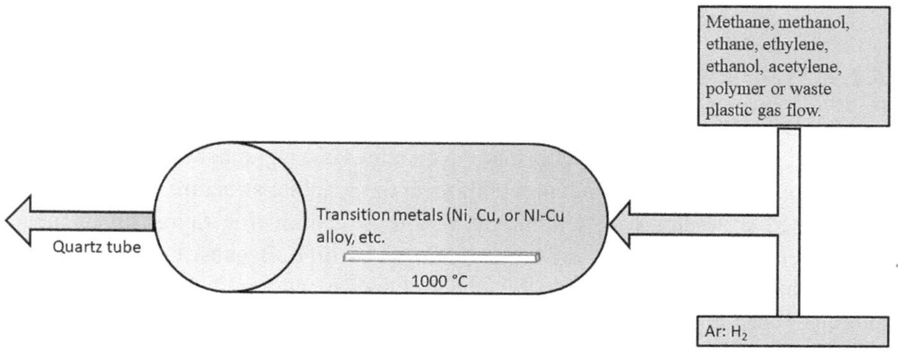

FIGURE 3.7 Chemical vapor deposition (CVD) process.

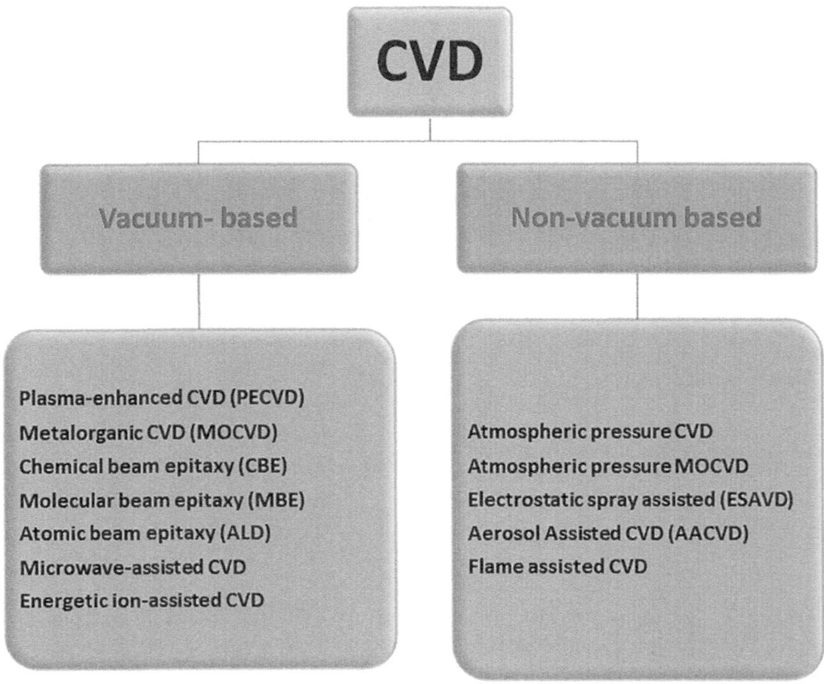

FIGURE 3.8 A variety of CVD procedures.

The most common materials used as CVD substrates are transition metals like crystalline Ni,[60] Cu,[61,62] or CuNi alloys, as well as Ir,[63,64] Ru,[65] Pt,[66] Co,[67] Ti,[68] and Re.[69] Crystalline Ni, Cu, or Ni-Cu alloys yield high monolayer graphene coverage and are incredibly affordable when compared to other substrates.[70–72]

3.2.2.2 Plasma-Enhanced Chemical Vapor Deposition (PECVD)

A well-known French synthetic chemist named Marcellin Berthelot designed the entire PECVD method in 1869 for the breakdown of gases such as CH_4 in glow discharge, a reaction that is the norm of PECVD today. In 1876, according to Ogier J., PECVD methods allowed Si:H and SiNx films to develop from SiH_4 and $SiH_4 + N_2$ precursors, respectively.[73,74] Because graphene can be created without a catalyst at a relatively low temperature, this method is more useful for large-scale industrial applications.[75] Even when gas-phase precursor materials are utilized, the cost is high. This is how graphene sheets were first produced.[76] On a number of substrates, such as Si, SiO_2, Al_2O_3, Mo, Zr, Ti, Hf, Nb, W, Ta, and 304 stainless-steel, PECVD is utilized to produce monolayer and thin layers of graphene. Instead of using thermal energy, as in the thermal CVD process, the PECVD technique uses electrical energy to create plasma. Neutral species, ions, and electrons are all mixed together to form plasma. Non-isothermal and isothermal plasma are two categories of plasma.[74]

The gas, plasma generator, and vacuum heating chamber make up the three basic components of the PECVD experimental arrangement, as shown in Figure 3.9 PECVD growth also makes extensive use of other configurations that pair plasma

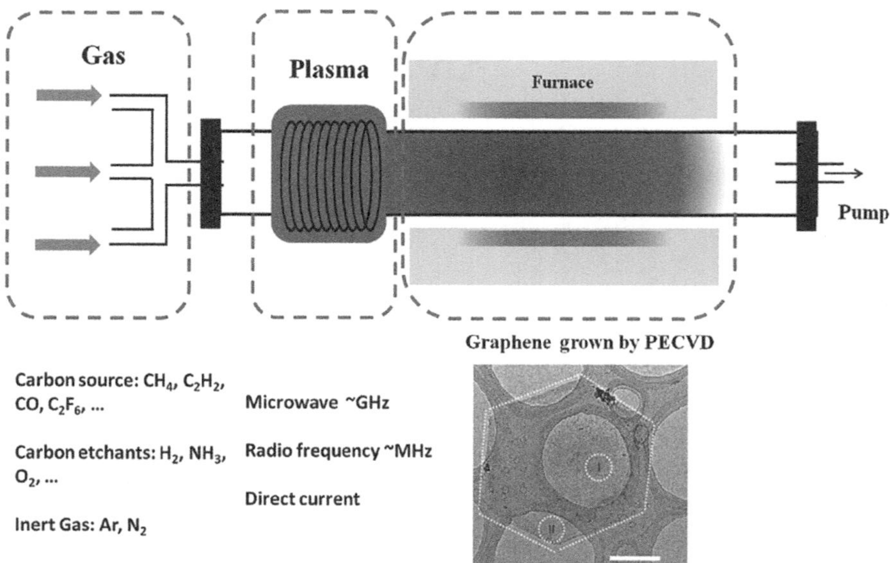

FIGURE 3.9 Schematic of experimental setup for PECVD, including gaseous system, plasma generator system, and vacuum heating system. (Reproduced with permission from Menglin Li et al.[77] Copyright 2016, Wiley.)

generators with growth chambers. The PECVD's central component, the plasma generator, can be broadly categorized into three groups based on the power source used to generate plasma: microwave (MW) plasma (typically operating at 2.45 GHz), radio frequency (RF) plasma (typically operating at 13.56 MHz), and direct current (DC) plasma. High frequency electromagnetic energy in the GHz range is what the MW plasma is. Graphene and its byproducts, including diamond films, nanowalls, and CNTs, have all been extensively synthesized using MW-PECVD. With a domain frequency in the MHz region, RF plasma is another well-liked source. The electromagnetic evanescent, wave propagation and electrostatic modes are the three basic ways that an RF generator's energy is connected to the plasma. With a higher energy density and a larger plasma volume, electromagnetic mode inductively coupled plasma (ICP) offers the benefit of producing high growth rates. In electrostatic mode, on the other hand, the comparatively low energy of capacitively connected plasma prevents its usage as a standalone plasma generator. DC plasma, another commonly utilized source because of its straightforward setup. Designs for DC glow plasma using parallel plates and pin-to-plate geometry can create sources that are uniform or nonuniform, respectively.[77]

Even when gas-phase precursor materials are utilized, the cost is high. This is how graphene sheets were first produced.[76] On a number of substrates, such as Si, SiO₂, Al₂O₃, Mo, Zr, Ti, Hf, Nb, W, Ta, and 304 stainless-steel, PECVD is utilized to produce monolayer and thin layers of graphene. Graphene and its byproducts, which may be divided into three functional groups, must be created from gaseous species: (i) Carbon radicals are produced via plasma-enhanced reactions from gaseous precursors that include carbon, which is then used to build graphene. (ii) To manufacture superior graphene and its byproducts, gases like H_2 and O_2 are combined as amorphous carbon etchants. (iii) As-grown graphene is frequently doped with gases like N_2 and NH_3, which allows for precise control of the material's electrical characteristics.[77]

The growth parameters have been optimized using simulations of the plasma process. On the basis of various gas systems, numerous plasma models have been created.[78–80] These models take into account a large number of species (ions, electrons, neutrals, and radicals) and reactions. Eight neutrals, eleven ions, and five radicals have been taken into consideration for the CH_4 or CH_4/H_2 plasma in a 1D fluid model.[81] This model also includes 27 electron processes, seven ion-neutral events, and twelve neutral-neutral reactions. Based on 1D fluid models, it has been discovered that the densities of radicals and ions in the plasma change with distance. A 2D fluid model is proposed that extends to the radial directions.[81] The simulation findings demonstrate that the electrode fringes, where the potential V changes drastically, are where the electron density reaches its greatest. As a result, electron-related events take place more frequently in these areas, resulting in the creation of high-energy ions and neutral densities. The power, gas mixture ratio, gas flow, and pressure all have an impact on the ion and radical densities. To comprehend the temperature-dependent growth kinetics of PECVD graphene production on cobalt substrate, two complimentary modeling approaches (1D and 2D) were established by Hinkov and colleagues. By utilizing gas-phase and surface reaction mechanisms, which comprise of 15 gas species, 43 gas reactions, 10 surface species, and 34 surface reactions, the gas temperature and species concentrations are predicted for various procedural parameters. It is determined how the growth temperature, microwave power, and CH_4 flow rate affect the mole fractions of the different gas species in the reactor and the surface coverage. The computer results unmistakably showed that hydrogen atoms are crucial to the development of graphene in microwave plasma systems. Then, in order to comprehend the fundamental processes that result in the movement of reactants across the boundary layer by gas diffusion from the main gas stream, which offers understanding regarding the formation of graphene on the cobalt substrate, a global sensitivity analysis was carried out.[82]

Vertically orientated (VG) graphene has stimulated significant interest for many years, although the development mechanism is not yet completely understood. Although there is indirect evidence that the electric field is involved, it is still unclear exactly what part it plays. At a methodical investigation, Sun et al.[83] discovered that VG development preferentially takes place in regions of plasma-enhanced chemical vapor deposition if there is a larger local field, such as at GaN nanowire tips. As opposed to developing perpendicular to the substrate on almost spherical nanoparticles, the VG expands along the field direction, which is perpendicular to the local surfaces of the particles. Persuasively, the sheath field was subjected to varying degrees of screening, and a direct link between field strength and VG growth was found. According to numerical analysis, as graphene grows, the field aids in the charge accumulating process, which finally transforms the cohesive graphene layers into distinct 3D VG flakes. The field also assists in attracting charged precursors to protrusions in the substrate, where they are sharpened and transformed into VG.

The study has served as a foundation for further research into the mechanisms driving vertical two-dimensional (2D) material growth.[83]

3.2.2.3 Epitaxial Growth

Epitaxial heat growth on a single crystalline silicon carbide (SiC) surface is among the most celebrated methods for making graphene. The prefix "epi" means "over" or "upon," while the word "taxis" denotes "order" or "arrangement" in the Greek word "epitaxy." When a single crystalline layer is applied on a substrate made of just one kind of crystal, epitaxial growth takes place, creating an epitaxial film. The primary process, in this instance, is the desorption of atoms from a SiC sample's annealed surface. The silicon atoms desorb at high temperatures because carbon has a far lower vapor pressure than silicon does, leaving the carbon atoms behind to form graphitic layers, also known as few-layer graphene (FLG).

When using other heating methods, such as resistive heating or e-beam heating, the result is the same. The heating procedure is done in a vacuum to prevent contamination. The molar densities indicate that at least three bilayers of SiC are necessary to free enough carbon atoms to produce one layer of graphene. As a result, it produces high-crystalline graphene on substrates made of single-crystalline SiC. Both homo-epitaxial growth and hetero-epitaxial development are common epitaxial growth mechanisms, depending on the substrate. A hetero-epitaxial layer has a film on a substrate made of a diverse material then a homo-epitaxial layer, which has a film on a substrate made of the same material.[11]

3.2.2.4 Pyrolysis

The Solvo thermal process was used to create graphene chemically in a bottom-up manner. Throughout this heat reaction, in a closed vessel, the molar ratio of sodium to ethanol was 1:1. Sheets of graphene can be easily separated by pyrolyzing sodium ethoxide with sonication. As a result, graphene sheets up to 10 m in size were created. A large D-band, a G-band, and an IG/ID intensity ratio of 1.16 were all visible in the resulting sheet's Raman spectroscopy, all of which are signs of flawed graphene. High-purity, functionalized graphene was easily made using this technique at low temperatures. However, the graphene's quality was unsatisfactory, due to a number of defects.[84] Another important method for producing graphene is the thermal breakdown of silicon carbide (SiC). Silicon desorbs at high temperatures, leaving carbon behind to form a thin coating of graphene.[85] This approach has been greatly enhanced by the continuous production of mm-scale graphene sheets at a temperature of 750 °C on a thin film of nickel plated on SiC substrate.[86] The benefit to this approach is the continuous synthesis of graphene films throughout the entire SiC coated surface. However, with this approach, a large-scale synthesis of graphene is not feasible. At 1000 K, ethylene thermally decomposes in a similar manner. One advantage of this synthesis method is its capacity to synthesize high purity graphene mono-layers.[87]

3.2.2.5 Unzipping Method

A carbon nanotube (CNT) is unzipped, employing chemical and plasma-etched methods (CNT). A long, thin graphene strip with sharp edges is known as a "graphene nano ribbon". The transition from a semimetal to a semiconductor electronic state depends on the nanotube's width.[88] Multi-layer or single-layer graphene can be produced depending on whether the starting nanotube has many walls or just one. The width of the resulting nanoribbons is determined by the diameter of the precursor nanotubes. Ammonia and lithium combine to form multi-walled carbon nanotubes (MWNTs) (NH_3). Unzipping methods that have been used to study the production of graphene and by products from CNT, include chemical assault,[89] plasma etching,[90] intercalation and exfoliation,[91] and metal-catalyzed cutting.[92] The first steps in the chemical unzipping of CNT to rupture the axial-directional C–C bond are H_2SO_4 treatment and $KMnO_4$ oxidation.[89] H_2SO_4 treatment initially breaks the axial-directional C–C bond of the CNT, and then $KMnO_4$ oxidation follows.[89] This method destroyed a significant quantity of precursor while producing a low yield of graphene.[93] MWCNT was intercalated with oxalic acid prior to chemical unzipping in order to boost yield. Because their size (0.34 nm), oxalic acid molecules can easily intercalate in the area (0.38 nm) between two MWCNT layers.[94] In another study, a two-step electrochemical method was used by Shinde and colleagues, Figure 3.10, to turn CNTs into nanoribbons made of a few layers of graphene.[95] The MWCNT working electrode in 0.5 M H_2SO_4 was first given a standard

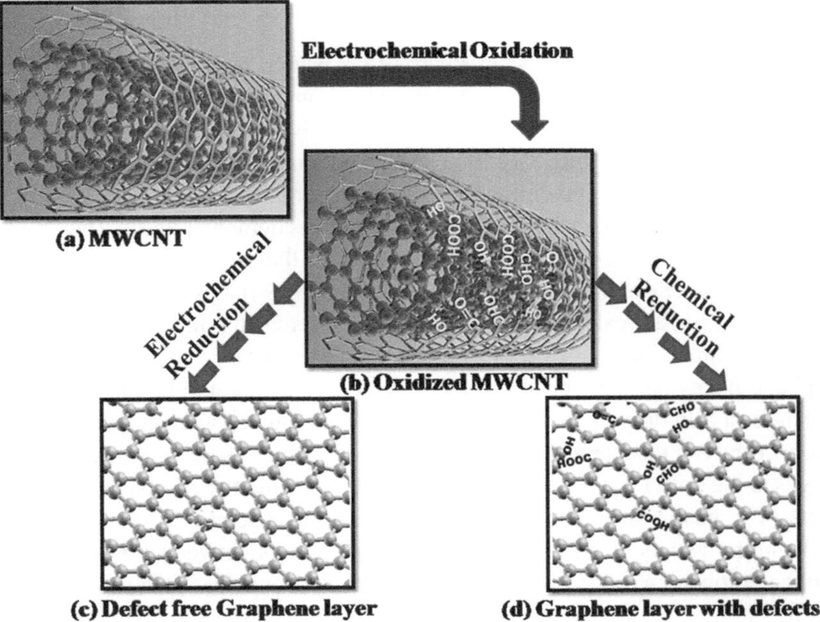

FIGURE 3.10 Diagrammatic Representation of the Electrochemical Transformation of GNRs from MWCNTs: (a) Pristine MWCNT; (b) MWCNT Deposited on Glassy Carbon Electrode after Oxidation to Generate Functional Groups on Edges under Controlled Potential So That It Gets Broken; (c) Electrochemical and (d) Chemical Reduction to Graphene Layers. (Reproduced with permission from Shinde et al.[95] Copyright 2011, American Chemical Society.)

anodic potential of 0.70 V vs mercury|mercury(I) sulfate (MMS) electrode, wherein an applied electric field started the dissociation of sp^2 carbon bonds, possibly at the tip of the MWCNT. Along the longitudinal axis, this persisted. Due to the stress created by the curved surface, fragmented MWCNTs in a horizontal plane were forced further apart, which may have contributed to their amazing conversion into graphene oxide layers. To fully explain the intermediates generated following C–C breakage, this hypothesized theory, however, requires additional experimental support. The researchers were able to construct thin nanoribbons using the same electrochemical unzipping procedure in single-walled carbon nanotubes (SWCNTs), and yet their following disentanglement was more challenging. It was discovered that this type of "unzipping" of CNTs under the influence of an interfacial electric field had special advantages with regards to the alignment of CNTs, which may enable the fabrication of graphene nanoribbons with controllable widths and fewer flaws. Although the electrochemical approach to graphene nanoribbons documented had the benefit of allowing for the control of edges and planes along the length and the tuning of orientation during the first oxidation step, some of the electronic properties may be impacted by intercalation or the adsorption of cations, anions, and solvent molecules on the defect site. However, this research offered fresh avenues for producing high-quality graphene in a good yield, and it had significant ramifications for a number of applications.[95]

3.2.2.6   ARC Discharge Method

Arc discharge mostly occurs when electricity is passed between two graphite electrodes and has been employed to produce carbon nanoparticles. This technique allows for the controlled production of graphene sheets.[96] A typical arch discharge method is depicted in Figure 3.11. In a study by Kim and colleagues, a method used in the experiments for fragmenting 2D multi-layer graphene sheets (MLGs) employing toluene droplets via an arc discharge method was demonstrated.[97] They revealed that the interaction between toluene and water, or the incorporation of the MLGs inside the toluene droplets. The outside of the granules (having a diameter of about 600 nm) that were created, appeared wrinkled, whereas the hollow structure of the spherical granule was made up of overlapping edges across a thin, consistent layer. Graphene flakes had also been produced with no aid of catalytic processes by arc discharge in a hydrogen atmosphere of modest pressure. Layer graphene has been generated by arc discharge utilizing a variety of buffer gases.[98] Because the hydrogen gas in the buffer gas can rupture the bonds connecting the carbon atoms together, sheets of graphene are unable to roll up. Rao et al. discovered that graphite rods can be employed as electrode materials with potentials between 100 and 150 A to produce graphene nanosheets in a mixed H_2/He environment.[99] Then, Wang et al.[100] created a more effective arc discharge technique for creating graphene nanosheets in the air in replacement for employing a mixed H_2/He atmosphere. High pressure was found to favor the synthesis of graphene nanosheets, whereas low pressure favored the synthesis of additional carbon nanostructures such as carbon nano-horns and nanospheres.

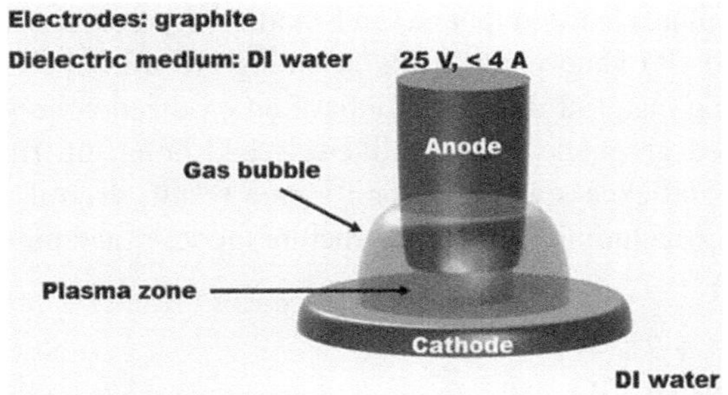

FIGURE 3.11 The aqueous arc discharge process was developed to produce a controlled number of graphene layers and degree of oxidation-including functional groups, as well as pore formation by simply adjusting the arc discharge power. Interestingly, the resulting graphene sheets showed enhanced water filtration performance in terms of a decrease in the ion diffusion and adsorption of organic dyes. (Reproduced with permission from Kim et al.[101] Copyright 2016, Elsevier Science Ltd.)

3.3 GO

The substance GO is two dimensional (2D). It is graphene in its oxidized state, with O functional groups adorning the sp^2 C basal plane.[102] By merely removing the functional groups from its surface, all of its physical properties can be adjusted from those of fully oxidized GO to, roughly, those of graphene. Through this procedure, it can change from an insulating substance to a semi-metal. In contrast to graphene, which is hydrophobic, since GO contains oxygen, it is hydrophilic and may be distributed in an aqueous solution.[15] A few nm to mm can be used to regulate and vary the size of GO flakes. This might impede research and development for real-world applications, particularly for large-area conductive thin films. By pre-exfoliating the graphite using microwave heating recently, Suet al.[103] successfully produced enormous mono-layer GO sheets (up to 2000 m^2), which gives rise to the possibility that additional oxidation processes aid in the exfoliation of layers of graphite flakes. GO is a desirable product due to the chemical composition and flake size tunability.[104] It exhibits noticeably more chemical activity, does not absorb visible light, and has a very low electric conductivity in comparison to graphene.[104]

GO, in line with the most popular Lerf-Klinowski (LK) model,[105,106] is composed of two randomly positioned domains: domains with sp^3-hybridized and oxidized carbon, as well as pure graphene with sp^2-hybridized carbon atoms. GO has more oxygen-containing functional groups on its sheet layer than graphene, altering its structure and changing its properties.

Evidently, the carboxyl and carbonyl groups were added close to the border of the single layer, while the hydroxyl and epoxy groups were unevenly positioned across the GO monolayer. There were two ring regions in GO: a benzene ring that is not an oxidized area and an oxidized aliphatic six-membered ring region. The ratio of these two zones' diameters is dependent upon the level of oxidation and uneven distribution. In contrast, the L-K structural model relies on a number of assumptions and disregards the impact of oxidants, raw graphene, and oxidation processes. Erickson and colleagues examined GO nano-plates using scanning electron microscopy (SEM) and discovered that the material did not only have an oxidized zone with high disorder and an unoxidized graphene area but likewise had hole imperfections caused by lamellar peeling and excessive oxidation.[107] As a result, several models have been published recently, including the binary structure model[108] and the dynamic structure model (DSM).[109,110]

3.3.1 Synthesis of GO

The most current techniques for producing GO rely on the apparent intercalation capacity of layered graphite. It increases the distance between flat carbon layers of crystalline graphite, allowing active metal atoms and several kinds of oxidizing agents to pass through, while also changing the exterior of the layers with chemically bound functional groups. Lastly, the complete disassembly of the graphitic crystal into distinct carbon monolayers with chemically changed surfaces is brought about by the action of suitable oxidizing agents. By Brodie, Staudenmaier, and Hoffman, potassium chlorate and strong acids (nitric and/or sulfur) are used in the synthesis of GO. The Hummers—Offeman method, which is the most often used, combines concentrated H_2SO_4, $NaNO_3$, and $KMnO_4$. Moreover, it yields well and takes a shorter period of time than earlier techniques. The finished product of the GO synthesis procedure is an aqueous solution of GO that is yellow in color.[111] Modern developments in GO procedures are built on Hummers' oxidation and exfoliation of graphite.[112] Known techniques for generating graphite oxide are examined in detail in Singh et al.[113] It is crucial to keep in mind that low (10 μm) starting material particle sizes are often required for the chemical oxidation of graphite.[114] Other synthesis methods are summarized in Table 3.1.

Chemical reactions typically require vigorous stirring or even ultrasonication to be completed. These elements make it more challenging to find large-sized GO flakes. Because of this, the average particle size in commercially available GO does not go above a few microns. The mean size of the flakes, however, has an important effect on the primary characteristics of GO and the collaborative characteristics of macroscopic structures created by GO (such as lamellar films). As a result, it is still difficult for technology to increase the size of produced GO particles. To create large and extra-large sized GO platelets (LGO and ULGO), a recent study has suggested improving the GO preparation process.[115] The purpose of the procedure was to divide the GO production process into two steps: first, a liquid medium was utilized to gently and chemically exfoliate the initial graphite into graphene flakes, and then, in moderate conditions, the resulting product was exposed to oxygen. At room temperature, efficient exfoliation of graphite was possible, in particular when crystalline graphite, chromium trioxide, and hydrogen peroxide were mixed.[116] The process conditions encouraged maintaining the purity of manufactured graphene flakes. Thus, vast graphene flakes can be produced uniformly over a large area by further oxidizing the obtained material using potassium permanganate at a lower concentration (compared to the conventional Hummers technique).

TABLE 3.1

Techniques for Preparing GO

Methods	Carbon Source	Oxidants	Reaction Time for GO	Temperature (°C)	Features
Brodie, 1859[14]	Graphite	$KClO_3$, HNO_3	3–4 days	60	Earliest method
Staudenmaier, 1898[118]	Graphite	$KclO_3$, HNO_3, H_2SO_4	96 h	RT	Improved efficiency
Hummers, 1958[119]	Graphite, ~44 μm	$KmnO_4$, $NaNO_3$, H_2SO_4	<2 h	<20-35-98	Water free, less than 2 h processing
Fu, 2005[120]	Graphite	$KmnO_4$, $NaNO_3$, H_2SO_4	<2 h	35	Validated $NaNO_3$ unnecessary
Shen, 2009[121]	Graphite colloidal, ~10 μm	Benzoyl peroxide (BPO)	10 min	110	Fast and non-acid
Su, 2009[103]	Sonicated graphite, <3000 μm	$KmnO_4$, H_2SO_4	4 h	RT	Large size GO
Marcano, 2010 & 2018[112]	Graphite, ~150 μm	H_2SO_4, H_3PO_4, $KmnO_4$	12 h	50	Bi-component acids, high yield
Sun, 2013[122]	Expanded graphite	$KmnO_4$, H_2SO_4	1.5 h	RT-90	Size-confined high yield, safe
Eigler, 2013[123]	Graphite, ~300 μm	$KmnO_4$, $NaNO_3$, H_2SO_4	16 h	10	High-quality GO
Chen, 2015[124]	Graphite, ~3–20 μm	$KmnO_4$, H_2SO_4	<1 h	<20-40-95	High yield
Panwar, 2015[125]	Graphite	H_2SO_4, H_3PO_4, $KmnO_4$, HNO_3	3 h	50	Tri-component acids, high yield
Peng, 2015[126]	Graphite, >10 μm	K_2FeO_4, H_2SO_4	1 h	RT	High-yield, less pollution
Rosillo-Lopez, 2016[127]	Defective arc-discharge carbon	HNO_3	20 h	RT	Nano-sized GO
Yu, 2016[128]	Graphite, ~44 μm	K_2FeO_4, $KmnO_4$, H_2SO_4, H_3BO_3	5 h	<5-35-95	Less manganite impurity, less acid, high yield
Dimiev, 2016[129]	Graphite	$(NH_4)_2\ S_2O_8$, 98%H_2SO_4, fuming H_2SO_4,	3–4 h	RT	Lightly oxidized, 25nm thick, ~100% conversion
Pei, 2018[130]	Graphite foil	H_2SO_4	<5 min	RT	Electrochemistry support; high efficiency and high yield
Ranjan, 2018[131]	Graphite	H_2SO_4, H_3PO_4, $KmnO_4$	>24 h	<RT-35-95	Cool the exothermal reaction to keep safe.

Source: Reproduced with permission from L. Sun,[117] Copyright 2019, Elsevier Science Ltd.
Note: RT: abbreviation of room temperature.

3.4 MOLECULAR SKELETON OF SINGLE LAYER GO

Before going on to a study of its properties, it is essential to provide a description of GO based on current research methodologies. Both the basal (hydroxyl and epoxy) and edge planes of the carbon monolayer that makes up GO comprise (essentially) sp^2- and (partially) sp^3-hybridized carbon atoms that have functional groups that contain oxygen (carboxyl, carbonyl). While sp^3-hybridized carbon atoms can be thought of as oxidized regions with covalently attached oxygen-containing functional groups, sp^2-hybridized carbon atoms can be thought of as unoxidized regions. High-resolution transmission electron microscopy (HRTEM) was used to directly observe lattice atoms and topological defects in a GO monolayer, and it was shown that these sp^3-hybridized carbon clusters are positioned slightly above or below the plane of sp^2-hybridized carbon atoms.[132–134]

In order to suspend the GO above the vacuum for increased contrast and atomic resolution imaging, samples were heated using a platinum coil identical to that shown in Figure 3.12a.[135] Slits were then made in the silicon nitride thin film employing a concentrated ion beam, as seen in Figure 3.12b. GO is wrinkled and appears evenly

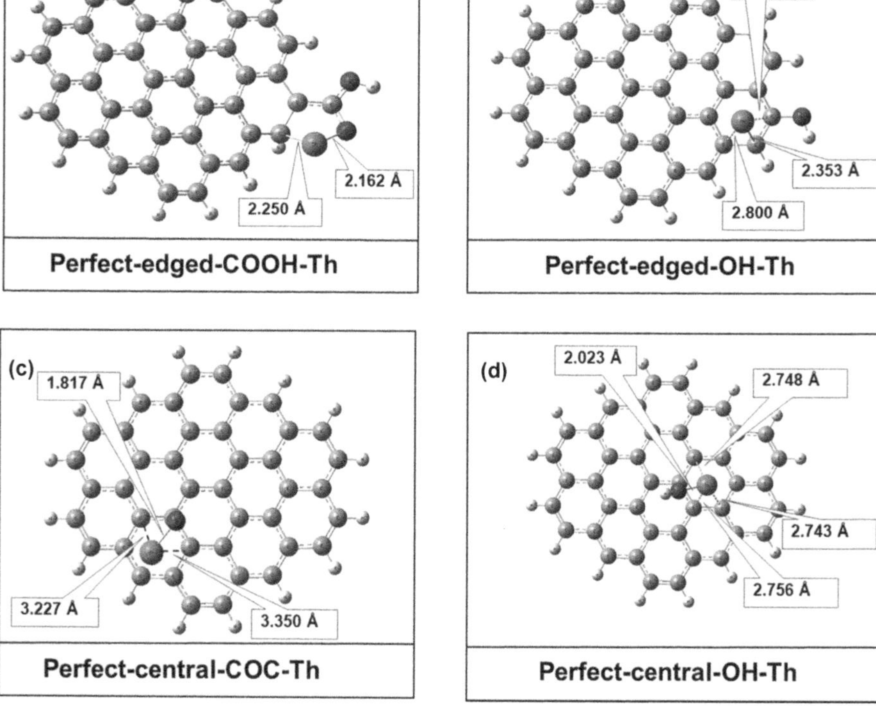

FIGURE 3.12 Different structures of the oxygen functional group on the perfect graphene surface. (a) Perfect-edged-OH-Th⁴⁺, (b) perfect-edged -COOH-Th⁴⁺, (c) perfect-central-COC-Th⁴⁺, and (d) perfect-central-OH-Th⁴⁺. (Reproduced with permission from Gao et al.[135] Copyright 2022, Elsevier Science Ltd.)

clustered across numerous samples, despite the occurrence of monolayer sections, as shown in Figure 3.12c and d. The hexagonal lattice's direct imaging is not possible at ambient temperature or in the high vacuum of the TEM due to the existence of an amorphous substance covering the GO surface, which response quickly under the beam's energy. When the sample is heated to 500 °C, it crystallizes more and reveals the GO atomic structure. In Figure 3.13a, a portion of GO was demonstrated in false color to

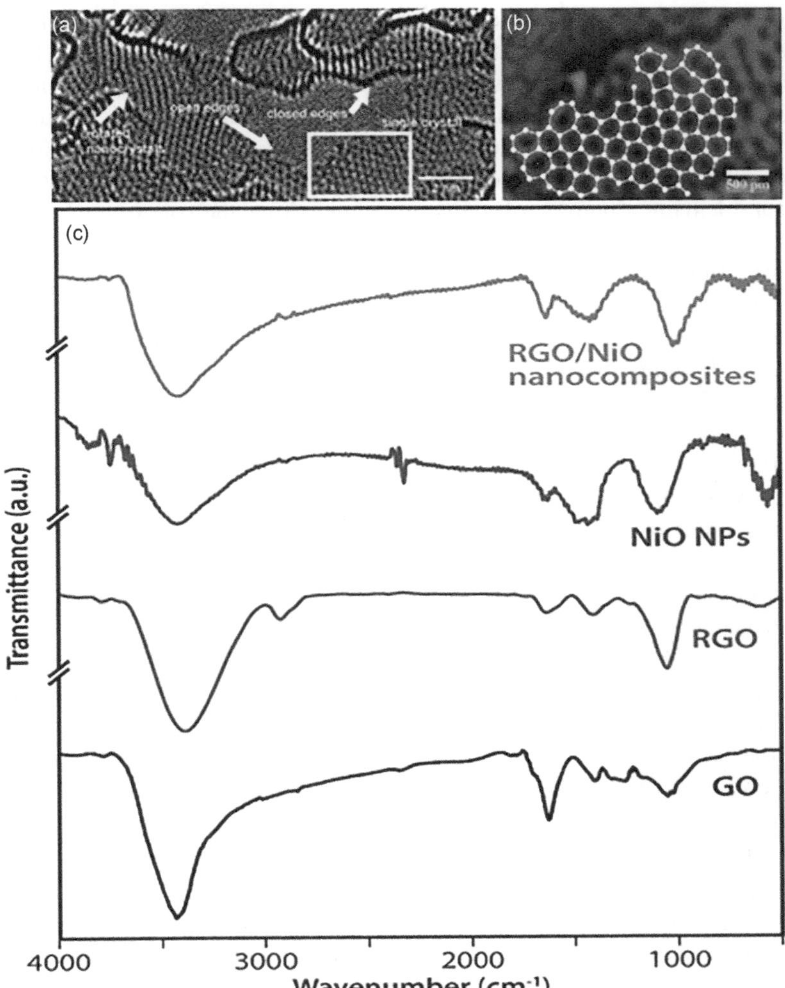

FIGURE 3.13 Atomic structure can be resolved using in situ heating with AC-TEM at 700 °C with features shown in (a). Sufficient resolution is achieved to identify defects in the bond structure (b). In (b), not all of the atomic structure could be accurately resolved, likely due to structural changes occurring during image acquisition, and only the directly resolved atomic structure is indicated by white dots and lines,[136] and (c) FT-IR spectrum of GO, RGO, NiO NPs, and RGO/NiO nanocomposites. (Reproduced with permission from Sadhukhan et al.[137] Copyright 2020, Elsevier Science Ltd.)

increase contrast. The margins of the (folded over bilayer, high contrast) structure were apparent as well as the regions of disorder that isolated the nano-crystallites from one another. Due to their stark contrast, the closed edges may be distinguished from heavier atoms that might have contaminated the sample. The atomic composition of a patch of monolayer GO is depicted in Figure 3.13b at 700 °C with enough clarity to demonstrate the hexagonal sp^2 lattice's flaws (pentagons and heptagons).[136]

In another report, isolated graphene sheets were floating freely in a vacuum or air on a microfabricated scaffold. Despite being a single atom thick, these membranes still exhibited long-range crystalline organization. The study using TEM has also shown that these suspended graphene sheets are not completely flat; rather, they have tiny roughening that causes the surface normal to vary by several degrees and out-of-plane deformations to approach 1 nm. An extended 2D carbon "honeycomb" lattice with sp^2 bonds makes up graphene. The hydrocarbon adsorbates that are frequently observed during TEM imaging desorb during heating and subsequent electron beam exposure, leaving a pure graphene sheet behind. After the parent graphene is oxidized, the resultant GO material is discovered to be extremely uneven in terms of structure. Holes, graphitic regions, and high contrast disordered regions, indicating areas of high oxidation, were present. These three features, with approximate area percentages of 2%, 16%, and 82%, respectively, are the three main features. GO is known to develop holes as a result of the severe oxidation and sheet exfoliation, which[107] typically are found to be less than 5 nm^2 in size.

GO samples have not yet been the subject of any high resolution electron microscopy research, although a high resolution electron microscopy with aberration correction examination of the reduced version of GO has found nearly identical holes.[138] With the maintained honeycomb structure and 1.4 nm atomic spacing of graphene, graphitic areas covering 1 to 6 nm^2 show partial oxidation of the basal plane.[134] The GO sheet is covered in a continuous network of the disordered, highly contrasted oxidized areas of the basal plane.[107]

In a study of the thinnest suspended graphene membrane, TEM images of the produced membranes' central portions typically showed homogenous, featureless zones, but the edges of the membranes typically scroll (Figure 3.14a). Additionally, folded areas where a graphene sheet separated partially from the scaffold during microfabrication were noticed.[139] For the quantity of graphene layers, these folds produce a TEM signal that is extremely clear. In the case of monolayer graphene, the only indication of a folded sheet is a solitary black line that is spatially parallel to the electron beam, which is comparable to the TEM images of one-half of a single-walled carbon nanotube (SWNT).[139] By using electron diffraction, this sample's two layers were determined. These membranes offered no discernible absorption in the bright-field TEM pictures and were only discernible as phase contrast at a significant enough defocus.[140]

A single layer graphene membrane is demonstrated in Figure 3.14b. According to electron diffraction, the areas shown by the arrows are distinct layers of carbon atoms that are only connected at their edges and hung in space. The incident beam and objective aperture in this dark-field TEM image were adjusted to select only electrons that were dispersed by a modest angle by tilting the primary beam to just

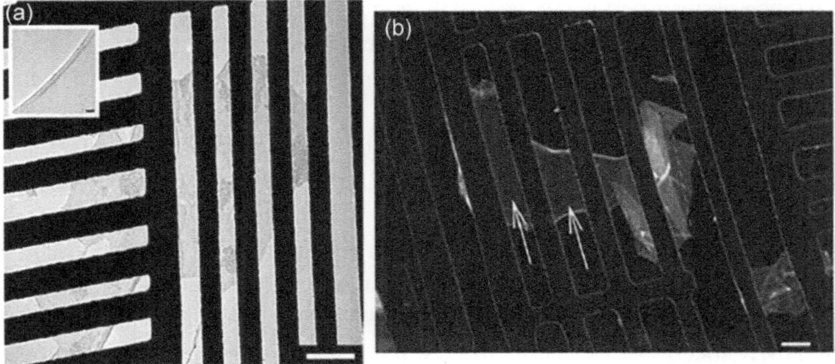

FIGURE 3.14 (a) TEM image of suspended graphene supported by a microfabricated metal grid (darker gray areas) (black lines). An edge scroll is visible in the inset. Scale bar is 1 m, and the inset is 20 nm. (b) Small angle dark-field TEM image of a single-layer graphene membrane. The dark grey area indicated by the white arrows is the single-layer region, as proved by electron diffraction measurements. Under these imaging conditions, the intensity is proportional to sample thickness. The right part of the flake is folded, and indeed, the recorded intensities in the folded areas are precisely integer multiples of the intensity in the single-layer area. (Reproduced with permission from Meyer et al.[140] Copyright 2007, Elsevier Science Ltd.)

outside the aperture. An image showing the thickness was obtained for a single element sample, as no Bragg reflections were chosen in this manner. As a result, the folded areas' gray levels are integer multiples of the area with a single layer. The imaging method is likewise highly sensitive to surface adsorbates.[140]

The chemical makeup of GO and functional groups that contain oxygen can be ascertained using spectroscopic research techniques such solid-state nuclear magnetic resonance (NMR), Raman spectroscopy, X-ray photoelectron spectroscopy (XPS), Fourier transform infrared spectroscopy (FTIR) and ultraviolet-visible spectroscopy (UV-Vis). With the help of FTIR (Figure 3.15a), it is possible to identify

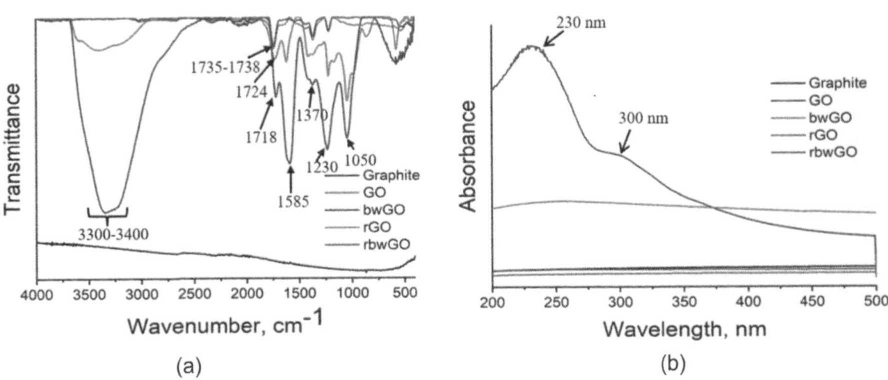

FIGURE 3.15 (a) FTIR and (b) UV-Vis spectra of GO, base-washed GO, and their reduced forms. (Reproduced from Aliyev et al.[141] Copyright 2019, the Authors.)

the existence of absorbed water, carboxyl, and hydroxyl groups (broad band at 3000–3800 cm^{-1}), aliphatic groups (at 2700–2980 cm^{-1}), carbonyl groups (at 1650–1750 cm^{-1}), C=C (1500–1630 cm^{-1}), hydroxyl groups (at 1300–1480 cm^{-1}), and epoxy groups (at 1000–1280 cm^{-1}).[141]

UV-Vis studies (Figure 3.15b) show that the primary optical absorption of GO takes place at a wavelength of about 230 nm. It is a part of the C=C chromophore peak and nanometer-scale sp^2 cluster of the π-π* plasmon. The auxochromes carboxyl, other carbonyl, and hydroxyl have an *n*-π* plasmon peak that is connected to the shoulder peak at 300 nm. The oxygen functions of reduction and substitution provide an explanation for the absence of peaks in the remaining samples.[141]

The NMR spectra of GO show three main lines: a signal at around 60 ppm allocated to carbon atoms coupled to epoxy functional groups, a signal at about 70 ppm belonging to carbon atoms attached to hydroxyl functional groups, and a signal at roughly 130 ppm attributed to sp^2 carbon. Three further faint signals at 101, 167, and 191 ppm are caused by ketone, carbonyl, and lactone groups.[141]

The XPS technique can clearly identify the type of bonds between carbon and oxygen in their many forms, including carbon atoms that have not been oxidized (sp^2 carbon), C–O–C, C=O, and COOH (Figures 3.16 and 3.17). Five separate chemical components that make up the C1 s signal of GO ascribed to sp^2 carbon atoms (286.79 eV), carbon atoms connected to hydroxyl (C–OH, 286.93 eV), epoxy (COC, 286.55 eV), carbonyl (C=O, 288.64 eV), OH (532.80 eV), CO–OH (532.10 eV), and carboxyl groups.[141,142]

A prominent G peak may be found at about 1594 cm^{-1} in the Raman spectra of GO and all sp^2-hybridized oxidized carbon. The spectra also show D, G, and D + G bands at 1334, 1594, and 2928 cm^{-1}, that are brought on by structural imperfections in functional groups containing oxygen on the basal carbon plane.[137]

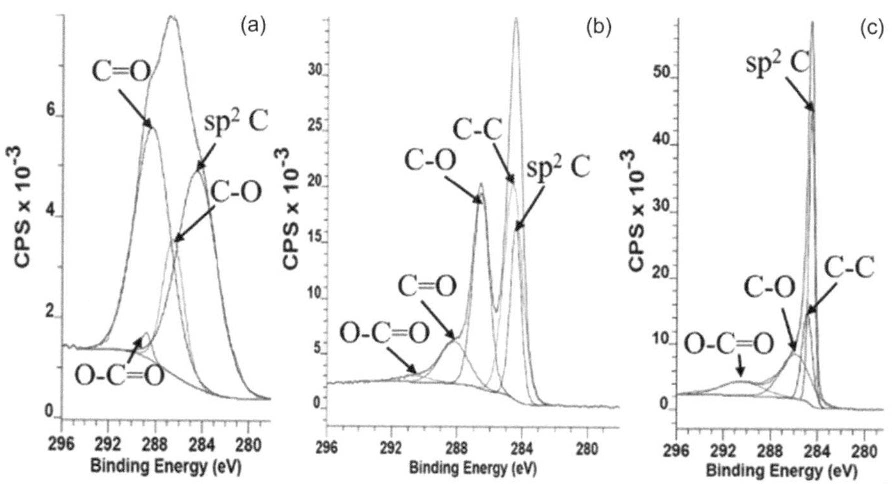

FIGURE 3.16 C1s XPS spectra of various GO samples, including (a) GO, (b) bwGO, and (c) rGO, and base washed GO. (Reproduced with permission from Aliyev et al.[141] Copyright 2019, the Authors.)

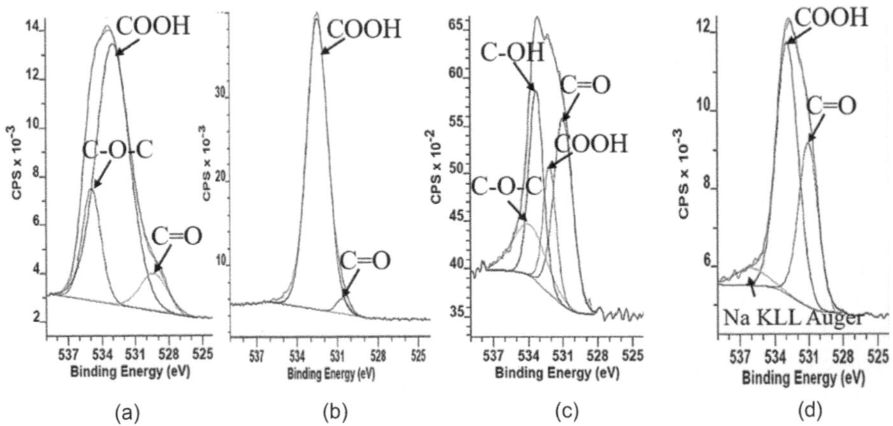

FIGURE 3.17 O1s XPS spectra of various GO samples, including (a) GO, (b) bwGO, (c) rGO, and (d) base washed and reducedGO. (Reproduced with permission from Aliyev et al.[141] Copyright 2019, the Authors.)

There is much debate over the chemical structure of GOs because of their extensive number of structural defects, non-stoichiometric organization, variation in oxygen-containing functional groups, and inconsistent GO synthesis methods. As a result, during the course of the last 80 years, various structural (chemical) models of GiO and/or GO have been put forth.[105,110,143] The Nakajima-Matsuo, Lerf-Klinowski, Dekany, and Ajayan models, along with the Hofmann, Ruess, Scholz-Boehm, and others, are well-known models for GO structure.[144]

3.5 rGO

Initially, it was only believed that GO could be utilized for low-cost mass production of graphene. The most crucial goal of the conversion GO was therefore the whole restoration of the structure and characteristics of graphene. The structure and chemical makeup of GO become unstable, with extreme levels of oxidation. The major fraction of functional groups bearing oxygen can be readily detached from the basal plane of the GO particle as a result of its strong surface reactivity. Most strategies for GO restoration are provided by several studies.[71,111,145] The existing GO to graphene conversion methods can potentially be categorized into three: physical (thermal), chemical, and hybrid, which simultaneously involve both physical and chemical aspects.

Thermal procedures frequently include heating GO either directly or by irradiation (microwave, infrared, visible, or ultraviolet (UV)) in a vacuum, inert, or reducing atmosphere. GO is transformed into rGO across a broad temperature range. The thermal reduction has the obvious benefit of eliminating oxygen-containing groups while also thermally annealing GO to restore its structure. It encourages the recovery of the GO basal plane's oxidation deficits via carbon atoms' sp^3-sp^2 rehybridization.[114]

Chemical reduction involves the employment of various reactants. Hydrazine hydrate for GO reduction is a frequently employed reagent. A few well-known reactants that successfully convert GO back into graphene are Fe powder, sodium borohydride, hydroquinone, hexamethylenetetramine, hydroiodic acid, sodium, and potassium alkaline solutions. This might be done in an aqueous or gaseous medium at a mild or even ambient temperature.[145] Catalyst-assisted photoreduction techniques may fall under the heading of chemical processes. As an illustration, the existence of titanium dioxide particles effectively helps to reduce GO in an aqueous medium when exposed to UV irradiation.

To increase the degree of reduction attained by chemical procedures, the solvothermal method, which combines thermal and chemical treatment under supercritical conditions, can be applied.[146] Better nitrogen doping of rGO is made possible by the same hybrid method. The ability of ULGO platelets to be controlled perforated under gentle UV irradiation in an argon atmosphere has been demonstrated. This approach allows for concurrent modification of the hole edges by carboxyl groups.[147]

3.6   GRAPHENE MATERIALS IN ADSORPTIVE WASTEWATER REMEDIATION

Even though graphene's electrical and electronic properties are the ones that are most frequently researched, its large theoretical specific surface area (SSA, 2630 m^2/g) and the existence of two basal planes of its single layer, make it a suitable nanoadsorbent for water treatment. Graphene's hydrophobic surface also draws organic pollutants with significant adsorption affinities, including dyes, phenols, and other harmful organic pollutants.[148] As a result, graphene has undergone numerous different functionalizations and composite formations. Moreover, the ability of several composites and functionalized graphite in various forms to purify water has also been investigated. Despite the fact that surface functional groups make GO and rGO good water purifiers, pure graphene is hydrophobic, which restricts its usage due to the low level of dispersion in water treatment. The GO and rGO's negative surface charges aid in the effective eradication of cationic pollutants such as heavy metals (HMs). It is common knowledge that the remediation of cationic HM ions is accomplished through electrostatic contact. Since GO changes over time, its use in water treatment is now somewhat restricted. However, due to its enormous surface area, rGO is outperforming GO as a material for water filtration. Since As^{3+}, As^{4+}, Se^{4+}, and Cr^{5+} are anionic dyes, rGO effectively removes these contaminants because it has a little negative charge.[149]

Surface functional groups, crystal structure, optical property, and faulty structure are some of the distinctions between GO and rGO that may be seen. Strong acidity and significant adsorption of cations and basic substances are characteristics of GO. In contrast, graphene exhibits a hydrophobic surface and has a high capacity for chemical adsorption thanks to a strong π-π interaction. In order to manufacture various types of nanocomposites while also improving the efficiency of separation and adsorption capability, graphene or GO can be modified in conjunction with organics or metal oxides.[150] Owing to their high SSA, controllable morphological characteristics, adjustable porosity, intrinsic hydrophobicity, and oleophilicity, chemical manipulation, tunable surface functionality, efficient synthesis, and excellent thermal, environmental, and resistance to chemicals, attention is now being placed on GO and rGO as the best materials for wastewater treatment.[151] Thus, functionalized graphene and graphene nanocomposites, which have developed outstanding water purifying applications in comparison to bare graphene, GO and rGO. They are developed by embedding them with distinct functional groups and complementary nanomaterials in order to improve the performance of graphene materials in environmental remediation.[152]

3.6.1 Powders

3.6.1.1 Removal of Metal Ions from Water

Water pollution is a serious environmental issue, so it is necessary to find a practical approach to remove metal ions in their original state from wastewater. Reverse osmosis, chemical precipitation, solvent extraction, ion exchange, adsorption, electrochemical treatment, and membrane filtration are just a few methods that can be used to remove HMs from water. Many of these methods are expensive. Among various techniques, adsorption has been found to be an effective, inexpensive, and simple technique for removing these ions. Among many materials used, carbon-based materials have been investigated in the absorption of HMs.[153] The utilization of graphene and its derivatives have been broadly exploited by researchers for the adsorptive remediation of metallic contaminants from wastewater/water as presented in Table 3.2.

For instance, Young-Chul Lee and Ji-Won Yang[154] used a hybrid of GO and titanium dioxide (GO-TiO$_2$) to remove HM ions (Zn^{2+}, Cd^{2+}, and Pb$^{2+)}$ from water. They found that the flower-like TiO$_2$ on GO considerably increased the efficacy of HM removal. Notably, at pH 5.6, the GO-TiO$_2$ hybrid's HM ion adsorption capabilities were 44.8 ± 3.4 and 88.9 ± 3.3 mg/g for removing Zn^{2+}, 65.1 ± 4.4 and 72.8 ± 1.6 mg/g for removing Cd^{2+}, and 45.0 ± 3.8 and 65.6 ± 2.7 mg/g for removing Pb^{2+} correspondingly after 6 h and 12 h, respectively. Colloidal GO, on the other hand, demonstrated adsorption capabilities of 30.1 ± 2.5 (Zn^{2+}), 14.9 ± 1.5 (Cd^{2+}), and 35.6 ± 1.3 (Pb^{2+}) mg/g given similar experimental conditions. They explained that as the hydrothermal treatment duration at 100 °C doubled from 6 hrs. to 12 hrs., TiO$_2$ blooms were noticeably produced onto GO. Prolonged treatment periods enhanced the surface area of the GO-TiO$_2$ composite, increasing its potential to remove HMs.[154]

In a study by Zare-Dorabei and colleagues, Pb^{2+}, Cd^{2+}, Ni^{2+}, and Cu^{2+} ions were simultaneously adsorbed from aqueous solutions using GO that had been chemically modified with 2,2'-dipyridylamine (GO-DPA). Concurrent with the adsorption procedure, ultra-sonification was employed, and it was shown that ultrasonic power contributed to the acceleration of the ion adsorption period by facilitating the dispersion of adsorbent. They further discovered that Pb^{2+}, Cd^{2+}, Ni^{2+}, and Cu^{2+} ions had maximum adsorption capacities of 369.75, 257.20, 180.89, and 358.82 mg/g, respectively. Moreover, the efficacy of adsorbent for use in industrial applications was demonstrated by the adsorbent removal efficiency on actual wastewater samples.[155]

In order to remove pollutants from wastewater, magnetic nanoparticles have also been investigated since they are known to have superior physical and chemical features, such as superparamagnetism, high surface area, straightforward separation in the presence of external magnetic fields, and excellent adsorption capacity. In order to remove Pb^{2+} and Hg^{2+} from aqueous solution, Zhang et al.[156] created magnetic

TABLE 3.2

An Overview of the Elimination of Various Metal Ion Pollutants by Graphene and Its Derivatives

Graphene-based Absorbent	Metal Ion Pollutant	Adsorption Conditions	Adsorption Isotherm Model	Adsorption Kinetics	Maximum Adsorption Capacity	Ref.
GO rGO	Ni^{2+}	pH = 8 $T = 25\ ^{\circ}C$ $t = 24$ hrs	Freundlich isotherm model (FRIM)	Pseudo-second-order	90.8% 84.8%	157
Magnetic fungal hyphal/GO nanofibers (MFHGs)	Co^{2+} Ni^{2+}	pH = 6 $T = 50\ ^{\circ}C$ $t = 8$ hrs	Langmuir isotherm model (LM)	Pseudo-second-order	97.44 mg/g 104.34 mg/g	158
GO modified with 2,2′-dipyridylamine	Pb^{2+} Cd^{2+} Ni^{2+} Ni^{2+}	pH = 5 $T = $ Room temperature $t = 4$ min	LM	Pseudo-second-order	369.749 mg/g 257.201 mg/g 180.893 mg/g 358.824 mg/g	155
Aminoethyl rGO	Pb^{2+} Ni^{2+} Cu^{2+} Co^{2+} Zn^{2+}	pH = 5.5 $T = 25\ ^{\circ}C$ $t = 20$ min	LM	Pseudo-second-order	173.6 mg/g 46.2 mg/g 58.6 mg/g 41.2 mg/g 54.4 mg/g	159
GO modified with bis(2-pyridylmethyl)amino groups	Cu^{2+} Ni^{2+} Co^{2+}	pH = 7 $T = $ Room temperature $t = 24$ hrs	Jossens isotherm model (JIM)	Pseudo-second-order	3.464 mmol/g 3.254 mmol/g 3.054 mmol/g	160

Flower-like TiO_2-GO	Zn^{2+}	pH = 5.6	—	—	44.8 ± 3.4 and 88.9 ± 3.3 mg/g	154
	Cd^{2+}	$T = 100\ °C$ (hydrothermal treatment)			65.1 ± 4.4 and 72.8 ± 1.6 mg/g	
	Pb^{2+}	$t = 6$ hrs and 12 hrs			45.0 ± 3.8 and 65.6 ± 2.7 mg/g	
Poly(amidoamine) modified GO	Cu^{2+}	pH = —	—	—	0.1368 mmol/g	161
	Zn^{2+}	T = room temperature			0.2024 mmol/g	
	Fe^{3+}	$t = 24$ hrs			0.5312 mmol/g	
	Pb^{2+}				0.0513 mmol/g	
	Zr^{3+}				0.0798 mmol/g	
Magnetic GO modified with 1,2-diaminocyclohexanetetra acetic acid	Cr^{4+}	pH = 3 $T = 30\ °C$ $t = 24$ hrs	FRIM	Pseudo-second-order	80 mg/g	162
rGO/nickel oxide (NiO)	Cr^{4+}	pH = 4 $T = 25\ °C$ $t = 24$ hrs	LM	Pseudo-second-order	198 mg/g	163
Porous Fe_3O_4 hollow microspheres/GO	Cr^{4+}	pH = 4, 5 $T = 20\ °C$ $t = 10$ hrs	LM	Pseudo-second-order	32.33 mg/g	164
Magnetic cobalt ferrite $(CoFe_2O_4)$/rGO	Pb^{2+}	pH = 5, 3 $T = 25\ °C$ $t = 2$ hrs	LM	Pseudo-second-order	299.4 mg/g	156
	Hg^{2+}	pH = 4, 6 $T = 25\ °C$ $t = 2$ hrs			157.9 mg/g	
Magnetite Fe_3O_4-rGO/MnO_2	As^{3+}	pH = 7 $T = 25.5\ °C$ $t = 24$ hrs	LM	Pseudo-second-order	14.04 mg/g	165
	As^{5+}				12.22 mg/g	
Magnetite Fe_3O_4/rGO	As^{3+}	pH = 7 $T = 20\ °C$ $t = 2$ hrs	FRIM and LM	Pseudo-second-order	13.10 mg/g	166
	As^{5+}				5.83 mg/g	

(*Continued*)

TABLE 3.2 (CONTINUED)

An Overview of the Elimination of Various Metal Ion Pollutants by Graphene and Its Derivatives

Graphene-based Absorbent	Metal Ion Pollutant	Adsorption Conditions	Adsorption Isotherm Model	Adsorption Kinetics	Maximum Adsorption Capacity	Ref.
GO/iron nanohybrid	As^{3+} As^{5+}	pH = 7 $T = 22\ °C$ $t = 24$ hrs	LM	Pseudo-second-order	306 mg/g 431 mg/g	167
Sulfur/rGO nanohybrid (SRGO)	Hg^{2+}	pH = 6–8 $T = 23\ °C$ $t = 3$ hrs	LM	Pseudo-second-order	90%	168
Polyamine-modified rGO hydrothermal (HT-rGO-N)	Hg^{2+}	pH = 5 $T = 23\ °C$ $t = 3$ hrs	FRIM	Pseudo-second-order	63.8 mg/g	169
Chemical reduction Polyamine-modified rGO (CM-rGO-N)					59.9 mg/g	

cobalt ferrite ($CoFe_2O_4$)-rGO nanocomposites. At pH 5.3 and 25 °C, Pb^{2+} had the highest adsorption equilibrium, at 299.4 mg/g, while Hg^{2+}, at pH 4.6 and 25 °C, had the highest adsorption equilibrium at 157.9 mg/g. Equally important, the magnetic properties of $CoFe_2O_4$-rGO made it easy to retrieve the adsorbent composite from water using magnetic separations in just two minutes at a low magnetic field.

3.6.1.2 Removal of Organic Pollutants

Organic contaminants include substances that are used in our daily lives but become pollutants when they are discharged as waste into water bodies. These substances include colors, insecticides, pharmaceuticals/drugs, and other harmful chemicals.[170]

3.6.1.2.1 Pharmaceuticals

The last two decades have seen the emergence of pharmaceuticals as possible emerging contaminants (ECs). Their remnants are widespread in surface and ground waters, often at trace concentrations of ng/L to low µg/L, but also they can also be pervasive in drinking waters.[171,172] Concerns about the potential long-term harm they could do to aquatic ecosystems and human health have increased as a consequence of their presence in drinking water.[173] Drugs for humans and animals, veterinary medications, and food additives are the source of pharmaceuticals, which are released via excreta from people or animals, with little to no modification to their chemical composition.[174] Pharmaceutical traces in surface waters, groundwater, and partially treated water are typically 100 ng/L or less, whereas treated water typically has concentrations under 50 ng/L.[175] But there is mounting evidence that they have been found and are persisting in a variety of natural freshwater resources around the world. A number of medicines were found in aquatic habitats in both Africa and Europe at concentrations greater than their ecotoxicity endpoints, according to Fekadu et al.[176] Pharmaceuticals are often well described, and their effects are evaluated prior to commercialization, so their adverse effects are anticipated to lead to endocrine disruption in living things and the emergence of antibiotic-resistant bacteria and genes.[171,177]

Graphene nanoplatelets with a large surface area were used in a study by Al-Khateeb and colleagues to adsorb or eliminate three of the widely used pharmaceutical substances: aspirin (or acetylsalicylic acid), acetaminophen (N-acetyl-para-aminophenol), and caffeine (1,3,7-trimethyl-1H-purine-2,6(3H,7H)-di). The widely used non-steroidal anti-inflammatory drug is aspirin (ASA), whereas acetaminophen (APAP) is used to relieve pain and fever. Numerous analgesic drugs include caffeine (CAF), which is known to stimulate the central nervous system. They demonstrated that the majority of the ASA, CAF, and APAP could be eliminated utilizing 10.0 mg GNPs, for 10.0 min, at 23 °C, in 8.0 pH solutions. The adsorption process is exothermic, as evidenced by the fact that the capacities for adsorption declined in an inverse correlation to solution temperature. Intra-particle diffusion and liquid film diffusion (LFD) models were utilized to fit the experimental data in order to identify the step that determines the rate of removal; however, the fitting only occurred for a portion of the adsorption, proving that no model was able to fully regulate the adsorption process.[178]

Rostamian et al.[179] compared the adsorption of sulfamethoxazole onto graphene and GO nanosheets; and they discovered that the optimal adsorbent dosage for both systems was 0.010 mg, a starting pH of 6, and a contact period of 110 minutes. Through adsorption isotherms and nonlinear fitted curves in Figure 3.18, they showed that sulfamethoxazole's adsorption increased to a certain point and remained constant as concentration was raised further. Additionally, from 25 to 45 °C, the adsorption was lessened by a rise in temperature. They added that the weakening of the forces interacting between the sulfamethoxazole and adsorbents was caused by the process being exothermic, thus causing a further rise in temperature. The equilibrium adsorption data of sulfamethoxazole were best represented by the Redlich-Peterson (RPM)

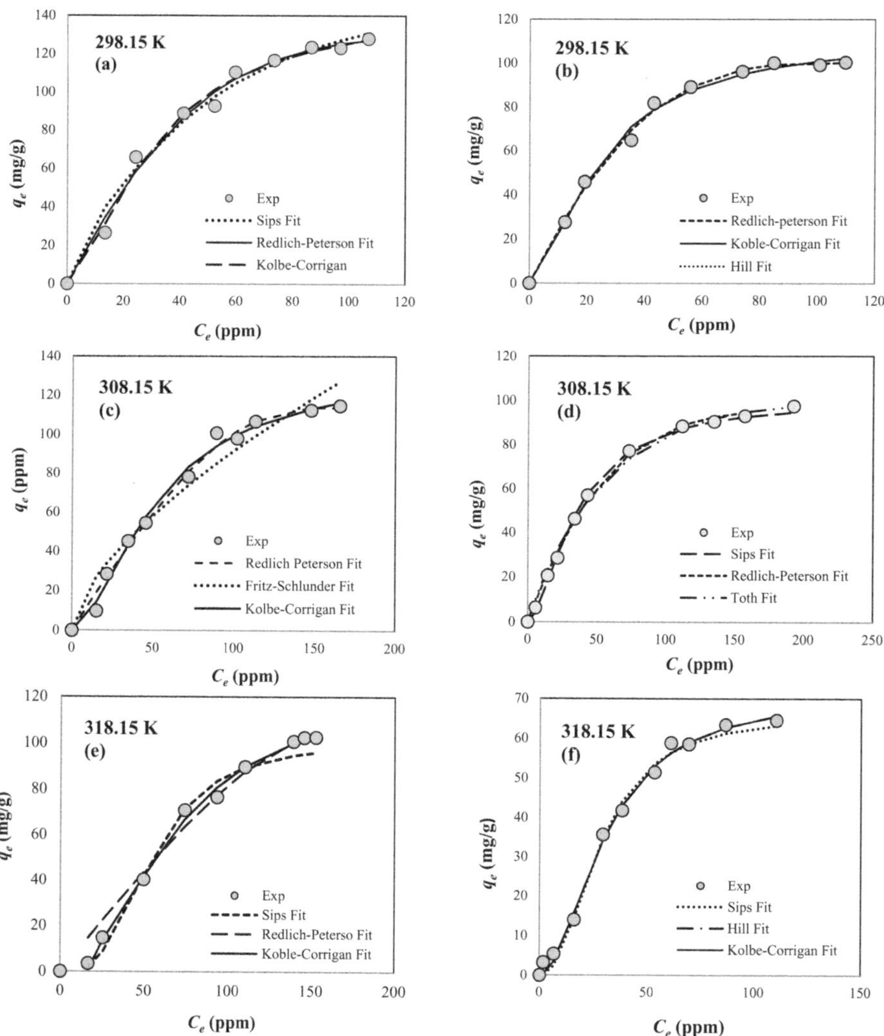

FIGURE 3.18   Equilibrium adsorption and the nonlinear fitted curve with different isotherms applied to SMX adsorption by GOS ((a), (c) and (e)) and GNS ((b), (d) and (f)) at different temperatures. (Reproduced with permission from Rostamian & Behnejad.[179] Copyright 2016, Elsevier Science Ltd.)

and Koble-Corrigan (KC) isotherm models, among other models, however, the kinetic experimental data on both adsorbents were well-suited by the pseudo-second-order model. The results of the study demonstrated that GO can be employed as a more effective adsorbent for the removal of sulfamethoxazole from aqueous solutions.

In another study, Brito and colleagues demonstrated that mixing GO with activated carbon improved adsorption efficiency for bisphenol A and paracetamol. They demonstrated that the maximum adsorption attained utilizing activated carbon alone was 234.7 and 200.9 mg/g for bisphenol A and paracetamol, respectively. However, when activated carbon was combined with GO, the adsorption capacity increased to 239.9 and 222.6 mg/g for bisphenol A and paracetamol, respectively. The effects of adsorbent quantity on the adsorption of paracetamol and bisphenol A after 2 h are shown in Figure 3.19(a) and (b). It was found that the dosage of the adsorbent had a direct proportional relationship with how well pollutants were able to bind to the

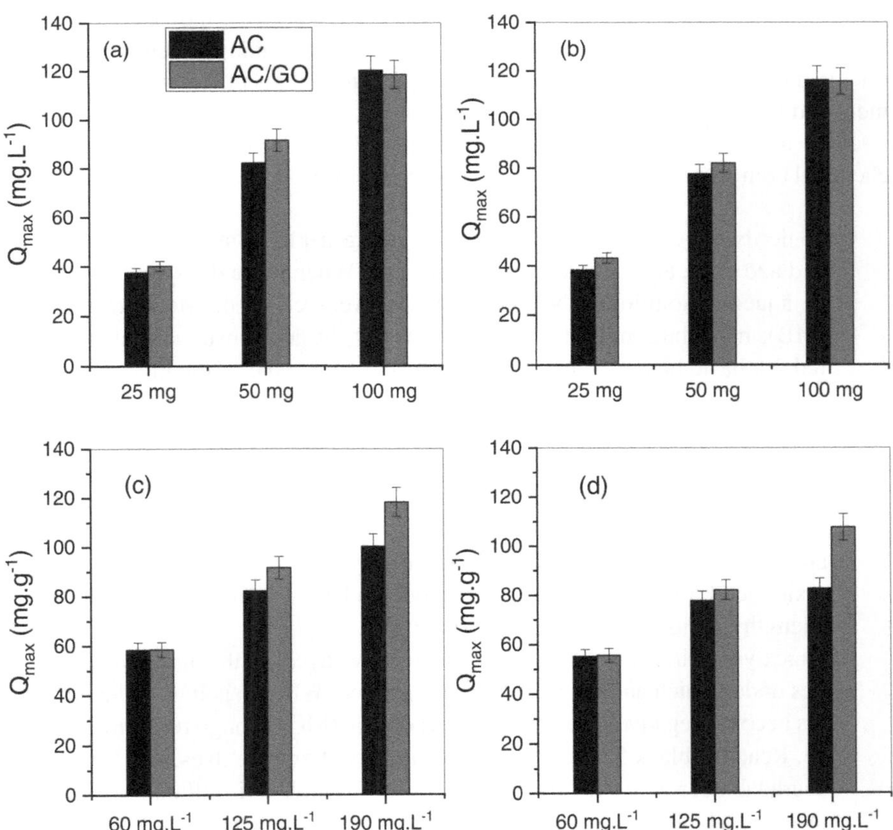

FIGURE 3.19 Effect of adsorbent dose for bisphenol A (a) and paracetamol (b) removal. Conditions: temperature, (25 ± 1) °C; pH, 8.3 (bisphenol A) and 5.7 (paracetamol); volume, 100 mL of contaminants at 125 mg·mL^{-1}; and contact time, 120 min. Effects of contaminant concentrations for bisphenol A (c) and paracetamol (d) removal. Conditions: temperature, (25 ± 1) °C; pH, 8.3 (bisphenol A) and 5.7 (paracetamol); contact time, 120 min. (Reproduced with permission from Brito et al.[180] Copyright 2023, Elsevier Science Ltd.)

adsorbent. With little adsorbent for both pollutants, the activated carbon/GO has higher adsorption in both pollutants. The effects of bisphenol A and paracetamol concentration on the adsorption process are shown in Figure 3.19c and d. This revealed that the overall method of adsorption was accelerated by the rise in the concentrations of both contaminants. The adsorption rate did not, however, rise linearly with concentration. The rapid saturation of the accessible active surface sites in the large concentration of pollutants may be the cause of this tendency.[180]

3.6.1.2.2   Dyes

In the textile, leather, and paper sectors, dyes are often used as coloring agents; nevertheless, after usage, they constitute a serious environmental risk to those around the industries. The employed colors are thrown into water sources, such as ponds and rivers. Dyes directly harm aquatic creatures as they enter rivers before they reach the food chain and start to harm other living things in the ecosystem. The bulk of the dyes have been labeled as being extremely harmful since they can change the Deoxyribonucleic (DNA). Because they are not biodegradable and remain for a long time in the environment, these dyes have a high potential to contaminate the environment and have a long-term negative impact on the ecosystem.[170]

Dyes are categorized into cationic, anionic, and non-ionic depending on their chemical composition, charge, color, and intended use.

Cationic dyes: All fundamental dyes, including triphenylmethane, xanthene, and azo dyes, are included in cationic dyes. When these dyes are dissolved in aqueous solutions, they become positively charged. Methylene blue (MB), malachite green (MG), crystal violet, basic fuchsin safranin, basic red 29, basic blue 159, basic yellow 28, etc., are some of the more popular cationic dyes. These dyes have a negative impact on the human liver, kidney, brain, and reproductive system in addition to being extremely mutagenic and carcinogenic. Additionally, these colors are not only harmful to aquatic life but also hazardous to people. Because of the elevated concentration of these dyes in seawater, photosynthesis in aquatic plants is hampered and light permeation is inhibited. Since they include aromatic amines and toxic metal ions that are either non-biodegradable or persistent in water for a lengthy period, azo dyes are poisonous.[170]

Anionic dyes: Direct dyes, acid dyes, and reactive dyes are the three subcategories under which anionic dyes can be separated. When in solution, these colors become negatively charged. Methyl orange (MO), Congo red, Direct Red 23, Reactive black 5, etc., are a few examples of anionic dyes. The less biodegradable reactive dyes are extremely water-soluble. It is difficult to remove the colorful effluent of reactive dyes because it easily flows through the treatment process. Its presence disrupts the ecosystem in numerous ways.[170]

Non-ionic dyes: Dispersed dyes are a type of non-ionic dye that is mostly utilized to create colored plastics, polyester, polyamide, etc. Non-ionic dyes like Direct Green 97 are typical examples. When colored material comes into contact with the dispersed dyes, it might cause in sensitization and allergies in the body. Additionally, these dyes pose a serious hazardous threat.[170,181]

MG dye adsorption on GO and rGO has been investigated by Robati and colleagues. They achieved optimal pH, contact time, temperature values of 3, 100 min, 25 °C and endothermic adsorption process. MG dye was absorbed onto the created adsorbents through an endothermic process, as seen by the effects of temperature. The pseudo-first-order kinetic model was discovered to be well-fitting and in strong agreement with the adsorption kinetic data of the MG dye on GO and rGO surfaces. With the highest adsorption capacity of 27.16 and 13.52 mg/g for GO and rGO, respectively.[182]

Nano GO was utilized as an adsorbent in a work by Soudagar and colleagues to eliminate the color MB from synthetic wastewater.[183] They discovered that the ideal GO dosage, temperature, and pH were 10 mg, 40 °C, and 8, respectively. Moreover, the equilibrium period of contact was 15 min. Furthermore, the endothermic and spontaneous nature of the adsorption process was noted, while the pseudo-second-order kinetic model of adsorption was used, and the highest amount of adsorption capacity was found to be 429.485 mg/g in the LM model.[183]

The effectiveness of an adsorbent made of GO/SiO_2 nanocomposite for removing MB from contaminated water was examined by Dan et al.[184] They showed that 555.50 mg/g was the highest adsorption capacity for MB determined by the LM model. Furthermore, the antibacterial activity of the GO/SiO_2 nanocomposite was investigated, and it was discovered that the nanocomposite had bactericidal action towards both gram-positive and gram-negative bacteria.

Yao et al.[185] explored the Azo dye removal from industrial effluent discharge utilizing immobilized Lipase enzyme on GO. They demonstrated that, until an optimum point, the enzyme activity improved with Lipase concentration, pH, and temperature. The best immobilization effectiveness and highest enzyme activity were achieved at a Lipase concentration of 6 mg/mL. The greatest Azo dye removal efficacy was measured at 240 min of contact time, and the 5 mg/L starting dye concentration was 89.47 percent. It was discovered after looking into the immobilized Lipase's reusability that it may be recycled up to four cycles. As a result, the adsorption of dyes by immobilized lipase on GO may have a substantial effect on a number of industrial sectors.

The adsorption capacity of four materials based on graphene, including graphite oxide (GrO), GO, chemical rGO (CrGO), and thermal rGO (TrGO), for the dyes MB and acid red (AR) were examined in a study by Paton-Carrero and co-workers. The adoption of MB is shown in Figure 3.20a, where there is evidence to suggest that raising the dosage of the adsorbent improved MB removal because doing so increased the surface area in contact with the dye, which in turn enhanced the number of active sites for adsorption.[186] With the exception of CrGO, the removal of MB in every material was close to 100% of the overall dye content. Only 30% of the MB in the solution was eliminated while using CrGO, thus less adsorption. This behavior was linked to the smaller number of oxygen groups present and the higher degree of stacking, which were both connected to the smaller surface area. The adsorption pattern with AR (Figure 3.20b) was different from that with MB. Sample TrGO was the best adsorbent in this situation because it had a high surface area and few oxygenated functional groups. Both the oxygenated functional groups and the anionic AR dye were negatively charged (in contrast to the cationic MB). Because of this, the adsorbents utilized were not favorable for RA adsorption since they either had a lot of oxygenated functional groups (GO, GrO) or they had a little surface area (CrGO).

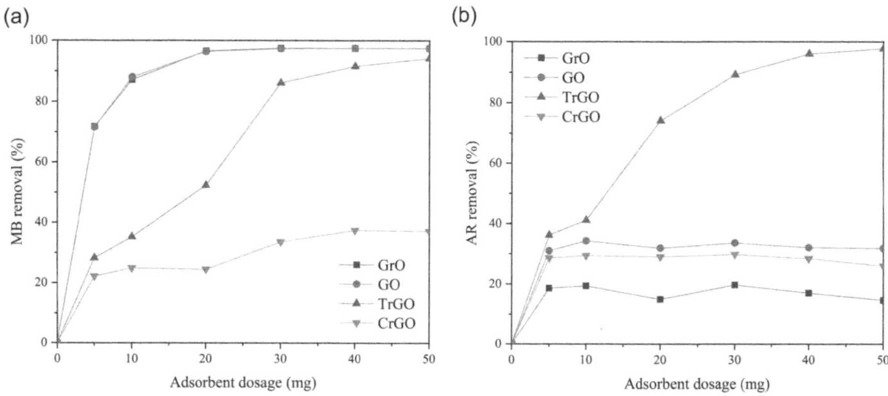

FIGURE 3.20   Dye adsorption vs. adsorbent dose. (a) methylene blue and (b) acid red. (Room temperature, t = 180 min, GO= 30 ppm). (Reproduced with permission from Panton-Carrero et al.[186] Copyright 2022, Elsevier Science Ltd.)

Since TrGO had a high surface area and a low percentage of oxygenated functional groups, the adsorption results for both dyes were identical. Focusing on MB adsorption, the adsorption capacities for GO, GrO, TrGO, and CrGO were 354.84, 327.99, 123.32, and 62.68 mg/g, respectively, with GO having the maximum adsorption capacity. Regarding AR, the materials' trends were as follows: TrGO had the maximum adsorption capacity, with values of 155.28, 58.99, 54.67, and 38.08 mg/g for TrGO, CrGO, GO, and GrO, respectively. Furthermore, almost all materials suit the LM model according to the adsorption isotherms, with the exception of GO for MB, which followed the FRIM model.[186]

3.6.1.2.3   Pesticides

Pesticides are chemicals made specifically for use in agricultural fields to safeguard crops from weeds and insects. Pesticides are thus a type of chemical designed primarily to exterminate insects or keep weeds from growing. They are insecticides, however, quite dangerous and non-biodegradable. It has been demonstrated that even a small amount of these substances can result in the development of cancer or other illnesses that are fatal.[170]

In order to eliminate phosphates from water samples, Vicente-Martinez and colleagues compared the effectiveness of GO and GO functionalized with silver nanoparticles (GO/AgNPs) as adsorbents. Under basic experimental circumstances, GO/AgNPs provided a maximum removal efficiency of 100% and instantly reached the equilibrium conditions. GO, on the other hand, had a maximum removal efficiency of 75%, required 9 minutes to reach equilibrium with 20 mg of adsorbent. Recovery percentages for GO and GO@AgNPs were 98 percent and 80 percent, respectively, for both adsorbents when they were both regenerated in an acidic medium.

The adsorption of phosphate ions in wastewater by chemically produced GO was investigated in another paper by Jun and colleagues.[187] A 70% removal efficiency for phosphate ions was discovered. With the use of iron nanoparticles in a chemically produced GO solution, phosphate removal efficiency was improved from

70% to 80%. At a temperature of 30 °C and an initial phosphate content of 100 mg/L, the capacity to remove phosphate reached a maximum of 89.37 mg/g. The equilibrium isotherms were described using the FRIM equation, and the isotherm constants were calculated.

A novel adsorbent for the removal of phosphate and nitrate ions from an aqueous solution was created by Sereshti et al.[188] The adsorbent was synthesized from magnetic Fe_3O_4 nanoparticles coated GO and modified with strontium nanoparticles (MGO-Sr). They investigated the adsorption process using the LM, FRIM, and Dubinin-Radushkevich (DR) isotherm models. According to the findings, phosphate and nitrate uptake adhered to the LM model's monolayer sorption process. The maximum adsorption rate of the phosphate and nitrate ions was 238.09 and 357.14 mg/g, respectively. They described an adsorption rate using pseudo-first-order, pseudo-second order, and intra-particle diffusion kinetic models. The results showed that phosphate and nitrate ions were removed using the pseudo-second-order model rather than intra-particle diffusion.

Researchers have extensively used graphene and its derivatives for the adsorptive remediation of organic pollutants from wastewater/water, as shown in Table 3.3.

3.6.2 MEMBRANES

Among the most effective and cost-effective techniques to provide drinkable water is membrane separation, a technique that is still in development.[208] The membrane water treatment process involves removing undesirable components from water through a semi-permeable barrier that lets some water molecules pass whilst excluding others (e.g., bacteria, proteins, ions, etc.). Depending on the application, these elements can be separated using sieving, size exclusion, Donnan exclusion, adsorption, or diffusion-based methods.[209] The principal membrane processes are categorized in Figure 3.21, along with the fundamental ideas and characteristics of each process.

Water, as well as various tiny molecules, can move more easily between the layers of GO membranes owing to pristine patches (non-oxidized) made of carbon atoms.[210,211] The interlayer distance is considered to rise in oxidized regions of GO, which are brought about by the existence of multiple oxygen rich functional groups,[210,212] impacting the water flux and the membranes' capacity to hold contaminants.[213] Consequently, the oxidized areas are very beneficial for molecular transport. Additionally, oxygenated groups can interact with ionic components via electrostatic interactions and hydrogen bonding due to their hydrophilic nature, which improves the selective penetration capabilities of GO membranes.[211]

Numerous research efforts have lately been drawn to forward osmosis due to its benefits over pressure-driven membrane technologies, including its minimal need for power and potent antifouling properties. Numerous methods have been established for the fabrication of GO-based membranes in order to synthesize varying membrane types with different features.[214] The application largely determines the production process and type selectin of GO-based membranes. These techniques include non-solvent-induced phase separation (NIPS), layer-by-layer assembly, interfacial polymerization, blending, and coating.[215] A summary of the production processes and kinds of the most recently explored GO-based membranes will be given in the next section.

TABLE 3.3
An Overview of the Elimination of Various Organic Pollutants by Graphene and Its Derivatives

Graphene-based Material	Pharmaceutical Pollutant	Adsorption Conditions	Adsorption Isotherm Model	Adsorption Kinetics	Maximum Adsorption Capacity mg/g or %	Ref.
Pharmacological residues						
Graphene nanoplatelets	Aspirin Acetaminophen Caffeine	pH = 8 T = 23 °C t = 1 hr	—	Pseudo-second-order	12.98 mg/g 18.07 mg/g 19.72 mg/g	178
GO	Tetracyclines	pH = 8.5 T = 25 °C t = 12 hr	LM	Pseudo-second-order	313 mg/g	189
GO	Doxorubicin hydrochloride (DOX)	pH = 3.4 T = 15 °C t = 12 hr	LM	Pseudo-second-order	1428.57 mg/g	190
rGO	Tetracycline (TC) Sulfamethazine (SMZ)	pH = 6 T = 25 °C t = 8 hr	LM	Pseudo-second-order	219.10 mg/g 174.42 mg/ g	191
GO nanosheet (GOS)	SMX	pH = 6 T = 25 °C t = 1.8 hrs	RPM	Pseudo-second-order	127 mg/g	179
Graphene nanosheet (GNS)		pH = 6 T = 35 °C t = 1.8 hrs			103 mg/g	
GO	Enrofloxacin (ENF)	pH = 7 T = Room temperature t = -	LM and FRIM	Pseudo-second-order	45.035 mg/g	192
Nitrogen-doped rGO/ Fe$_3$O$_4$ nanocomposite	Norfloxacin (NOR) Ketoprofen (KP)	pH = 6.8 T = 25 °C t = 24 hrs	LM	Pseudo-second-order	158.1 mg/g 468.0 mg/g	193

Maltodextrin/rGO	Diclofenac	pH = 7 T = 20 °C t = 50 min	LM	—	9.788 mg/g	194
	Amoxicillin	pH = 7.4 T = 20 °C t = 50 min			12.17 mg/g	
Maltodextrin/reduced graphene/copper oxide	Diclofenac	pH = 7 T = 20 °C t = 50 min			12.84 mg/g	
	Amoxicillin	pH = 7.4 T = 20 °C t = 50 min			526.31 mg/g	
Phosphates and Nitrates						
Graphene	Nitrate ions (NO_3^-)	pH = 7 T = 30 °C t = 12 hrs	LM	Pseudo-second-order model	89.97 mg/g	195
GO	Phosphate ions (PO_4^{3-})	pH = 7 T = 25 °C t = 12 hrs	FRIM	—	89.37 mg/g	187
Fe_3O_4/GO/MGO-Sr	Phosphate ions Nitrate ions	pH = 6 T = 25 °C t = 1.5 hrs	LM	Pseudo-second-order	238.09 mg/g 357.14 mg/g	188
Ceria nanoparticles decorated partially rGO (CeO_2-PRGO)	(PO_4^{3-})	pH = 4 T = Room temperature t = 2, 5 hrs	LM	Pseudo-second-order	45.8 mg/g	196
Zirconium-loaded rGO(RGO-Zr)	$H_2PO_4^-$	pH = 5 T = 25 °C t = 24 hrs	LM	Pseudo-second-order	27.71 mg/g	197

(Continued)

TABLE 3.3 (CONTINUED)

An Overview of the Elimination of Various Organic Pollutants by Graphene and Its Derivatives

Graphene-based Material	Pharmaceutical Pollutant	Adsorption Conditions	Adsorption Isotherm Model	Adsorption Kinetics	Maximum Adsorption Capacity mg/g or %	Ref.
Titania-functionalized GO (T-F GO)	$H_2PO_4^-$	pH = 6 T = —	LM	—	33.11 mg/g	198
GO		t = 24 hrs			1.68 mg/g	
GO		pH = 10 T = 25 °C t = 20 min	—	—	11.25 mg/g	199
GO-AgNPs		pH = 7 T = 25 °C t = 20 min	LM		1666.7 mg/g	
Dyes						
GO	MB	pH = 6 T = 25 °C t = 5 hrs	LM	Pseudo-second-order	243.90 mg/g	200
Magnetic β-cyclodextrin–chitosan (CS)/GO	MB	pH = — T = 25 °C t = 2 hrs	LM	Pseudo-second-order	84.32 mg/g	201
GO rGO	MG	pH = 3 T = 25 °C t = 2 hrs	Intra-particle diffusion	Pseudo-first-order	27.16 mg/g 13.52 mg/g	182
Exfoliated GO (EGO)	MB Methyl violet (MV) Rhodamine B, (RB) Orange G (OG)	pH = 3 T = 25 °C t = 20 min	LM	Pseudo-second-order	17.3 mg/g 2.47 mg/g 1.24 mg/g 5.98 mg/g	202

rGO	MG	pH = 3.7 T = 25 °C t = 7 hrs	LM	Pseudo-second-order	476.2 mg/g	203
Magnetic GO	MB	pH = 9 T = 30 °C t = 3 hrs	LM and FRIM	Pseudo-second-order	98%	204
GO	Basic yellow 28 Basic red 46	pH = 7 T = 20 °C t = 4 hrs	LM	Pseudo-second-order	68.5 mg/g 76.9 mg/g	205
Oil Products						
Graphene nanoplatelets Thermally reduced graphene (TRG)	Diesel	pH = — T = 23 °C t = 30 min	FRIM	Pseudo-second-order	1550 mg/g 805 mg/g	206
GO	Diesel	pH = — T = 30 °C t = 30 min	FRIM	Pseudo-second-order	1335 mg/g	207

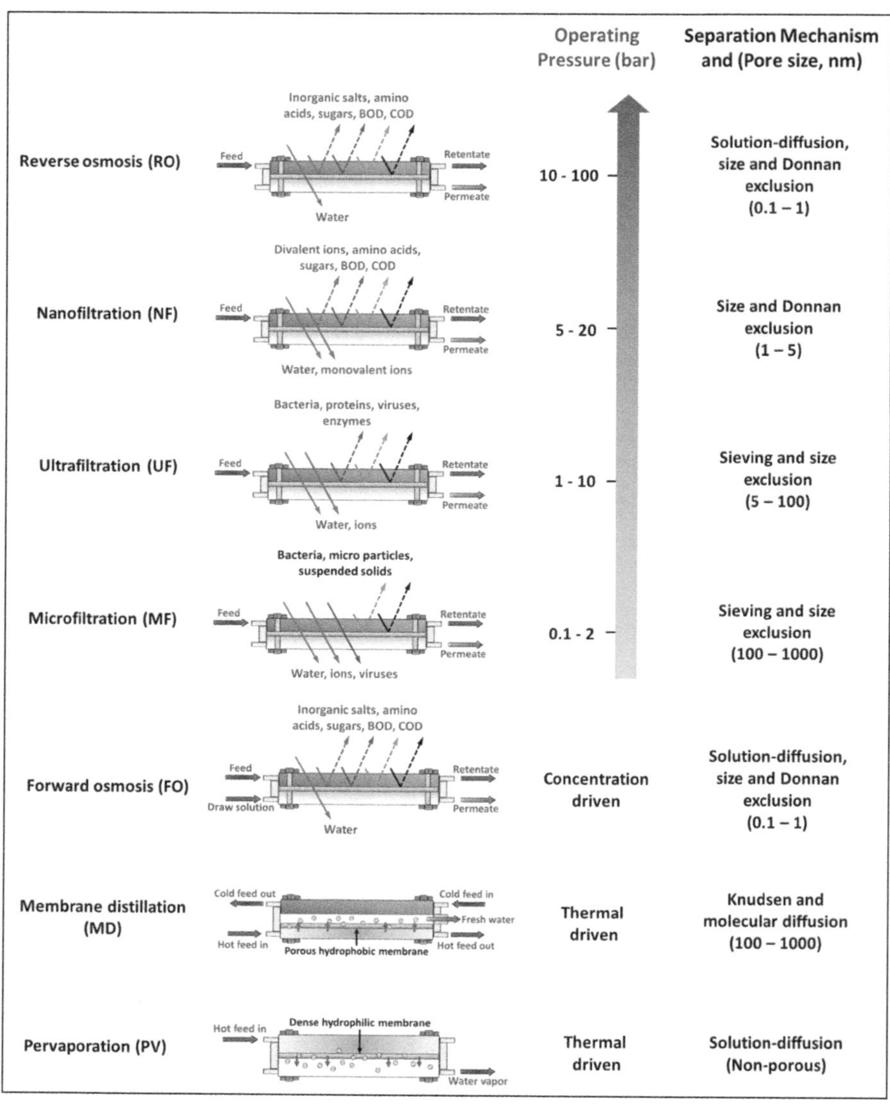

FIGURE 3.21   The basic concepts and classifications of membrane separation processes. (Reproduced with permission from Alkhouzaam & Qiblawey.[214] Copyright 2021, Elsevier Science Ltd.)

3.6.2.1 GO-Based Membrane Fabrication Techniques

3.6.2.1.1 NIPS

As a technique for creating polymeric membranes, NIPS is one of the most used.[216] The membrane in the NIPS technique is created by exchanging the solvent derived from the polymer solution with the non-solvent from a coagulant bath. The NIPS approach involves three components: polymer, solvent, and non-solvent. Polysulfone (PSF), Polyethersulfone (PES), Polyvinylidene Fluoride (PVDF), and Cellulose

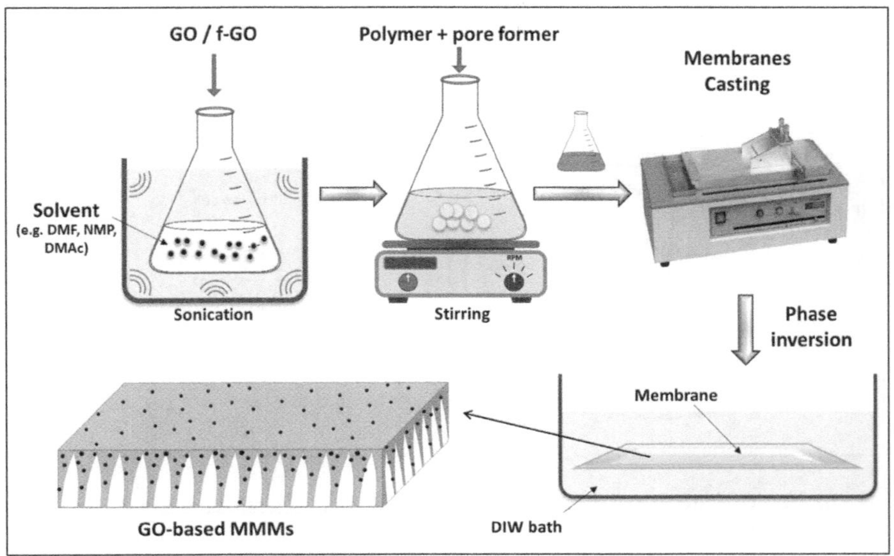

FIGURE 3.22 GO-based MMMs fabrication using NIPS technique. (Reproduced with permission from Alkhouzaam & Qiblawey.[214] Copyright 2021, Elsevier Science Ltd.)

Acrylate (CAs) were only a few of the polymer materials that were used for membrane manufacturing using the NIPS method. By inserting the GO (a useful filler) into the polymer matrix, NIPS can be utilized to create GO-based mixed matrix membranes (MMMs) in addition to producing perfect polymer substrates. This can boost the membrane's flow, rejection, and antifouling capabilities while also improving the pore structure.[217] The synthesis of GO-MMMs using the NIPS method typically begins with sonicating the p-GO (or f-GO) nanoparticles in the solvent, casting the dope solution after thoroughly combining them with the polymer to create a uniform mixture, and then dipping into a coagulant bath, the asymmetric membrane is created by precipitating the polymer with water.

The fabrication of GO-based MMMs using the NIPS approach is shown in Figure 3.22. The production of GO-based MMMs frequently uses the solvents N-Methyl-2-pyrrolidone, Dimethylformamide (DMF), and Dimethylacetamide (DMAc). Other additives, such as polyvinylpyrrolidone (PVP) and polyethylene glycol (PEG), are typically incorporated into the polymer solution in addition to the GO to serve as pore-forming agents.[218] GO-based MMMs can also be created through straightforward casting and drying (without dipping in a coagulant bath). This technique is frequently utilized to create thick, impermeable membranes for pervaporation applications (PV).[219,220]

3.6.2.1.2 Vacuum/Evaporation/Pressure-Assisted Self-Assembly (VAS, EAS, and PAS)

Both free-standing[221] and GO-based selective layer (SL) membranes[222] can be made using PAS, VAS, and EAS approaches.[222] In order to create lamellar structures

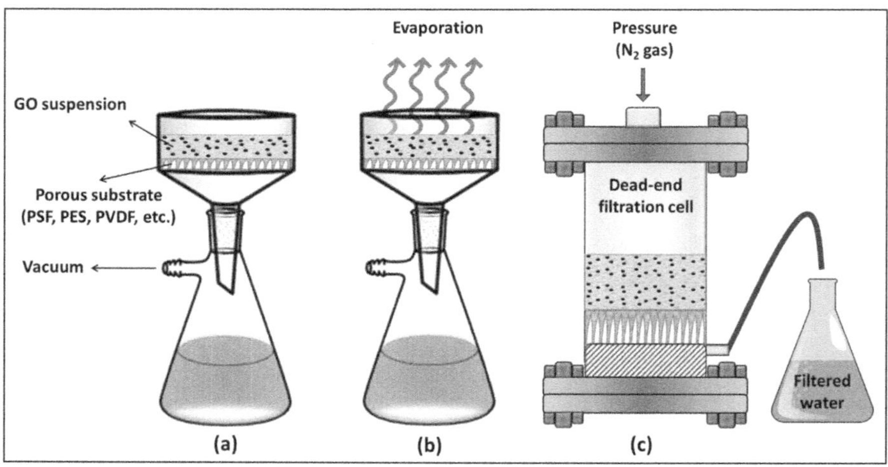

FIGURE 3.23   Illustration of the GO composite membranes fabrication using (a) VAS, (b) EAS, and (c) PAS methods. (Reproduced with permission from Alkhouzaam & Qiblawey.[214] Copyright 2021, Elsevier Science Ltd.)

with thin, single layers of GO nanosheets, a GO slurry is deposited and filtered via a porous substrate, often an ultrafiltration (UF) or microfiltration (MF) substrate. This method is used by PAS, VAS, and EAS. The setup of the production of membranes based on GO using VAS, PAS, and EAS approaches is shown in Figure 3.23. GO nanosheets' interlayer distance (d-spacing) creates nanochannels that, in contrast to the holes of traditional membranes, can permit water permeation while obstructing the passage of other pollutants. In order to accomplish the desired separation performance, a reasonable amount of research has been carried out in the past years to develop the lamellar structure of GO-SL membranes. A layered GO membrane with a d-spacing of ~3 permits the penetration of a water single layer. However, in aqueous conditions, the d-spacing can be increased up to a value of ~9, at which point some hydrated ions can pass, lowering the selectivity.[223] GO laminates' d-spacing can be adjusted by changing the content of GO,[224] modifying the deposition rate and method,[225,226] modifying the degree of oxidation of GO nanoparticles,[223] crosslinking and intercalating with other materials[227,228] by GO-coated membranes that have been post-functionalized.[229] According to Xu and colleagues,[226] the VAS technique was used to slowly deposit GO nanosheets, which resulted in a minor narrowing of the interlayer spacing and the creation of a preferred nanostructure with rapid water transport nano-channels. The rapid deposition of GO nanosheets, on the other hand, led to a less homogeneous nanostructure because the GO laminates' water permeability was hampered by their haphazard stacking.

As a result, membranes created through delayed deposition had fluxes that were about 2.6 times larger than those created through quick deposition, along with a marginal improvement in salt rejection. In another study, Tsou et al.[225] investigated how the assembly method affected the microstructural architecture and functionality of modified polyacrylonitrile (mPAN)/GO composites using the PAS, VAS, and EAS

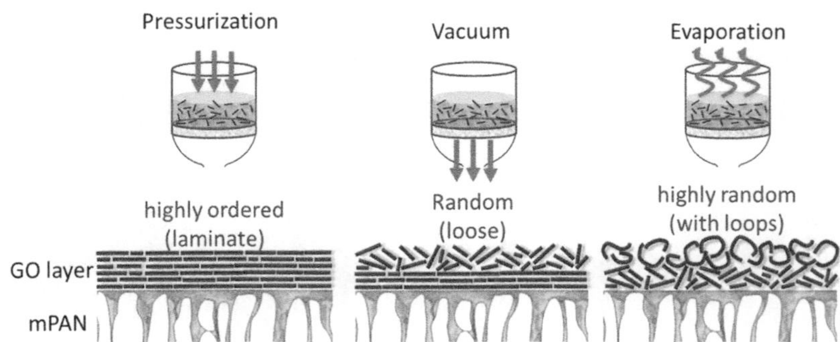

FIGURE 3.24 Schematic diagrams of composite GO/mPAN membranes fabricated through PAS, VAS, and EAS. (Reproduced with permission from Tsou et al.[225] Copyright 2015, Elsevier Science Ltd.)

techniques, Figure 3.24. It was discovered that composites made using the PAS approach displayed excellent pervaporation separation performance together with excellent stability at elevated temperatures. This was due to the fact that GO membranes generated using the PAS approach had a denser (Figure 3.25a, d and g) and more highly organized laminate structure than those prepared using the VAS and EAS methods. Furthermore, the GO layer is heterogeneous with a loop architecture when constructed utilizing the EAS approach, led to an interlayer spacing of 11.5 Å that was reduced using VAS and PAS to 9.7 and 8.4 Å, respectively.[214]

Most often, intercalating GO with different substances and crosslinkers is utilized to regulate the interlayer spacing. GO laminates with different d-spacing can be made using materials with various chain lengths and structural configurations. Sun et al.[230] used the PAS method to build GO nanoparticles on a mixed cellulose ester substrate after crosslinking them with poly (vinyl alcohol) (PVA) at various concentrations. Glutaraldehyde (GA) was used to further crosslink the covered membranes. According to the authors, altering the crosslinker concentration (PVA) can customize the distance between layers in GO laminates, which in turn affects how well pervaporation desalination performs. The membranes coated with a 5:1 PVA:GO ratio, with d-spacing increased from 8 Å to 12 Å, had the best pervaporation desalination performance.[214]

To learn more about how different aliphatic terminal diamines affect the lamellar structural architecture of constructed GO membranes and the efficiency of polydopamine (PDA) modified α-Al_2O_3 supports, Qian and colleagues modified GO nanoparticles.[231] In order to create diamine modified GO nanosheets, a variety of aliphatic terminal diamines with varying kinetic diameter length, including 1,2-diaminoethane (A2), 1,4-diaminobutane (A4), 1,3-diaminopropane (A3), 1,5-diaminopentane (A5), 1,7-diaminoheptane (A7), 1,6-diaminlhexane (A6) and 1,8-diaminooctane (A8). FESME pictures for membranes of GO, A2-GO, A3-GO, and A4-GO produced on tubular -Al_2O_3 substrates are shown in Figure 3.26. As demonstrated in Figure 3.26a–d, there were no observable morphological differences between the membranes of GO, A2-GO, A3–GO, and A4–GO. All of the membranes had the normal wavy creases and were free of any visible flaws.

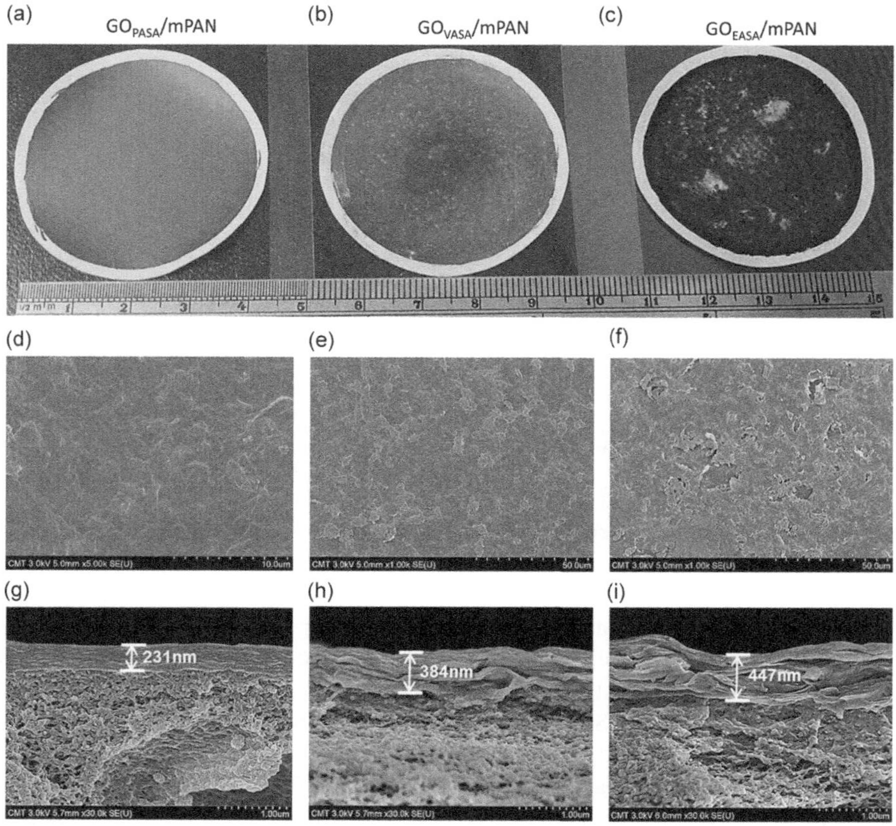

FIGURE 3.25   Photographs and SEM images of composite GO/mPAN membranes fabricated through PASA, VASA, or EASA. (a), (d) and (g): GO$_{PASA}$/mPAN; (b), (e) and (h): GO$_{VASA}$/mPAN; (c), (f) and (i): GO$_{EASA}$/mPAN. (Reproduced with copyright permission from Tsou et al.[225] Copyright 2015, Elsevier Science Ltd.)

The cross-sectional morphology of the membrane of GO, A2-GO, A3-GO, and A4-GO is shown in Figure 3.26e–h. All membranes having a lamellar structure were securely incorporated on the PDA-modified framework. Nevertheless, as a result of the weak action force between the support surface and GO layer that was not changed with PDA, a clear gap was detected amongst the GO membrane and the -Al$_2$O$_3$ support. Additionally, the thickness of the GO membrane was around 412 nm, with the same loading amount of 1.3393 g/m^2. The Ax-GO membranes' thicknesses rise to 535 nm, 688 nm, and 792 nm, respectively. This increased interlayer separation of Ax-GO is what causes the Ax-GO membranes to become thicker. Since diamines are intercalated between GO sheets, the modified GO nanosheets' d-spacings varied between 8.46 Å and 12.28 Å. Furthermore, the best membranes that displayed the highest PDA performance and exceptional stability for up to 7 days of intense heat,

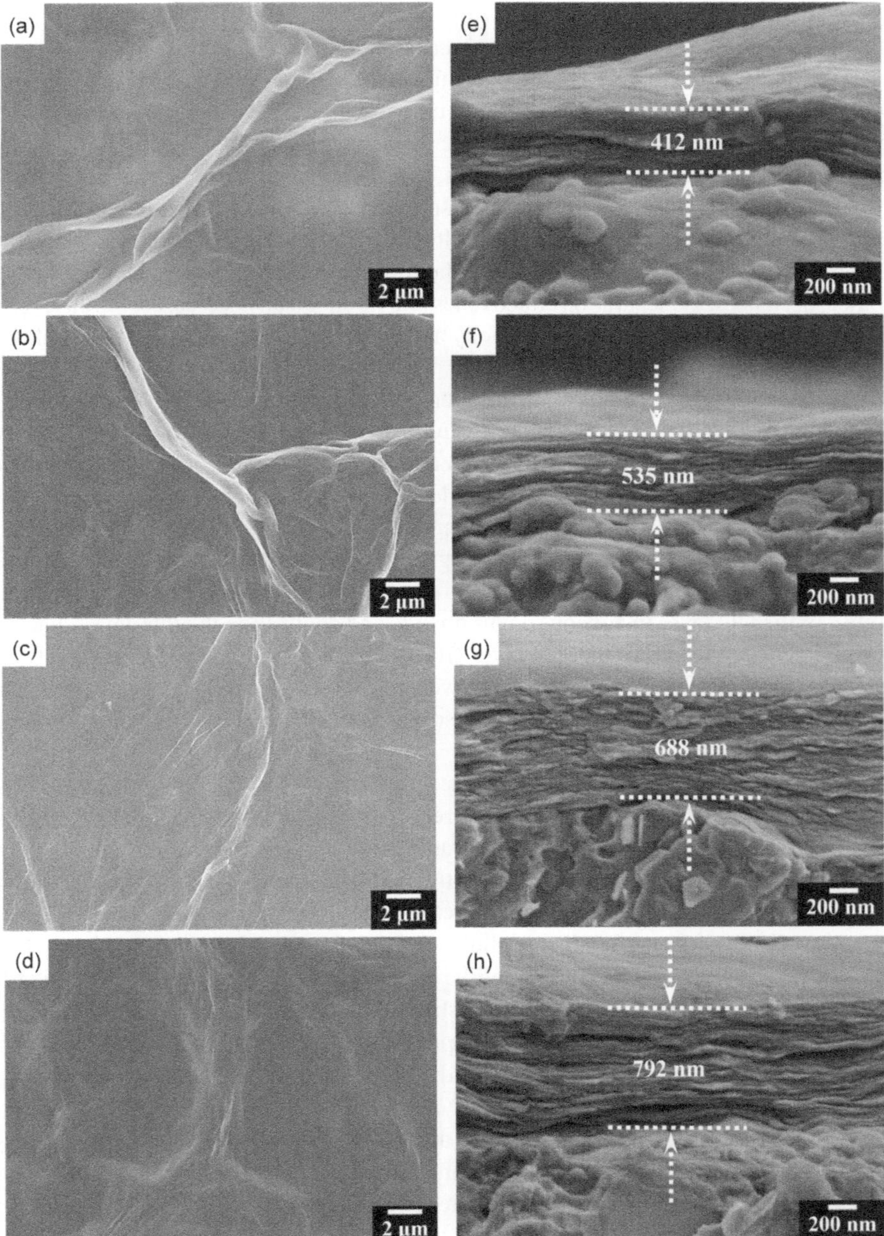

FIGURE 3.26 Top view (a, b, c, d) and cross-section (e, f, g, h) FESEM images of GO (a, e), A2-GO (b, f), A3-GO (c, g) and A4-GO (d, h) membranes on the PDA-modified α-Al$_2$O$_3$ tubes. (Reproduced with permission from Y. Qian et al.[231] Copyright 2019, Elsevier Science Ltd.)

were made from GO cross-linked with 1,4-diaminobutane. Surfactants were also employed to regulate the GO-SL membranes' interlayer spacing.

Lian and his coworkers used cetyltrimethylammonium bromide (C16TAB) as a d-spacing modifier to deposit a GO layer utilizing the VAS method.[232] The molecular length and breadth of C16TAB, a cationic surfactant, were roughly 1.99 and 0.2 nm, respectively. According to the researchers, the interlayer gap was increased by about five times, which resulted in a 13-fold rise in water flux. Nevertheless, the enlarged pores led to a considerable reduction in salt rejection. Although the permeability of the membranes can be increased by intercalating or crosslinking GO nanosheets with other substances, the separation effectiveness was dramatically decreased. Furthermore, to other characteristics like hydrophilicity and charge, the intercalant's or crosslinker's molecular size was taken into account.

Kang et al.[229] created GO composites, unstructured nanofiltration (NF) membranes with low negatively charged surfaces using one-step co-deposition. GO-membranes were post-functionalized to improve both their performance and microstructure. The VAS approach was used to deposit a GO layer, which was then crosslinked with tannic acid/Ni solution to create loose NF membranes. Because tannic acid/Ni-crosslinked membranes' interlayer spacing was marginally greater than that of uncross-linked membranes, more dyes were rejected, and there was higher flux. Enhancing the durability of the GO layer in aqueous environments and controlling the interlayer spacing can be done by reducing the GO layer (rGO).[223] The rGO films' hydrophobicity as well as the closer interlayer spacing, however, may result in a large decrease in the water flux. This problem can hence be solved by additional functionalizing of the membrane of rGO.

While the methods for fabricating free-standing GO membranes are the same, each study has a different technique for removing the GO-film from the support. A gratis forward osmosis GO membrane was created by Liu et al.[223] utilizing the VAS procedure by filtering a GO mixture through a mixed cellulose ester substrate. After that, hydriodic acid vapor was used to decrease the GO film to rGO. The rGO film was then removed from the mixed cellulose ester substrate by submerging it in water until it floated freely on the surface of the water (Figure 3.27). GO suspension is filtered on a permeable nitrocellulose substrate; Chen and colleagues created a free-standing GO membrane using the VAS approach.[221] However, by sequentially dipping the gratis film in N-methylpyrrolidone and ethyl ether, the film was removed.[221] Zhao et al.[233] used the VAS technique to filter a film made of GO and GO-Palygorskite (GO-PGS) composite across a cellulose acetate membrane. After that, acetone was used to dissolve the CA membrane in order to create a gratis MF GO membrane. Gratis GO membranes typically have great separation efficiency, but their lack of stability in comparison to other membrane types prevents them from being used under a lot of pressure (e.g., nanofiltration and reverse osmosis).

3.6.2.2 Separation Mechanisms

The three main parts of the GO membranes' material separation process are interaction with functional groups that include oxygen, interlayer spacing, and endospore (pore size exclusion for large-volume molecules). When truly purifying water, a variety of techniques can be used to effectively separate all contaminants.[234]

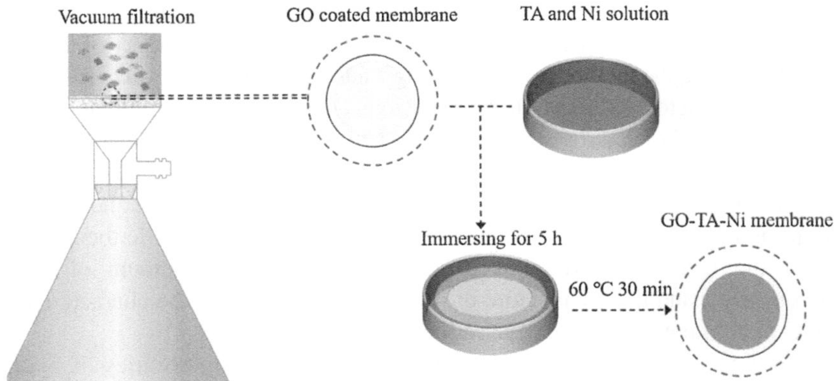

FIGURE 3.27 Illustration for the preparation of GO-TA-Ni membrane. (Reproduced with permission from Kang et al.[229] Copyright 2020, Elsevier Science Ltd.)

3.6.2.2.1 Separation Mechanism of Metal Ions

Human endeavors, including mining, the release of waste gases, irrigation using sewage, as well as usage of heavily loaded items, are the main causes of HM contamination. HM removal from water is a pressing issue that needs to be resolved because HMs seriously impair people's health. Among the popular techniques for removing the HM ions is membrane separation. GO can be utilized as a separation material through surface complexation, electrostatic adsorption, cation-π interaction, and ion exchange (as shown in Figure 3.28) due to its large SSA and abundant functional group content.[234] At the same time, it demonstrated good HM ion adsorption capacity.

GO was initially used as an adsorbent material, and it was discovered to have a superb ability for Cu^{2+} adsorption.[234,235] Following that, a GO membrane was developed to separate and adsorb Cu^{2+} more effectively, and the molecular dynamics simulations demonstrated that the good ability of metal ions to bind to GO surfaces was caused by their complexation and electrostatic attraction.[236] Other ions besides Cu^{2+}

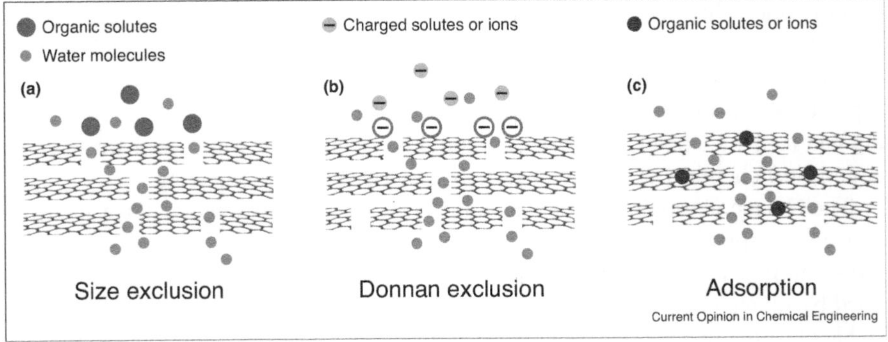

FIGURE 3.28 Schematic diagrams of the purposed separation mechanisms of GO membranes: (a) size exclusion, (b) Donnan exclusion and (c) adsorption. (Reproduced with permission from Yu Zhang & Chung.[238] Copyright 2017, Elsevier Science Ltd.)

have also had their absorption characteristics examined. For instance, the membrane created by Yu is capable of detecting and separating Hg^{2+}, and a variety of adsorption mechanisms give the membrane effective Hg^{2+} adsorption and removal capabilities.[237]

Numerous factors influence how metal ions adsorb on GO membranes. With an increase in pH, the electrostatic interaction between negatively charged GO sheets and metal ions becomes more powerful. The initial pH controls the types of metal ions in the solution and impacts the surface potential of GO. The rate and capacity of adsorption are accelerated by rising temperature and pressure, while the concentration of the feed, the variety of metal ions involved, the amount of oxygen-functionalities, and the architecture of the membrane all affect the characteristics of the GO membrane's selective penetration.[239]

According to Sun and colleagues, the varied penetrating abilities of hydrated metal ions (Na^+ > Mn^{2+} > Cd^{2+} > Cu^+) are caused by variable contact intensities (i.e., electrostatic attractions and chemical interactions) between the GO functional groups and metal ions.[240] The coordination of the metal ions with the oxygen-functionalized groups on a free-standing GO membrane restricts the penetration of metallic ions across the membrane.[240] More metal ions were adsorbed through coordination due to the increased oxygen-functionalized groups on the GO membrane, and the performance of the removal was somewhat diminished by increasing the feed concentration and pressure.[239]

Another crucial factor is the membrane's structure. The layer-by-layer (LbL) GO-modified Torlon hollow fiber membrane was created by Zhang et al.[241] in order to separate the metal ion and obtain superior rejection. Prior to repeating the ethylenediamine (EDA) and GO deposition processes and performing an amine-enrichment modification with polyethylenimine (HPEI), the base layer was first cross-linked with HPEI. The combination of the Torlons support and the GO framework layer can produce typical integrally skinned NF membranes with less polymer usage, while still efficiently sealing the flaws of the narrowly spaced pore size distribution in composite membrane (as shown in Figure 3.29). The hollow-fiber membrane (as depicted in Figure 3.30a–f) had superior mechanical strength and thermal resistance characteristics than its layer membrane equivalent. Zhang's membrane was utilized for ten cycles and continued to operate well. The GO/Torlons composite membrane had exceptional water permeability of 4.7 L m^{-2} h^{-1} bar^{-1} and rejections of Pb^{2+}, Ni^{2+}, and Zn^{2+} greater than 95%. During a 150-hour nanofiltration test, the membrane additionally demonstrated outstanding long-term performance stability. While the efficacy of membrane separation was significantly influenced by the thickness of the GO layer and the resulting nano-sieving channels, waterways could be changed for wastewater treatment by adding stiff chemical groups or flexible polymer chains as the spacing.[242]

The GO membrane's structural instability is known to prevent its industrial usage; hence a complexing agent's function is essential to increase the membrane's structural stability. The isophorone diisocyanate (IPDI)/GO framework membrane developed by Zhang et al.[243] had high water permeability in addition to being structurally stable and effective at removing dyes and metal ions. The rejection of metal ions was indeed decreased by the crosslinking of GO and IPDI, which also lowered the charge on the membrane's surface. However, the rejection of metal ions was further enhanced by raising the pH of the feed solution. In short, it has been demonstrated that the GO membrane rejects HM ions (Table 3.4) better than the bare membranes. For instance,

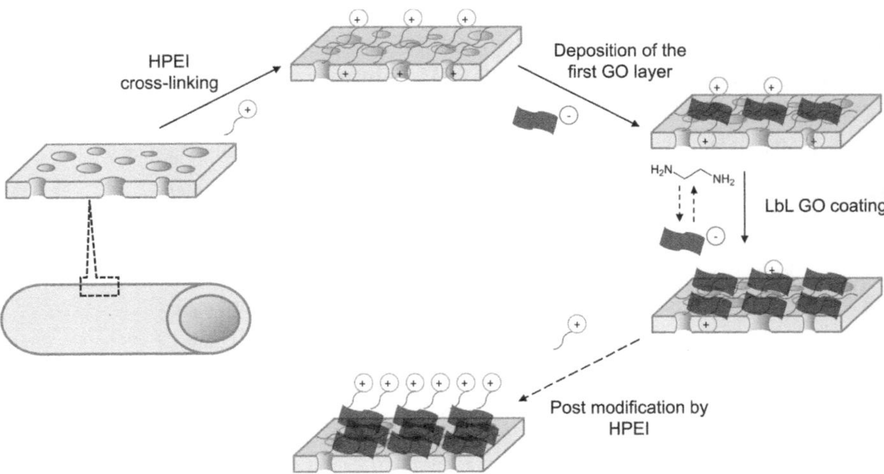

FIGURE 3.29 Schematic diagram of the procedure to construct the LbL GO framework membrane (Reproduced with permission from Yu Zhang et al.[241] Copyright 2016, Elsevier Science Ltd.)

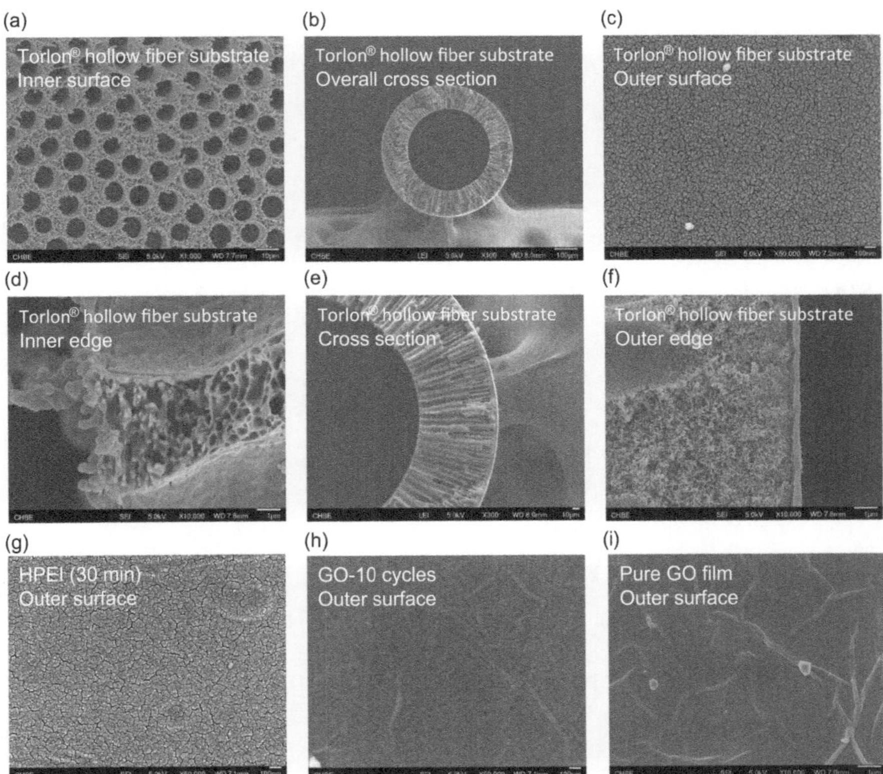

FIGURE 3.30 Morphology of the Torlons hollow fiber substrate (a–f), HPEI (30 min) (g), GO-10 cycles composite membrane (h) and a typical GO film formed by the vacuum-assisted process (i) (Reproduced with permission from Yu Zhang et al.[241] Copyright 2016, Elsevier Science Ltd.)

TABLE 3.4

Summary of HM Ions from Water GO Membranes

Graphene-Based Material	Pollutant	Maximum Adsorption Capacity mg/g or %	Ref.
PLGO	As^{3+}	140 mg/g	245
GO/CNTs	Co^{2+}	37 mg/g	246
	Ni^{2+}	40 mg/g	
	Cu^{2+}	50 mg/g	
	Zn^{2+}	42 mg/g	
	Cd^{2+}	48 mg/g	
	Pb^{2+}	98 mg/g	
$GO-NH_2$	Co^{2+}	116.35 mg/g	247
PVDF-g-4VP/GO-CTPy/PTFE	Cu^{2+}	625 mg/g	248
GO Nanosheets	Cd^{2+}	106 mg/g	249
	Co^{2+}	68.2 mg/g	
rGO-Ag	Cr^{5+}	100%	250
GO-Ag	Cr^{3+}	96.7%,	250
	Pb^{3+}	100%	
	Cu^{2+}	99.9%	
$SPES-GO-MnO_2$	Ni^{2+}	67.4%,	251
	Cu^{2+}	87.1%	
GO	Uranium ions	—	252
SGO@UiO-66-TFN	Cu^{2+} and Pb^{2+}	>99.4%	253
GO-AuNCs@BSA	Hg^{2+}	90.45%	237
CS/GO	Pb^{2+}	461.3 mg/g	254
	Cr^{5+}	310.4 mg/g	
$PAN/GO/Fe_3O_4$	Pb^{2+}	799.4 mg/g	255
	Cr^{5+}	911.9 mg/g	
HPEI-modified GO/EDA framework membrane	Pb^{2+}	95.7%	256
	Ni^{2+}	96.0%	
	Zn^{2+}	97.4%	
	Cd^{2+}	90.5%	
Torlon hollow fiber modifies LbL GO framework membrane	Pb^{2+}	95.88%	257
	Ni^{2+}	99.74%	
	Zn^{2+}	98.07%	
PSF + GO + DMF mixed matrix membrane	As^{3+}	~83%	258
PSF + GO + DMF mixed matrix membrane	$Cr^{2+}Cu^{2+}$, Cd^{2+}, and Pb^{2+}	~90%	259

(*Continued*)

TABLE 3.4 (CONTINUED)
Summary of HM Ions from Water GO Membranes

Graphene-Based Material	Pollutant	Maximum Adsorption Capacity mg/g or %	Ref.
PSF/PDA/IRMOF- 3/GO-1	Cu^{2+}	46.2%	243
GO-IPDI membrane	Pb^{2+}	66.4%	
	Cr^{2+}	71.1%	
	Cd^{2+}	52.8%	
rGO-K	Cr^{3+}	99.4%	260
	Fe^{3+}	99.8%	
cGO/SAA	Ni^{2+}	88%	261
	Pb^{2+}	92%	
GO	Pb^{2+}	95.7%	262
	Ni^{2+}	96%	
	Zn^{2+}	97.4%	
	Cd^{2+}	90.5%	

compared to the virgin PSF/PDA membrane, the rejection of HM ions via the GO-incorporated PSF/PDA membrane was greater.[244]

3.6.2.2.2 *Separation of Organic Pollutants*

3.6.2.2.2.1 Pharmaceuticals The ringed hydrocarbons, such as naphthalene, phenanthrene, and pyrene, are examples of organic pollutants like polycyclic aromatic hydrocarbons that are produced when fossil fuels, biomass, and mineral extraction are burned.[263] These chemicals are extremely hazardous as they affect the kidneys, lungs, brain, and heart, leading to tumors and chronic cardiac problems. In addition to polycyclic aromatic hydrocarbons, monocyclic aromatic chemicals are especially risky because they interfere with nerve cell function. Another category of organic pollutants, including phenols, bisphenol A, phenyls, and biphenyls, threatens humankind and harms marine life. Even at very low concentrations, these organic contaminants are slow to biodegrade. Adsorption is one of the simplest methods for extracting organic chemicals from wastewater among the different waste removal strategies employed to date, and compounds produced from GO have shown promise for the restoration of water sources that are polluted. Research on the capacity of graphene and its derivative to adsorb different organic molecules is extensive.[264] Adsorption is frequently used in water pollution removal, sugar purification, soap cleaning processes, and chromatographic operations. In wastewater cleanup, adsorbents such as GO capsules, zeolites, nanocomposite, and iron oxide composites are widely utilized. A few illustrative examples of graphene and its by-products' used in the removal of dyes are mentioned in the following paragraphs. Table 3.5 summarizes the adsorption capacities of the different organic pollutants on graphene-based membranes.

TABLE 3.5

Summary of Organic Pollutants from Water Using GO Membranes

Graphene-Based Material	Pollutant	Efficiency and Mechanism of Pollutant Removal	Ref.
Dyes			
GO/ginger extractive (GE) membrane	RB	99%	274
GO/PDA/PES-SPES membrane	RB	87.03 mg/g	268
PVA-GO-sodium alginate (PVA-GO-NaAlg) nanocomposite hydrogel (HG)	Lanasol blue 3R dye (LB3R)	>83% L; NF	275
PAN/GO hybrid	MB	100%; NF	270
Zirconium-based MOFs (UiO-66)/ PAN–PDA–GO	Congo red (CR)	99.6%	276
	MO	94.8%	
	RB	95.5%	
	MB	100%	
PES/3-aminopropyltriethoxysilane (APTS)/ functionalized GO	Sunset yellow (SY)	>95%; NF	277
	Acridine orange (AO)	>95%; NF	
rGO@MoO$_2$	RB	100%; NF	278
	MB	100%; NF	
	Evans blue (EB)	100%; NF	
	MB	90%; NF	
	Fuchsine acid (FA)	95%; NF	
rGO@WO$_3$	RB	100%; NF	
	MB	100%; NF	
	Evans blue (EB)	100%; NF	
	MB	85%; NF	
	FA	98%; NF	
Poly (ethylene imine)-600 (PEI-600) grafted GO sheets	DR	>99%	279
	MB	>94%	
PEI-10000 grafted GO sheets	DR	>99%	
	MB	>99%	
Pharmaceuticals			
HPEI-GO/PES ultrafiltration membrane	BSA	92.0%; UF	266
Polyamide (PA)/GO	Norfloxacin	53.32%; NF	267
Graphene, graphene oxide, carbon nanotubes and C60	Steroid hormones	GP-AO2: 6.6 mg/g; adsorption	265
Zirconium-based MOFs (UiO-66)/ polyacrylonitrile (PAN)–PDA–GO	Tetracycline-hydrochloride	95.5%	276
	Oxytetracycline	94.8%	
	Ciprofloxacin	98.6%	
SiO$_2$–GO nanohybrid/PSF membrane	BSA	98%; UF	280
Modified GO and deposition of silver carbonate (Ag$_2$CO$_3$) on the membrane surface by suction filtration was used to prepare polyvinylidene fluoride (PVDF) (PGA)	BSA	79%; UF	281

(*Continued*)

TABLE 3.5 (CONTINUED)
Summary of Organic Pollutants from Water Using GO Membranes

Graphene-Based Material	Pollutant	Efficiency and Mechanism of Pollutant Removal	Ref.
GO@UiO66/PVDF	BSA	93%; UF	282
Multilayered GO	Ibuprofen	100%	283
rGO$_{3h}$/CNTs	Sulfamethoxazole	>77%	284
Phosphates and Nitrates			
CS/GO	20 types of pesticides	95% removal of rifampicin; adsorption	272
Zirconium functionalized NCS-GO composite (NCT@GO/Zr)	Phosphate Nitrate	172.41 mg/g 138.88 mg/g	273

Among organic pollutants, pharmaceuticals such as antibiotics, steroid hormones, anti-acids, anti-inflammatory, lipids-lowering drugs, etc., have been observed in wastewater. In a work by Nguyen and colleagues, they compared the ability of graphene, GO, CNT, and C60 nanoparticles to adsorb steroid hormones such as estrone (E1) and 17-estradiol (E2) for eventual inclusion in membrane composites.[265] Graphene showed high adsorption capacities for hormones undeterminable in isotherm studies (over 10 mg/g). Figure 3.31 illustrates the adsorption capacity of the different nanoparticles with a change in pH. Second-grade graphene (GP AO-2, 20–30 monolayers) and first-grade graphene (GP AO-1, 3 monolayers) adsorbed less hormones at a pH higher than 10, which was explained by the electrostatic repulsion between negatively charged particles and E2. They went on to explain that there were two factors that might have controlled GO's adsorption of E2. On the one hand, electrostatic repulsion amongst the negatively charged GO surface and E2 prevented

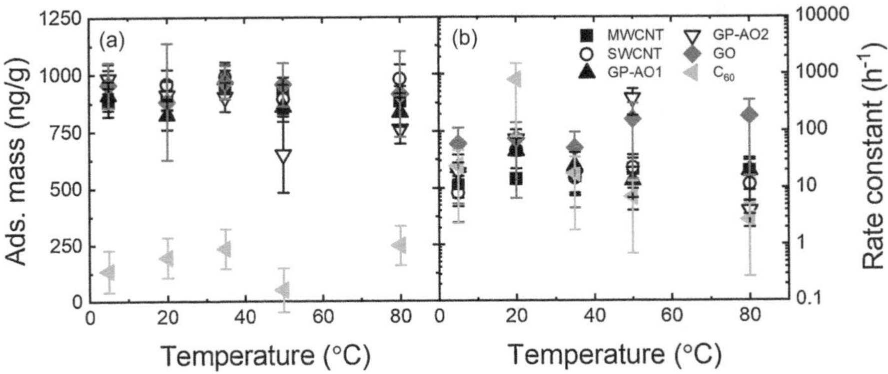

FIGURE 3.31 Hormone removal as a function of pH (100 ng/L E2, 1 mM NaHCO$_3$, 10 mM NaCl, 20.0 ± 0.5 °C, pH 2–12). (Reproduced with permission from Nguyen et al.[265] Copyright 2021, Elsevier Science Ltd.)

adsorption, much like it did in the situations of GP-AO1 and GP-AO2. On the other hand, it was proposed that GO stacked sheets were exfoliated at high pH as a result of increased repulsion forces between the hydrophilic functional groups. At high pH, E2 would consequently be able to penetrate the interlayer gaps, interfering with adsorption. The GP-AO1 under investigation was the most effective nanomaterial for steroid hormone adsorption and ought to be applied to composite membranes. GO and GP-AO2 were less appropriate due to their subpar adsorption capability because of their quick adsorption kinetics and low SSAs.[265]

Hyperbranched HPEI-GO/PES hybrid ultrafiltration membranes were made using the traditional phase inversion approach in another study by Yu et al.[266] by dispersing HPEI-GO nanosheets in the PES casting solution. They discovered that the hybrid membranes had excellent mechanical properties and that the incorporation of HPEI-GO enhanced the antifouling and antibacterial performance of the hybrid membranes in their prepared state. Figure 3.32 shows the Bovine serum albumin (BSA) adsorption amount on the different PES membranes. It is evident that there was no appreciable decline in the capacity of HPEI/PES hybrid membranes to bind proteins. However, compared to pure PES membranes, HPEI-GO/PES hybrid membranes showed less protein adsorption. It can be noted that the pure PES membrane's protein adsorption capacity was 61.11 g/cm^2. However, with an increase in HPEI-GO content, the protein adsorption amount of the HPEI-GO/PES hybrid membrane declined dramatically, and when the concentration of HPEI-GO reached 5%, the adsorption amount dropped to 25.89 g/cm^2. According to the findings, the HPEI/PES hybrid membrane and pure PES membrane are not as resistant to protein adsorption as the hybrid membrane incorporating HPEI-GO. Due to HPEI's high solubility in water, its poor capacity to resist protein adsorption of the HPEI/PES membrane may be partially the result of HPEI being leached during the phase inversion procedure. However, due to their flat structure and bigger surface area, HPEI-GO nanosheets

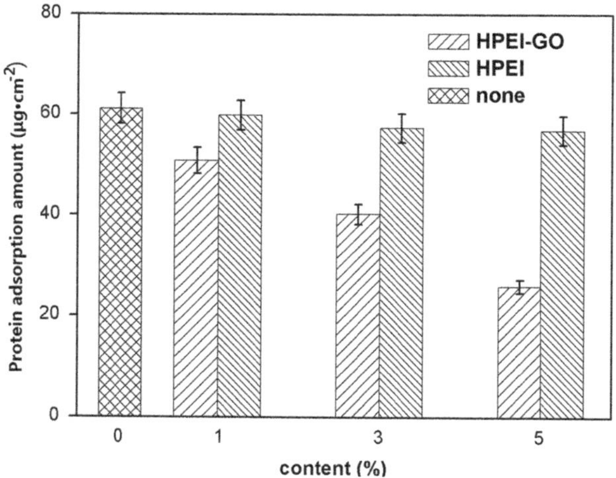

FIGURE 3.32 BSA adsorption amount on the different PES membranes. (Reproduced with permission from L. Yu et al.[266] Copyright 2013, Elsevier Science Ltd.)

were difficult to leach out, which improved the interaction between PES molecules and HPEI-GO molecules and made it necessary for HPEI-GO nanosheets to be wound and fixed in a polymer matrix. They further eluded that the hybrid membrane had potential uses that go beyond just separating proteins and peptides or purifying water.[266]

To be able to increase permeability/selectivity and boost anti-fouling qualities in several water treatment processes, Wang et al.[267] have investigated the integration of nano-materials in thin film composite membranes. Through an interfacial polymerization reaction involving the piperazine (PIP) and trimesoyl chloride (TMC) monomers, GO was incorporated into the semi-aromatic polyamide active layer of a thin film composite membrane. Additionally, employing these thin film composite membranes with various GO loadings, the rejection of three target pharmaceuticals and personal care products compounds—paracetamol, norfloxacin, and sulfamethoxazole—was examined. The addition of GO significantly increased the rejection of pharmaceuticals and personal care products as a result of the beneficial changes in the hydrophilicity, surface shape, and charge of the membrane. Wang et al.[267] further presented the normalized water flows of both the pure PA thin film composite membrane and the GO-modified PA thin film composite membrane during the foulant rejection test. Within 40 min, the flow of both unmodified and GO-modified PA thin film composite membranes marginally decreased before stabilizing over time. Observed water flow decreases for the PA/GO-0, PA/GO-4, PA/GO-8, and PA/GO-16 membranes, respectively, were 6.95, 2.76, 4.63, and 1.11 percent. The GO-modified membranes, on the other hand, consistently demonstrated higher flow throughout the fouling test, demonstrating improved anti-fouling performance with the GO-containing membrane, which may be explained by the enhanced hydrophilicity resulting from the hydrophilic groups of GO. The membrane's adsorption qualities could be changed by the surface hydrophilicity, improving the membrane's hydrophilicity and to some extent, reducing fouling behavior.

The impact of the GO loadings on the target pharmaceuticals and personal care products compound rejection was also presented in the same study. Both pure PA thin film composite membrane and GO-modified PA thin film composite membrane demonstrated a comparatively low paracetamol rejection (5%), which was attributed to its lowest molar weight and maximum hydrophilicity, due to the phenolic group among the three pharmaceuticals and personal care products. The largest rejection was seen with norfloxacin, which among the three pharmaceuticals and personal care products, has the highest molar weight. Norfloxacin's greater rate of rejection was also influenced by the hydrophobic group on it in the interim. Furthermore, the norfloxacin rejection decreased gradually up to 40 min, after which it performed consistently over time. Norfloxacin was rejected by the PA/GO-4 membrane more frequently than the PA/GO-0 membrane (53.32 ± 1.78 percent vs 43.99 ± 1.44 percent at 60 min), which suggests that the PA/GO-4 membrane is more hydrophilic. However, when GO loading rose, membrane hydrophilicity also increased. This led to an improvement in water flux, which in turn had some impact on the rejection of pharmaceuticals and personal care products. The sulfamethoxazole rejection remained largely constant over the course of the 2 h operation. Sulfamethoxazole ($pK_a = 5.7$) mostly manifested as an electronegative species in the neutral state. According to Wang et al.,[267] the PA/GO-4 membrane had a greater rejection rate (41.85 ± 1.09 percent at 60 min) than the PA/GO-0 membrane (13.56 ± 1.02 percent at 60 min). This is because the embedding of GO increased the membrane's negative surface charge. Their rejection during the fouling test was mostly caused by the electrostatic attraction between the negatively charged sulfamethoxazole and the GO-modified PA thin film composite membrane. However, the PA/GO-16 membrane exhibits lesser sulfamethoxazole rejection, while having the highest negative charge. Many other factors can influence the rejection of pharmaceuticals and personal care products throughout the filtering process. These phenomena might be influenced by how the PA/GO-16 membrane's enhanced water flow affects the rejection of pharmaceuticals and personal care products.[267]

3.6.2.2.2.2   Dyes

This section focuses on the removal of organic pollutants using membranes made of graphene. Wang et al.[268] studied the fabrication of the GO-PDA/ PES-sulfonated PES (SPES) adsorptive membranes for the eradication of RB from aqueous solutions. They began with the preparation of the PES-SPES membranes using a non-solvent immersion precipitation technique. The proposed system of the step-by-step in-situ growth procedure used to create the GO-PDA/PES-SPES adsorptive membrane is shown in Figure 3.33. The catechol structure in the dopamine (DA) polymerized in-situ to generate the DA-preactivated membrane after the PES-SPES membrane was initially submerged in the DA solution. The GO dispersion was then added, and DA self-polymerization caused the GO sheets to develop on the DA-preactivated membrane. The membrane was then removed, dried, and fully polymerized at room temperature to create a solidified GO-PDA/PES-SPES adsorptive membrane.

Figure 3.34a and b showed the results for the membranes. Figure 3.34a showed that when looking at the non-woven fabric, it is clear that no visible adsorption has occurred on RB. In contrast, the GO-PDA/PES-SPES adsorption membrane has a much higher adsorption capacity (reaching 16.57 mg/g) than PES-SPES membrane (at 0.894 mg/g) and DA-preactivated membrane (2.576 mg/g). While Figure 3.34b indicated that the pH level of the solution had a substantial impact on the adsorption capacity of the GO-PDA/PES-SPES membrane. The adsorption capacity increased

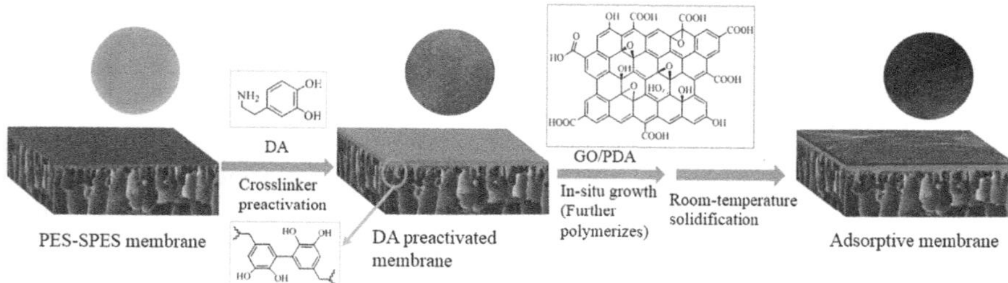

FIGURE 3.33   Schematic diagram of fabrication of GO-PDA/PES-SPES membrane by stepwise in-situ method. (Reproduced with permission from X. Wang et al.,[268] Copyright 2021, Elsevier Science Ltd.)

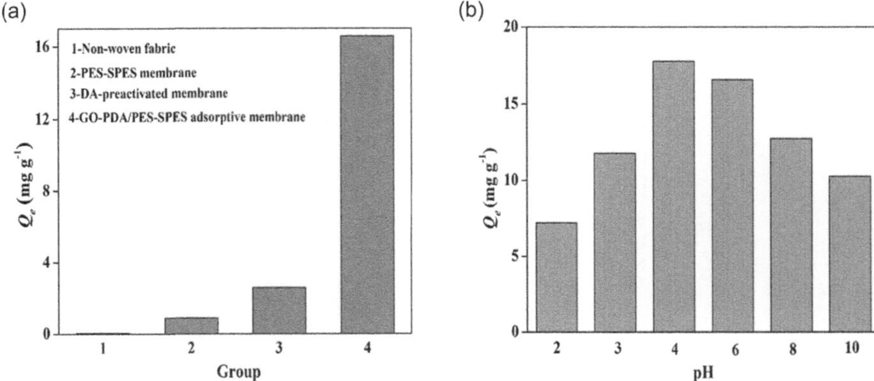

FIGURE 3.34 Static adsorption thermodynamics of GO-PDA/PES-SPES membranes. (a) Adsorption capacity of different membranes, (b) Effects of pH value on adsorption capacity. (Reproduced with permission from X. Wang et al.[268] Copyright 2021, Elsevier Science Ltd.)

from 7.196 to 17.77 mg/g with rising pH values (from 2.0–4.0), while the adsorption capacity decreased, in contrast, as the pH levels rose from 4.0 to 10.0. The findings indicated that the adsorption of RB onto the membrane was accurately predicted by the pseudo-second-order kinetic model and the LM Model, and the GO-PDA layer's maximum adsorption capacity is 87.03 mg/g.[268]

GO nanosheets are used in a revolutionary method to create a new kind of water separation membrane that allows water to circulate via the nanochannels between the GO layers while undesirable substances are rejected by charge effects and size exclusion. The GO membrane was created by layer-by-layer depositing GO nanosheets on a PDA-coated PSF substrate and cross-linking them with 1,3,5-benzenetricarbonyl trichloride. Not only did the cross-linking give the stacked GO nanosheets, but the GO membrane flux was discovered to be between 80 and 276 Liters per square Meter per Hour/MPa, or around 4–10 times more than that of the majority of commercial NF membranes. Although the GO membrane at this stage of development had a relatively low rejection of monovalent and divalent salts (6–46%), it showed an average rejection of MB (46–66%) and a high rejection of Rhodamine-WT (93–95%).[269]

In a different study, Qui et al.[270] found that utilizing the thermally induced phase separation approach, a robust 3- and 2-dimensional homogeneous polyacrylonitrile/ GO (PAN/GO) nano-porous membrane with selective separation properties is generated on a big scale. The successful operation of the membrane was made possible by the addition of 0.2 weight percent GO. The performance of the PAN/GO hybrid membranes in terms of dye rejection is shown in Figure 3.35. It is evident that all membranes exhibit higher rejection to methyl blue than AR 18. With 100% rejection for methyl blue and 99.8% for AR 18, PAN-0.2 has the highest level of rejection. Due to electrostatic repulsion and size sieving effects, GO is known to create a negative charge on membrane surfaces as a result of the acidic functional groups, which resulted in a significant rejection of negative dyes (both AR red 18 and methyl blue) due to these effects. They also stated that there was little doubt the expansion of the

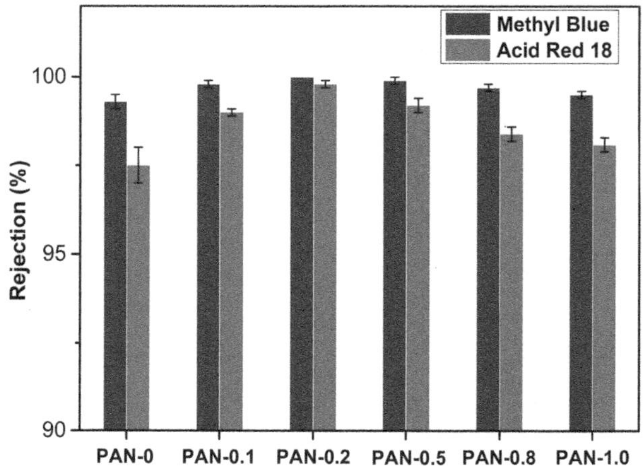

FIGURE 3.35   Dye rejection performance of the PAN/GO hybrid membranes. (Reproduced with permission from Qiu et al.[270] Copyright 2018, Elsevier Science Ltd.)

size of the pores in the separating layer caused a drop in the crystal size and average pore radius of the hybrid membrane, which rose with rising GO concentration. Additionally, the hybrid membrane presented a desirable option for the manufacture of unstructured NF membranes with performance improvement in a practicable manner.

3.6.2.2.2.3   Phosphates and Nitrates

Some of the most serious environmental contaminants that come from human activity in soil, atmospheric, and aquatic body contamination are pesticides. The increased need for food production and health-protection initiatives has resulted in an unchecked use of pesticides in recent decades, which has had negative effects on human health and the environment.[271] CS-GO composite membranes comprising various concentrations of GO have been created and employed as sorbents for solid-phase extraction of pesticides in a work by Silvestro and colleagues. Membranes made of CS GO were also glutaraldehyde surface crosslinked to increase their dimensional stability in a liquid medium. Each pesticide's recovery capacity was measured and reported in Figure 3.36 as a percentage ($R\%$). It was easy to confirm that the yields were dependent on the membranes' hydrophobicity by looking at the $R\%$ data. In overall, when using CS or CS-GO with low GO concentration, contaminants having a log K_{ow} between 0.8 and 2 (pesticides A–D; blue-colored) displayed low $R\%$ (1 percent and 5 percent). This might be because those composite membranes' higher hydrophilicity allowed for stronger interactions with the contaminants yet impaired the extraction process. Notably, the $R\%$ values in the CS-GLU sample were somewhat greater than those in CS-GO1 GLU and CS-GO5 GLU. The CS-GLU membrane's (Figure 3.36) limited permeability most likely allowed the pesticides to adsorb, mostly on the surface layers, making extraction that ought to be simple and very challenging. It's interesting to note that the recovery of those

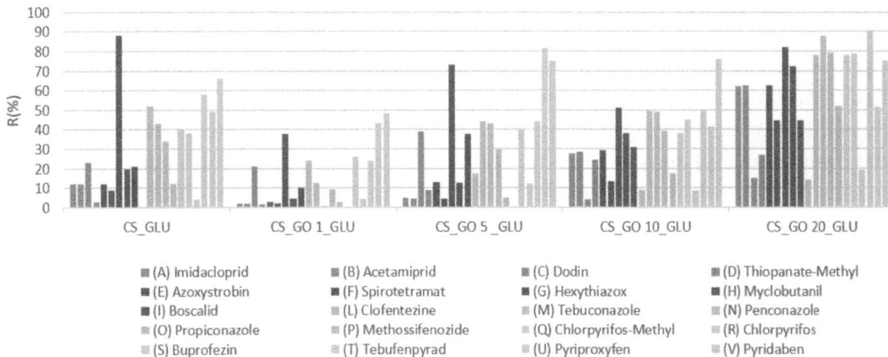

FIGURE 3.36 Recovery capacity of the CS_GLU and CS_GOX_GLU composite membranes. Pesticides are reported in ascending order of log K_{ow}. (Reproduced from Silvestro et al.[272] Copyright 2021, the Authors.)

pollutants was improved in cases when there was a lot of GO, like in the case of the CS-GO10 GLU and CS-GO20 GLU composites. A synergistic interaction between CS and GO was theorized to explain this behavior. In reality, a favorable interaction between insecticides and composite membranes with a low log K_{ow} was enabled by the concurrent presence of hydrophilic groups (NH_2 and OH groups of CS, together with OH and COOH groups of GO). Therefore, it became feasible to modify the CS adsorption characteristics to favor interactions with hydrophobic contaminants and minimize interactions with hydrophilic pollutants by using composite membranes with a high quantity of GO. Hexythiazox (G) is the one exception, whereby a high *R%* was achieved using all systems; the same pattern of behavior was observed for pesticides with a log K_{ow} value between 2 and 3 (pesticides E, G, H, and I; red-colored). For all composites, evaluating the pesticides with a log K_{ow} between 3 and 6 (green and yellow colors) revealed quite high R percent values. Intriguingly, the extraction of more hydrophobic contaminants was not hampered by the increase in matrix hydrophobicity; on the contrary, it increased along with the rise in GO concentration. Actually, the CS GO20 GLU sample worked best as an absorbing substance, producing the best results. This behavior implied that by adjusting the GO concentration in the CS-based membranes, the production of hydrogen bonds and hydrophobic interactions (-stacking) might be regulated, with the benefit of a superior pollutant extraction.[272]

The ability of nano chitosan-GO composite (NCS@GO) to eliminate nitrate (N) and phosphate (P) from aqueous solutions was examined in a different study by Salehi et al.[273] To make the NCT@GO composite selective for the adsorbate anions, high and low concentrations of zirconium (Zr) were added. A response surface methodology (RSM) was used to construct NCS@GO/H-Zr, which showed outstanding P and N uptake of 172.41 mg/g and 138.88 mg/g, respectively, with an acceptable pH range of 3 to 11, viable selectivity between the two adsorbates contestant anions, intended recycling potential and effectiveness of P and N's desorption, and retained 76 percent and 85 percent for P and N adsorption capacity, following 10 cycles.

3.6.3 Porous Materials

Nanopores are created across individual graphene sheets to create advanced 2D nanomaterials known as nano-perforated graphene (NPG) materials.[285] Despite the possibility of developing nanopores from isolated defective sites, careful design of the dimension and density patterns of the pores necessitates certain perforation techniques. To control surface properties and absorption mechanisms throughout the perforated materials and to take advantage of the altered properties at the nanoscale, the creation of pores in a highly controlled manner is essential.[90,286]

The uses of the NPG materials include separation, sensors, batteries, and super capacitators. The porous graphene (microporous and mesoporous interconnected structure), with a focus on separation, allows access to ion and molecule movement, opening up new possibilities for developing creative materials with increased adsorption capabilities and reusability. Furthermore, NPG is a more desirable option for the removal of emerging contaminants owing to its larger surface area, super hydrophobicity, simple water separation, recyclable materials, and chemical stability, especially when it is generated utilizing a low-cost, scalable synthesis technique.[148] NPG is typically synthesized via chemical processes (Figure 3.37), such as chemical etching, chemical vapor deposition, doping, templating, electron beam irradiation, and ion bombardment.[287] These methods, however, are exceedingly expensive and not practical for use since they require complicated processes like physical ion beam irradiation and organic synthesis in order to produce small, uneven amounts of porous structured graphene. Therefore, it is still extremely difficult to evenly inject pores into a graphene sheet. By converting chemically produced GO into porous rGO through reduction and thermal processing, porous graphene can be manufactured on a massive scale. Theoretically, graphene has a SSA of 2630 m^2 g^{-1}.[288] Perforation

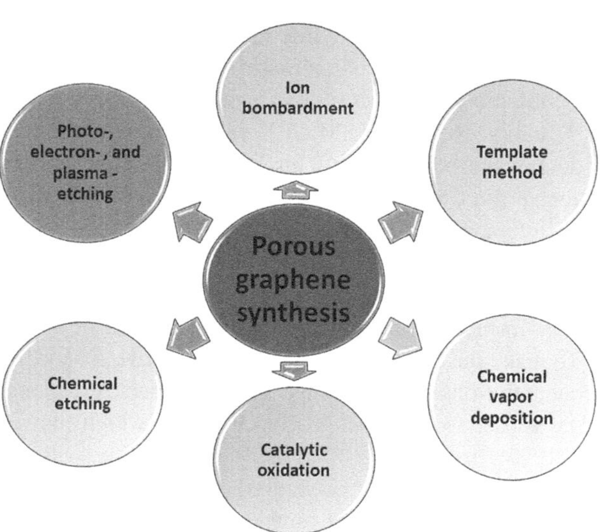

FIGURE 3.37 Synthesis methods for porous graphene.

techniques need to be developed and enhanced if materials made from NPG are to attain their maximum capacity in a variety of applications.[289–291]

3.6.3.1 NPG Synthesis Methods

3.6.3.1.1 Photo-, Electron-, and Plasma-Etching

With the use of light, electron beam, and oxygen plasma etching techniques, numerous endeavors in research have been made to create porous graphene by locally oxidizing or degrading flaws on the surface of graphene sheets.[292–294] Akhavan et al.,[295] for instance, described how to create graphene nano mesh (GNM) with 200 nm pores by photodegrading GO sheets with UV assistance at the point where ZnO nanorods' tips come into contact with the sheets (Figure 3.38). By constructing the sheets on the nanorods, The GO sheets were joined physically to the tips of the ZnO nanorods, and the generated pore diameters were consistent with the ZnO nanorods' diameter. Without doping, the resulting GNM with a macroporous structural architecture displayed p-type semiconducting properties. Additionally, the diameter of the ZnO nanorods can be used to alter the size of the holes on GNM, and in the future, the quantity of holes on GNM has to be raised.

Additionally, the nano-sculpting of the suspended graphene sheets by the electron beam may be controlled. With this method, various characteristics of the graphene sheets, such as gaps, slits, and holes at the nanoscale, can be seen in a matter of seconds. Applying the aforementioned process makes it simple to construct GNM in an eco-friendly manner, but it may be difficult to precisely regulate the size, shape, and distribution of the resulting porous structure.

It has been demonstrated that treatment using focused electron-beam irradiation in a Transmission Electron Microscopy (TEM) at room temperature may controllably nano sculpt suspended multilayer graphene sheets with few-nanometer precision. Moreover, the structures created by electron beam irradiation are steady and persistent over time. The graphene sheet does not experience any substantial long-range deformation as a result of the considerable removal of carbon. Particularly, during cutting, the sheets do not begin to distort, curl, fold, or fold away from the focal point.[296] To synthesize the NPG, Lemme and coworkers have etched graphene with gaps down to

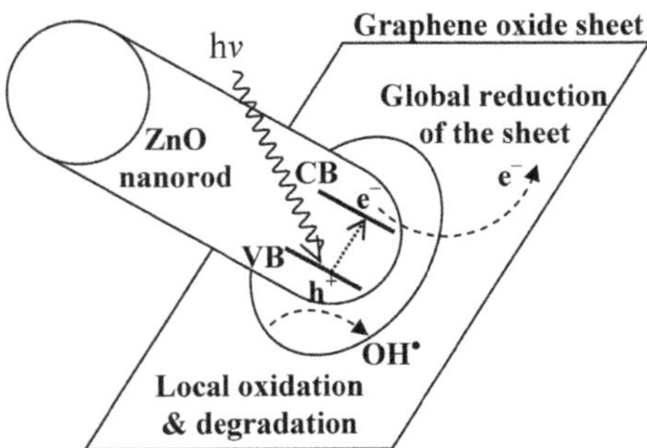

FIGURE 3.38 Schematic illustration of the mechanism describing the formation of graphene nanomeshes by using the photocatalytic property of the ZnO nanorods. (Reproduced with copyright permission from Akhavan et al.[295] Copyright 2010, American Chemical Society.)

about 10 nm utilizing the helium ion beam.[297] Sequential high-resolution imaging was used to He ion etch a suspended graphene with dimensions of about 150 nm in length and 1.5 nm in width. A perspective range of 2 m × 2 m and image with a size of 2048 × 2048 pixels were used to expose the graphene to the He ion beam, resulting in pixels that were spaced ~1 nm apart. With a duration of 50 µs, a reliable line dosage of 0.8 nC/cm was obtained. Such a high-resolution image is displayed in Figure 3.39b, which has been enlarged and clearly marked to separate the suspended graphene from the underlying SiO_2 and the chromium/gold interactions. Figure 3.39c displays a series of photos obtained under similar circumstances (No. 1–3, where picture 1 is the same as Figure 3.39b). The graphene flake's circle designates the area where the initial etching took place. The amount of material that was etched was found to increase with each scan with the He ion beam. The dwell time, and consequently the image quality, was enhanced to 500 s after 13 scans, which is equivalent to a line dosage of 8 nC/cm but remains insufficient to fully etch the graphene (Figure 3.39c, No.14). According to the images, it is preferable to remove edge atoms from the graphene crystal than

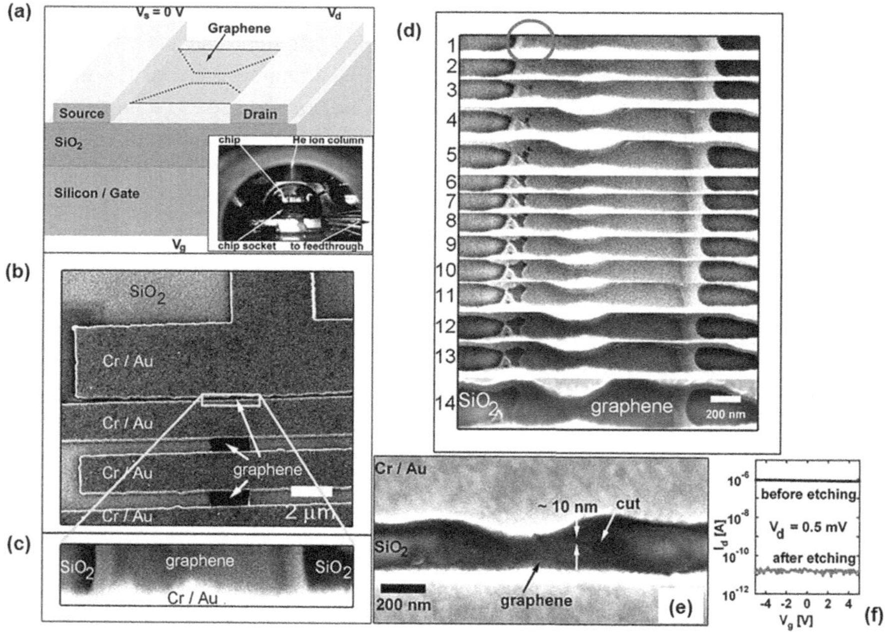

FIGURE 3.39   (a) Schematic of a graphene device. Inset: photograph of the microscope chamber with installed chip, (b) HeIM image of suspended graphene devices. The yellow box indicates the area that was subsequently imaged and etched in high resolution. (c) High-resolution image used to etch graphene, (d) HeIM image (with false color) of a suspended graphene device after etching with minimum feature sizes of about 10 nm (color online). The circle indicates the area where etching occurred initially (color online). (e) HeIM image (with false color) of a suspended graphene device after etching with minimum feature sizes of about 10 nm (color online) and (f) Electrical measurement of the device before and after etching. (Reproduced with permission from Lemme et al.[297] Copyright 2009, American Chemical Society.)

inside atoms. With a field of view of 100 to 10 nm, the residual graphene sheet was etched utilizing live scanning mode. The live screen image in this case was used to verify etching. The HeIM image in Figure 3.39d depicts a consequent cut with minimal features with 10 nm or smaller diameters.[297]

In a different study, Zeng et al.[298] documented the synthesis of GNM utilizing an etch mask made of an anodic aluminum oxide (AAO) membrane and a poly (methylmethacrylate) (PMMA), the layer of adhesion between the rGO sheets and the AAO template on the Si/SiO_2 wafer. The periodic and consistent pore size of GNM under O_2 plasma etching could be adjusted by utilizing several AAO templates. The GNM's pore size of about 67 nm and width of approximately 33 nm was formed when 8 nm thick PMMA was utilized as the adhesion layer, and O_2 plasma etching was carried out for 30 s. PMMA films with thicknesses of 45, 15, and 8 nm were prepared. A good-quality GNM with a large area of 100 μm^2 was produced by optimizing the thickness of PMMA at around 8 ca. nm. With 40 seconds of O_2 plasma etching, holes in rGO could be produced when the thickness of PMMA was increased to 15 nm. However, the quality of the etched rGO sheet and the density of the holes produced were poor. The hole density increased with a longer etching period (for instance, 60 s). While some regions of the rGO sheets were not etched, certain holes were expanded and fused together. This was attributed to the fact that the thicker PMMA layer required more time to punch through and because the thickness variation in PMMA causes uneven etching of rGO sheets. Etching with O_2 plasma is also an isotropic etching method, which means it happens both laterally and vertically. As a result, lateral etching in the PMMA layer became substantial with a longer etch time. The rGO sheets' holes grew and took on an uneven shape as a result. In fact, with the PMMA film thickness of 45 nm, many of the rGO sheets were totally destroyed, whereas other portions remained intact after 120 s of O_2 plasma etching. All of these findings indicated that optimizing the etching time of O_2 plasma and PMMA layer thickness was necessary to produce high-quality GNM.[298]

3.6.3.1.2 Template Method

An efficient technique for converting graphene into porous graphene is template synthesis. The transformation utilizes a variety of organic and inorganic structures as templates. The right template might be chosen based on the desired dimensions and morphology of pores. Both the soft-template approach and the hard-template method fall under the heading of this method.

3.6.3.1.2.1 Soft-Template Methods

Under moderate synthesis circumstances, the soft-template approaches employ a variety of amphiphilic molecules as structure-directing agents, including surfactants and copolymers. The fabrication of mesoporous materials with a 2D sandwich structure made of layers of graphene and mesoporous silica was done from the bottom up using the cationic hexadecyltrimethylammonium bromide (C16TAB). The negatively charged nature of GO sheets is caused by oxygen functionalized groups present. In an alkaline solution, C16TAB has the capacity to electrostatically adsorb and self-assemble on the surface of negatively charged GO. After the elimination of C16TAB and hydrolysis of the silicon precursor tetraethyl orthosilicate (TEOS), the GO-based silica hybrid was created.[299]

High-temperature thermal annealing produces graphene-based silica sheets. The statistics on adsorption show a high specific area of 980 m²/g. Another approach involves hydrothermally processing a suspension of graphene, C16TAB, and $SnCl_4$ to create a mesostructured graphene-based SnO_2 composite.

Diankov et al.[300] investigated how graphene formed on SiO_2 was affected by remote hydrogen plasma. For processes involving plasma species, strong monolayer selectivity was seen, as evidenced by the creation of isotropic holes in the monolayers' basal plane as well as etching from the sheet's borders. When compared to bilayers or thicker sheets, etch pit density on monolayers is about two orders of magnitude larger. The etch pit shape was also very different for bilayer or thicker sheets: uniformly sized hexagonal etch pits showed that etching is very anisotropic and starts from previously existing flaws instead of nucleating continually, as on monolayers. Regarding graphene, both monolayer and multilayer, the etch rate showed a clear relationship with sample temperature: it was very sluggish at ambient temperature, peaked at 400 °C, and was completely inhibited at 700 °C. When graphene is treated with hydrogen plasma on the significantly smoother mica substrate, the phenomenology is quite similar to that of the rougher substrate SiO_2, indicating that something besides the roughness of the substrate regulates the reactivity of monolayer graphene with hydrogen plasma species.

3.6.3.1.2.2   *Hard-Template Methods*

For the hard template method to create porous graphene, the template must first be prepared. To do this, the actual hard template must be prepared, and its surface must be functionalized to acquire the needed properties. Graphene or GO can then be applied to the template, depending on the need. In the final phase, the template is selectively eliminated without damaging its framework.[301]

In order to create foams with nano porous graphene having regulated pore size, high surface area, and ultra-large pore volume, Huang et al.[302] established a unique hard-template design strategy driven by hydrophobic interaction. It has been demonstrated that choosing hydrophobic silicon spheres as templates is essential for producing nanoporous graphene foams. Their graphene foams had the highest overall pore volume value of any porous graphene materials that have been disclosed (4.3 cm³ g⁻¹). According to their results, the surface area was 851 m² g⁻¹. The hydrophobic basal planes of GO interact with the hydrophobic surface of methyl group grafted silica spheres to produce self-assembling lamellar-like structures.[302]

Su et al.[303] used a straightforward, hard template-directed, ordered assembly method to make a 3D bubble-shaped macroporous graphene film, using a hollow graphene bubble to serve as a replicable building block. The precisely constructed PMMA spheres of the GO precursor and template were created using vacuum filtering. The template was effectively removed utilizing pyrolysis after GO was next converted into graphene during calcination. The free-standing film was developed.

3.6.3.1.3   *Chemical Etching*

Chemical etching or activation of graphene planes is a template-free method of creating 2D porous graphene. By effectively oxidizing C atoms with either catalysts or oxidants, graphene planes may be effectively etched.[304] Xu and colleagues etched graphene in a single step using H_2O_2 as an oxidant, and they noticed that the graphene self-assembled into a 3D network structure. In this procedure, a well-dispersed aqueous dispersion of GO was infused with a measured quantity of H_2O_2 aqueous solution. Mechanically strong monolithic (Holey Graphene Frameworks) HGFs were produced by autoclaving the mixture for six hours at 180 °C with Teflon lining. Hydrogels were created by reducing and self-assembling graphene sheets by using a hydrothermal technique. These hydrogels have an interconnected 3D macro-porous network, with pores that range in size from sub-micrometers to several micrometers. The GO defects with higher activity were partially oxidized and etched by H_2O_2 molecules at the same time, leaving voids in carbon that progressively evolved into nanopores. Seeing as the flawed carbon sites were often dispersed throughout the GO's basal plane, the etching process took place over the entire graphene sheet, producing a lot of in-plane pores that were a few nanometers in size. Control trials without the addition of H_2O_2 produce Graphene Frameworks (GFs) that are non-holey graphene sheets. Additionally, too much H_2O_2 would cause an aggressive etching that would enlarge the holey graphene's pore size and shatter it into tiny pieces. Due to this, the HGFs got to be exceedingly fragile and unsuitable. The freeze-dried HGF has a surface area of 830 $m^2\,g^{-1}$ that was precisely determined by Brunauer-Emmett-Teller (BET) method, which is significantly greater than that of non-holey GF, according to tests on nitrogen adsorption-desorption ($\sim$260 $m^2\,g^{-1}$). Utilizing a hydraulic press, the as-prepared HGF may be significantly compressed to create a compact and independent HGF film with an approximately 60-fold increase in packing density, thanks to its extremely porous structure and outstanding mechanical stability.[305]

In a work by Lui and colleagues, mono- and multiple-layer areas of graphene were both observed in the same sample at the same time (Figure 3.40), allowing for direct comparison of oxidation as a function of layer count. At various temperatures, samples were warmed up in an O_2/Ar gas flow.[306] Examining dozens of samples that

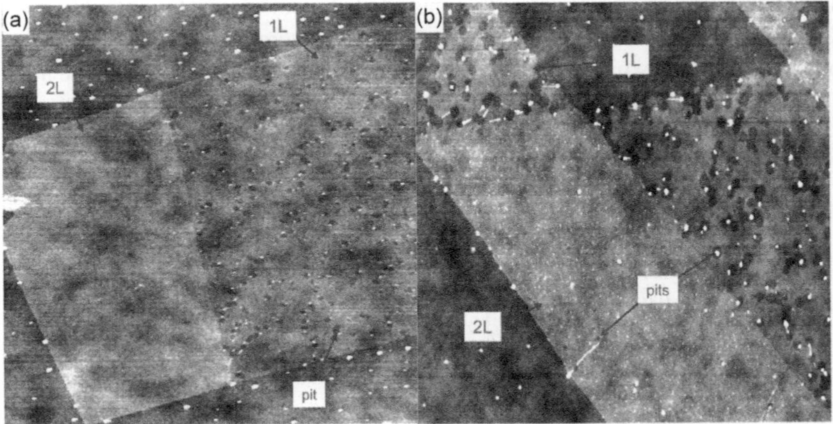

FIGURE 3.40 AFM images of oxidized single-layer (1L) and double-layer (2L) graphene. (a) (5.06 × 5.06 μm²) Oxidized at 500 °C for 2h (P(O_2)) 350 torr. (b) (6.95 × 6.95 μm²) Oxidized in reduced O_2 pressure at 600 °C for 40 min (P(O_2)) 260 torr. (Reproduced with copyright permission from L. Liu et al.,[306] Copyright 2008, American Chemical Society.)

had undergone oxidation at various temperatures, it was found that single layers underwent oxidative etching more quickly than multiple layers. In single-layer graphene, at 500 °C, oxidation results in 20–180 nm in diameter etch pits, but not in double-layer sheets (Figure 3.40a).[306]

Another technique that is effective for producing porous graphene is acid-etching. Carbon erosion may occur if graphene is exposed to particular acid/oxidizer solutions. When GO is exposed to the acid/oxidizer solution while being sonicated or exposed to microwave radiation, this reaction takes place. HNO_3 and $KMnO_4$ are two frequently utilized acids and oxidizers for etching graphene.[307] As shown by Fan et al.,[308] $KMnO_4$ can be used as an oxidant to prepare porous graphene when subjected to microwave radiation (Figure 3.41). With a surface area of 1374 m^2 g^{-1}, the graphene with pores, as a result, was distinguished by pores that were roughly 3 nm in size. As illustrated below, the reaction between carbon and $KMnO_4$ produced the pores in the graphene:

$$4MnO_4^- + 3C + H_2O \leftrightarrow 4MnO_2 + CO_3^{2-} + 2HCO^{3-} \tag{3.1}$$

In another study, Romanos and colleagues showed that nanospace engineering of KOH activated carbon is possible in the case where a powerful base, such as KOH, is used to create porous graphene by regulating the level of carbon consumption and metallic potassium intercalation into the carbon lattice during the activation process.[309] They were able to quantitatively control the creation of high SSA, porosities, and sub-nanometer (1 nm) and supra-nanometer (1–5 nm) pore volumes by combining KOH concentration and activation temperature. The technique frequently resulted in a bimodal distribution of pore sizes, a substantial, generally a fixed, number of sub-nanometer pores, and a fluctuating number of supra-nanometer pores. They showed the relationship between porosity, KOH:C weight ratio, and activation temperature; they proved that the porosity plateaus at 900 °C, above KOH:C = 3. In reality, the amount of metallic potassium that penetrated the carbon matrix increased as the KOH:C weight ratio increased. The carbon matrix grew larger and became more porous. The lattice expanded further as the temperature increased, and as the intensity of carbon consumption increased, so did the porosity.

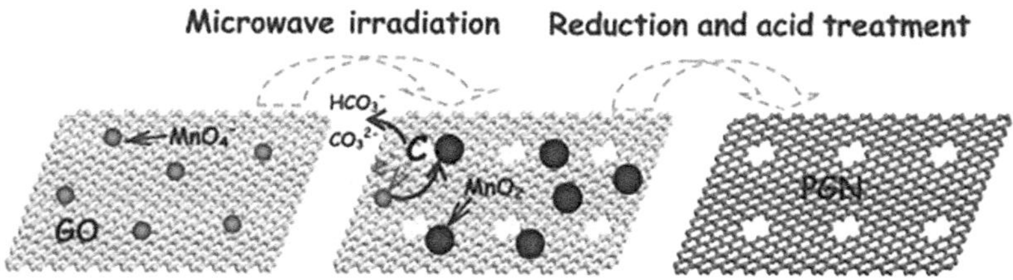

FIGURE 3.41   Illustration of the formation of porous graphene material with pores on the surface of sheet. (Reproduced with permission from Z. Fan et al.,[308] Copyright 2012, Elsevier Science Ltd.)

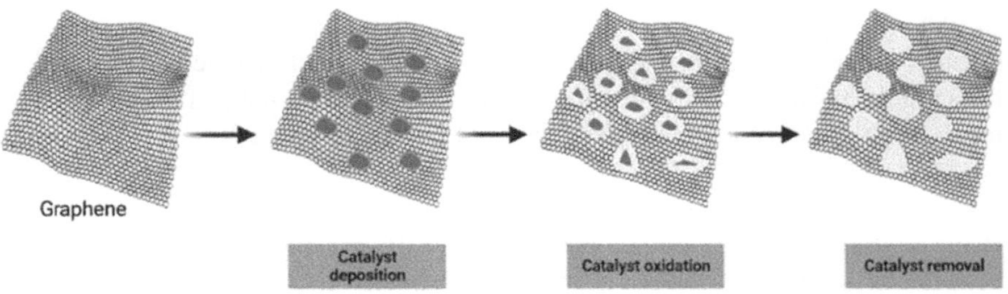

FIGURE 3.42 The three-step procedure for making holey graphene.

3.6.3.1.4 Catalytic Oxidation

Graphene can also be catalytically oxidized with metal, metal oxide, or nonmetal oxide to produce GNM. For example, Lin et al.[310] explored a straightforward and easy to control technique for producing holey graphene structures, in which the graphene surfaces were etched with holes through silver (Ag) nanoparticle-catalyzed graphitic carbon oxidation that had been placed (Figure 3.42). In this method, graphene was first decorated with catalytic Ag nanoparticles, resulting in graphene with an Ag nanoparticle surface (Ag-G). MWNTs were used as the supporting material in a solvent-free deposition procedure to accomplish this. In the second phase, an open-ended tube furnace was used to heat the Ag-G samples and subject them to well-managed air oxidation. On the initially unbroken graphene surfaces, a sizeable number of holes afterwards emerged. The majority of holes had Ag nanoparticles adhering to them. Under the conditions present, some holes also showed up as tracks that looked to be connected to the directional motion of the attached Ag nanoparticles. In Step II (i.e., oxidation temperature at 300 °C) slightly oxidized Ag-G samples were refluxed in diluted (2.6 M) nitric acid in the third phase after the Ag catalysts had been removed. After a meticulous water treatment, the solid was dried. Given that nitric acid converted metallic Ag into Ag^+ during this process, each and every catalytic Ag nanoparticle was entirely eliminated from the samples (i.e., $AgNO_3$). With repeated washings, the Ag salt could be successfully removed because it was soluble in the aqueous dispersion. The method was found not only flexible because it allowed for regulated hole sizes on the graphitic surface, but it was also simple to scale up in comparison to the methods for producing holey graphene. Furthermore, they postulated that this method could increase the practicality of using holey graphene in a variety of applications that require huge quantities.

3.6.3.1.5 Chemical Vapor Deposition (CVD)

A template CVD technique developed by Ning and colleagues prompted the gram-scale fabrication of nano-mesh graphene with a tightly controlled structure.[311] The nano-mesh graphene's porous structure had pores smaller than 10 nm and surface corrugations and the restriction of the number of graphene layers was less than two. Both contributed to stopping the agglomeration of the sheets of graphene; it possessed substantial SSAs up to 1654 m^2 g^{-1} and substantial total pore volumes up to 2.35 cm^3 g^{-1}. The nano-mesh graphene demonstrated exceptional electrochemical capacitance (up to 255 F g1), cycling stability, and rate performance because of its distinctive porous structure and high surface area. This synthesis method is beneficial for applications that require porous graphene as a bulk material and offers prospects for the high-volume, low-cost manufacture of porous graphene.[311]

3.6.3.1.6 Ion Bombardment

Using a straightforward method, O'Hern et al.[312] investigated the ionic transport behavior of regulated, a high density, small diameter holes over large expanses of mono-layer graphene produced by CVD. They used ion bombardment to first create reactive, localized flaws in graphene, and then oxidative etching to grow those defects into permeable pores. They showed control of mono-layer graphene membrane selectivity at the sub-nanometer length scale using potassium chloride and an organic dye diffusion measurement. In scanning transmission electron micrographs (STEM)-observed ion-bombarded and etched graphene, they demonstrated the appearance of many pores of sub-nanometer size (Figure 3.43d–f). No pores in the graphene were seen when ion bombardment was not present (Figure 3.43a–c).

In a related study, Celebi et al.[313] created 2D membranes using a simple, dependable procedure. To generate graphene with few defects and good grain connection and avoid the occurrence of unfavorable cracks, this method used CVD. It was possible to produce freestanding graphene layers with a thickness of less than 1 nm by cleanly transferring dual graphene layers in succession onto a SiNx frame with

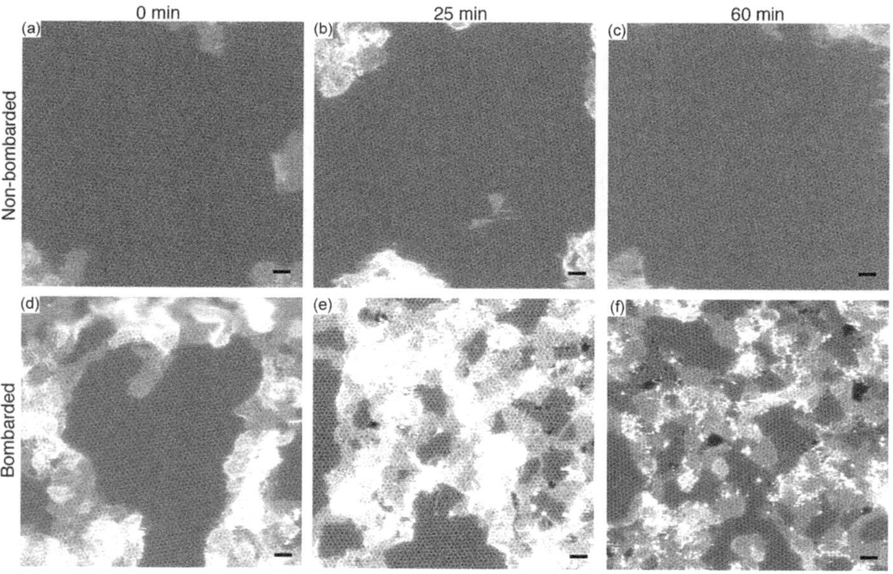

FIGURE 3.43   STEM of pores created in graphene membranes. Comparison between graphene etched in acidic potassium permanganate with (d–f) and without (a–c) ion bombardment demonstrates that both bombardment and etching are necessary for pore creation. Etch times are 0-, 25-, and 60-min. Scale bars are 1 nm. (Reproduced with permission from O'Hern et al.[312] Copyright 2014, American Chemical Society.)

49 pores that were each 4 μm in diameter. It was discovered that the freestanding graphene was strengthened by the double transfer and was prevented from leaking due to random flaws. Because grain boundary flaws, polymer residues, or dust particles might promote fracture development while perforating the graphene, it has been discovered that the cleanliness and quality of the graphene are critical during this graphene transfer procedure. Water and gases were still unable to pass through the double-layer graphene film. Then, to create porous membranes, nanopores were created using a focused ion beam (FIB). For drilling nanopores smaller than 10 nm in diameter, a He-based FIB was employed, while a Ga-based FIB was used to perforate apertures between 14 nm and 1 mm in diameter. Fast and precise drilling was made possible by low exposure doses (5×10^{-6} to 5×10^{-5} pA/nm^2 for Ga$^+$ ions and 6×10^{-3} pA/nm^2 for He$^+$ ions), which produced well-defined pore diameter distributions. Additionally, they demonstrated that their porous graphene membranes absorb water considerably more quickly than ultra-filtration membranes made of acrylic, cellulosics, and PSF, hence requiring less pressure.[313]

Helium ions may be precisely used in a modified helium ion microscope for lithography to cut and design graphene. The foundation of helium ion microscopy is the field ionization of helium ions using a cryogenically cooled tungsten tip that has been shortened by an atom trimer. The gun is positioned in the middle to enable imaging using the ion emission from a single trimer atom. The imaging gas pressure can be changed to alter the beam current, with typical operation occurring between fA and pA. Interaction between the sample and the ion beam affects both the device's resolution when used as a microscope and the smallest size of an etched line when used as a lithography tool. According to Bell et al.,[314] graphene on Si/SiO$_2$ substrates can be patterned, aligned, and exposed in accordance with a design with computer-controlled exposure. It was also noted that, once appropriate beam doses were established, the sample would suffer only minor visible damage or doping, whereas, in the case of a gallium ion beam, extensive deterioration and ionic contamination would develop under similarly accelerated voltage due to the higher atomic mass of the gallium ion.

In a more recent study, Wei and workmates reported the establishment of an effective method for concurrent defect engineering and in-situ reduction of GO membranes utilizing carbon ion beams.[315] It was initially revealed that GO membranes' laminar structure may be controlled in order to reduce interlayer space and customize nanoscale holes. They discovered that the GO structure exposed to an ion beam was significantly influenced by the GO layer's thickness and the ion fluence used. Further, they revealed that the ion beam-induced method could be successfully used on GO membranes that had been modified with ethylenediamine (EDA) for improved selectivity, which suggests that it could have far broader applications.[315]

3.6.3.2 Removal of Metal Ions

Metal ions have been adsorbed using divers graphene-based porous graphene materials as per literature (Table 3.6). In a study by Zhou et al.[316] porous graphene was used to remove Pb^{2+} from an aqueous solution. They studied the effect of utilizing different hydrothermal treatment periods on the removal of Pb^{2+} and revealed that the adsorption capacity of Pb^{2+} decreased with an increasing hydrothermal treatment

TABLE 3.6
Summary of HM Ions from Water Using Porous Graphene and Its Derivatives

Graphene-Based Material	Metal Ion Pollutant	Maximum Adsorption Capacity, mg/g	Ref.
Porous GO/carboxymethyl cellulose (CMC) monoliths	Cu^{2+}	82.93 mg/g	319
	Pb^{2+}	76.70 mg/g	
	Cd^{2+}	46.13 mg/g	
	Co^{2+}	59.99 mg/g	
	Ni^{2+}	72.04 mg/g	
Porous graphene	Pb^{2+}	150 mg/g	316
Porous graphene nanomembranes	Cu^{2+}	84.76 mg/g	320
	As^{3+}	7.92 mg/g	
Porous GO-maize amylopectin composites	Pb^{2+}	84.76 mg/g	318
	Mn^{2+}	7.92 mg/g	
	$Cr_2O_7^{2-}$	13.6 mg/g	
	Cd^{2+}	17.64 mg/g	
	Cu^{2+}	30.56 mg/g	
	Nd^{3+}	25.52 mg/g	
	La^{3+}	12.48 mg/g	
	Y^{3+}	16.96 mg/g	
	Yb^{3+}	23.32 mg/g	
	Er^{3+}	30.32 mg/g	
Novel porous graphene-like carbon hydrogel	Ag^{+}	234.32 mg/g	321
Sulfur, nitrogen dual-doped porous graphene nanohybrid	Hg^{2+}	803 mg/g	317
	As^{3+}	99%	287
Porous GO based inverse spinel nickel ferrite nanocomposite	As^{3+}	42%	322
	As^{4+}	79.2%	
Reduced porous GO based inverse spinel nickel ferrite nanocomposite	As^{3+}	53%	
	As^{4+}	99.7%	

period. Furthermore, they explained that porous graphene with a shorter hydrothermal treatment period had more graphitic structural defects and oxygen-containing functional groups, which contributed to increased adsorption capacities.

Doping of porous graphene material has also been investigated by Tain and colleagues.[317] They proposed the dual doping of porous graphene with sulphur and nitrogen; and investigated it for the adsorption of Hg^{2+}, Pb^{2+}, Cd^{2+} and Cu^{2+} ions. They found that the sulphur and nitrogen-doped porous graphene nanohybrid material exhibited outstanding selectivity and ability to separate Hg^{2+} ions in an aqueous solution with other metals. Furthermore, the maximum adsorption capability was 803 mg/g with high removal ability of greater than 99% at any pH between 2 to 10.

Zhao et al.[318] studied the adsorption-desorption of properties of porous graphene-maize amylopectin composites for various metal ions, then found that the maximum adsorption capacities of Pb^{2+}, Mn^{2+}, $Cr_2O_7^{2-}$, Cd^{2+}, Cu^{2+}, Nd^{3+}, La^{3+}, Y^{3+}, Yb^{3+} and Er^{3+} were 84.76, 7.92, 13.6, 17.64, 30.56, 25.52, 12.48, 16.96, 23.32 and 30.32 mg/g, respectively.

3.6.3.3 Removal of Organic Pollutants

3.6.3.3.1 Removal of Pharmaceuticals

Porous graphene is considered to be a more desirable alternative for ECs (including pharmaceuticals) removal because of its larger surface area, super hydrophobicity, ease of separation from water, recyclability, and chemical stability, especially when it is created using a low-cost, scalable synthesis technique.[287,323] A summary of the adsorption of pharmaceuticals using porous graphene-based materials is depicted in Table 3.7. Atenolol (ATL), carbamazepine (CBZ), ciprofloxacin (CIP), diclofenac (DCF), gemfibrozil (GEM), and ibuprofen (IBP) are six commonly used pharmaceuticals. Khalil and coworkers evaluated the effectiveness of porous graphene (PG) synthesized at relatively low temperatures as a potential candidate for the removal of these drugs from their aqueous solution.[148] In addition, they demonstrated quick kinetics and adsorption capacities above 100 mg/g for some of the tested contaminants, as well as significant removal efficiencies for trace concentrations of all investigated emerging contaminants (>99 percent) at a low dose of porous graphene (100 mg/L).[148] Furthermore, in another study, Khalil and associates[324] investigated the viability and effectiveness of an innovative supported (reduced) graphene oxide-boron nitride foam (GO/BNF), which was made using a straightforward foam-gel casting process, as a highly reactive filter medium for removing gemfibrozil from tap water and wastewater. The removal rate of GEM medication from aqueous media was then accomplished using the GO/BNF, which had the fastest adsorption kinetics of all the evaluated graphene-based materials (GO and PG). The highest adsorption capacity of GO/BNF observed was 55.1 mg/g. One potential method of GEM adsorption by the graphene-based composite foam was discussed in terms of chemisorption. High correlation coefficient values led to the conclusion that the LM was followed during the GEM's adsorption equilibrium with the foam material. The LM is frequently the best representation of a chemical adsorption process in which the adsorbent and adsorbate produce ionic or covalent chemical interactions between their reactive groups. The kinetic data's good fit to the pseudo-second-order kinetic model contributed more credence to the chemisorption theory.

Through the use of a nucleophilic aromatic substitution process, porous GO adsorbents were successfully developed by joining GO sheets with decafluorobiphenyl (DFB). According to Shan et al.,[325] the increased surface area and pore size of the GO-based adsorbents as they were created made them ideal for the diffusion and adsorption of common pharmaceuticals such as CBZ, sulfamethoxazole (SMZ), sulfadiazine (SDZ), IBP, paracetamol (PCT) and phenacetin (PNT). The adsorption capacities of GO-DFB20 for CBZ, SMZ, SDZ, IBP, PCT and PNT were 340.5, 428.3, 214.7, 224.3,

TABLE 3.7

Summary of Organic Pollutants Removed from Water Using Porous Graphene and Its Derivatives

Graphene-Based Material	Pollutant	Maximum Adsorption Capacity, mg/g	Ref.
Dyes			
Porous G-like carbon hydrogel	MB	13381.62 mg/g	321
Porous GO-based carbon	MB	1100 mg/g	329
Highly porous graphene/CNT hybrid beads enhanced by carbonized polyacrylonitrile	MB	521.5 mg/g	330
Porous reactive graphene nano-adsorbents (APG)	MB	821.05 mg/g	331
Porous graphene microspheres	MB	596 mg/g	326
	MO	65 mg/g	
Porous graphene nanosheets	RB	99%	287
	MB	99%	
Nitrates and phosphorous			
Porous graphene nanosheets	Nitrate	99%	287
Porous bio-templated GO/MgMn-layered double hydroxide composite	Phosphate	244.08 mg/g	328
Pharmaceuticals			
Porous GO by connecting GO sheets with decafluorobiphenyl (DFB)	Sulfamethoxazole (SMZ)	749.6 μmol/g	325
	Paracetamol (PCT)	663.9 μmol/g	
Porous graphene hydrogel	Ciprofloxacin	235.6 mg/g	332
Porous graphene (PG)	Atenolol (ATL)	1455.36 mg/g	148
	Carbamazepine (CBZ)	191.67 mg/g	
	Ciprofloxacin (CIP)	409.15 mg/g	
	diclofenac (DCF)	82.74 mg/g	
	Gemfibrozil (GEM)	160.86 mg/g	
	Ibuprofen (IBP)	110.30 mg/g	
Novel porous composite foam-supported graphene	Gemfibrozil (GEM)	55.1 mg/g	324

350.6 and 316.1 μmol/g, respectively. The adsorption kinetics and isotherms of SMZ and PCT on the GO-DFB20 are depicted in Figure 3.44. Since SMZ and PCT are widely used pharmaceuticals and have distinct molecular sizes and hydrophilicity, they were selected as model pollutants. Figure 3.44a depicts the SMZ and PCT on the GO-DFB20 adsorption kinetics. On the GO-DFB20, PCT showed quicker absorption than SMZ, removing nearly 90% of its equilibrium amount in only 12 hours, as opposed to SMZ, whose adsorption equilibrium took roughly 48 hours to attain. The molecular sizes of SMZ and PCT were consistent with their varied adsorption times. The bigger molecule SMZ took longer to reach the adsorption sites in the pores of the GO-DFB20 due to the slower diffusion rate because of the porous structure. Based on

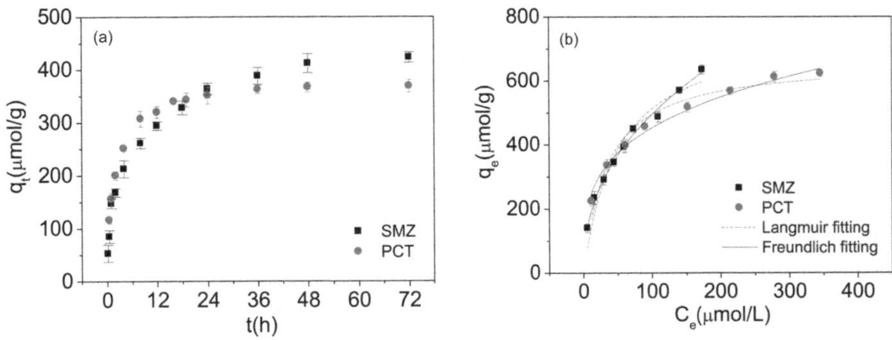

FIGURE 3.44 Adsorption kinetics (a) and isotherms (b) of SMZ and PCT on GO-DFB20. (Reproduced with permission from Shan et al.,[325] Copyright 2017, Elsevier Science Ltd.)

the relatively quick adsorption, it suggests that the generated pore size in the GO-DFB20 should be suitable for tiny contaminants to diffuse into and adsorb on the pores. Figure 3.44b displays the sorption isotherms of SMZ and PCT on the GO-DFB20, as well as the fitting outcomes obtained using the LM and FRIM models. The maximal adsorption capacities of GO-DFB20 for SMZ and PCT were further determined by the LM fitting to be 749.6 μmol/g and 663.9 μmol/g, respectively.

3.6.3.3.2 Removal of Dyes

With a singular combination of large surface area, super hydrophobicity, consistent porosity architecture, optical transparency, great oxidation resistance, and chemical stability, porous graphene is made up of few-layered planar graphene sheets. They are particularly effective adsorbents of dissolved organic pollutants because these planar porous nanostructures give easy access, speed up the adsorption and desorption of contaminants, and subsequently boost the adsorption rate and capacity. To link the faults for contaminant uptake in porous assemblies, porous graphene could offer a sizable degree of robustness. Additionally, it may possess a high structural stability, which is necessary throughout the adsorption and desorption processes to enable simple collection and repeated usage.[287] A summary of the adsorption of dyes using porous graphene-based materials is presented in Table 3.7.

With the aid of highly porous graphene created by thermal processing without the need for a template or catalyst, RB and MB have been adsorbed by Tabish and collaborators. When they compared their porous graphene with standard graphene equivalents, the prepared porous graphene nanosheets exhibited a clear improvement in hydrophobicity, adsorption capacity, and recyclability, which made them the best candidates for use in water treatment. The UV-vis absorption assays were used to examine the RB's adsorption kinetics, and it has been demonstrated that the adsorption of the RB was completed in 60 minutes with a 99.9% elimination efficiency. After 60 minutes, the RB absorbance band became weak. The kinetic models were effective in modeling the adsorption isotherm, and the adsorption capacity value obtained from the pseudo-first-order kinetic model was found to be in good agreement with the experimental data; however, the pseudo-second-order value performed badly in comparison. Thus, the adsorption in the instance of RB followed the first kinetic model. On the other hand, the maximal adsorption capacity of MB was found to be 313 mg/g. The adsorption capacity values for MB were also higher than those reported for conventional adsorbents. The highest correlation coefficient, corresponding theoretical and experimental values of q_e, and pseudo-second-order kinetics model results showed that both internal and exterior mass transfers occurred when experimental and theoretical conditions were taken into consideration.

Another study by Li et al.[326] used a 3D-printed inkjet nozzle and the gas-liquid microfluidic approach to create graphene microspheres. By varying the gas pressure, it was possible to easily control the graphene microspheres' diameter between 0.5 and 3.5 mm. They studied the adsorption performance of the MB and anionic MO by using the porous graphene microspheres. The maximal adsorption capacity of porous graphene microsphere is 65 mg/g for MO and 596 mg/g for MB. One of the greatest reported adsorption capabilities for 3D-geometric graphene adsorbents, the graphene sponge (526 mg/g), is 13 percent less effective than MB microspheres' adsorption capability.[327] The maximal adsorption capacity of 596 mg/g for MB, was nearly 21% more than that of graphene reduced using a conventional hydrothermal process. Additionally, it was discovered in the research that, upon the use of rGOs, MB exhibits the maximum adsorption capability. However, MO's capacity was 65 mg/g, which was 62 percent less than graphene reduced using the conventional hydrothermal technique. The anionic dye MO's and the cationic dye MB's various adsorption values demonstrated the selectivity. The porous graphene microsphere adsorption property for the cationic dye was specifically improved, however, the adsorption property for the anionic dye was constrained. This phenomenon was utilized for the separation of anionic and cationic dyes, revealing a novel method for waste-sorting treatment. Isothermal adsorption research revealed that the adsorption procedure followed LM's isothermal adsorption, indicating that the MB molecules were evenly dispersed throughout the monolayer and adsorbed uniformly. The graphene microsphere's adsorption kinetics fit the pseudo-second-order model, which indicated that a chemical reaction was in charge of controlling the adsorption process. The intra-particle diffusion was not the only rate-regulating process, according to the research on diffusion processes; other mechanisms, like boundary-layer adsorption, were also at play. Overall, this study offered a novel approach for the large-scale mass manufacture of graphene functional materials with well-controlled microstructures and surface features, which demonstrated superior performance in applications such as selective separations and environmental remediation.[326]

3.6.3.3.3 *Removal of Phosphorus and Nitrates*

Some of the most common contaminants that are hazardous to living things include nitrates and phosphates. In a study by Lai et al.,[328] they investigated the adsorption of phosphates using eco-friendly adsorbent of MgMn-layered double hydroxide (MgMn-LDH) and (GO) grown *in situ* to obtain L-GO/MgMn-LDH, adopting Garcinia subelliptica leaves as an inexpensive and natural template. Following calcination, the composite displayed a layered porous structure and demonstrated selective phosphate recognition, resulting in a desorption rate of 85.8 percent and a significantly high and recyclable selective phosphate adsorption capacity of 244.08 mg/g. Figure 3.45a shows the time-dependent phosphate adsorption capabilities of

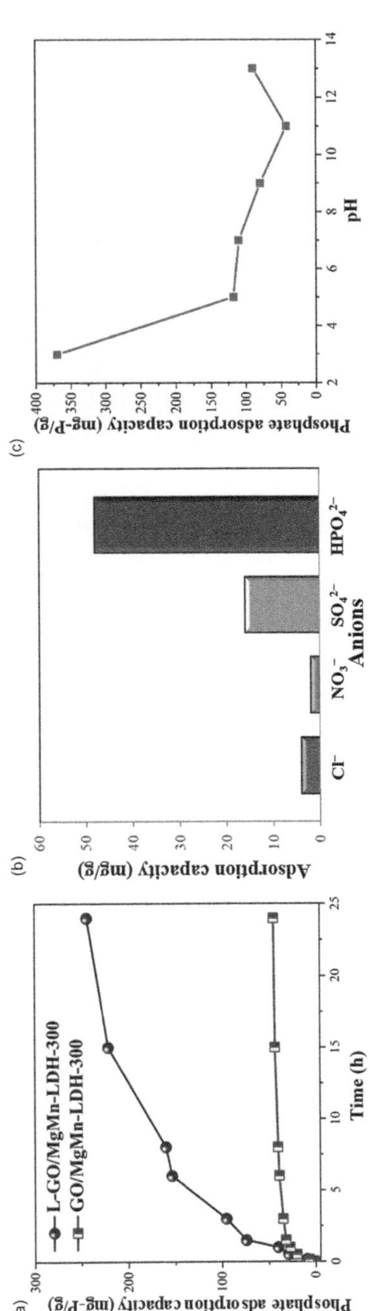

FIGURE 3.45 The practical phosphate sustainability by LDH composites in wastewater treatment. (a) The phosphate adsorption capacities of GO/MgMn-LDH-300 and L-GO/MgMn-LDH-300 as a function of time. With the introduction of leaf-template, the LDH composite shows 5.5 times the phosphate adsorption capacity. (b) The selectivity of L-GO/MgMn-LDH-300 toward different ions. (c) The pH dependence of phosphate adsorption capacity of L-GO/MgMn-LDH-300. (Reproduced with copyright permission from Lai et al.[328] Copyright 2020, Elsevier Science Ltd.)

the GO/MgMn-LDH-300 and L-GO/MgMn-LDH-300 composites. During the first 1.5 hours, the GO/MgMn-LDH-300 showed fast phosphate adsorption and, after 3 hours, it reached equilibrium. After 24 hours, it demonstrated an adsorption capacity of 44.50 mg/g. In complete contradiction, the L-GO/MgMn-LDH-300 not only displayed even faster adsorption in the first 1.5 h, but also eventually attained a high phosphate adsorption capacity of 244.08 mg/g, which is 5.5 times that of the GO/MgMn-LDH-300 composite after 24 hours. The phosphate adsorption processes of both adsorbents were dominated by chemisorption, as shown by the fact that LDH composites showed greater correlation coefficients (R^2) of the pseudo-second-order model. When the absorption of phosphate, was investigated in the presence of antagonistic ions, such as HPO_4^{2-}, Cl^-, SO_4^{2-}, and NO_3^-. The results (Figure 3.45b) demonstrated that L-GO/MgMn-LDH-300 had a better adsorption capacity than other competing ions because it had a strong selectivity toward phosphates. The findings supported earlier observations that higher-valence anions interacted with positively charged LDH sheets more strongly than monovalent anions did. The pH of the wastewater is a crucial factor in the phosphate adsorption process. In this study, the phosphate adsorption on L-GO/MgMn-LDH-300 was investigated at various pH levels ranging from 3.0 to 13.0. As the pH value increased from 3.0 to 11.0, it was clearly observed in Figure 3.45c that the capacity for phosphate adsorption diminished. Lower phosphate adsorption on LDH composites occurred as a result of increased competition between phosphate and OH, as well as the negatively charged surface as pH increased. About pH 11, the phosphate adsorption capacity with the lowest value was 42.2 mg/g. It was observed that phosphate adsorption increases as the solution pH approaches 13.0. Since phosphate is pH-related in solution, it occurred in several forms as $H_2PO_4^-$, HPO_4^{2-}, and PO_4^{3-} with pK1 = 2.12, pK2 = 7.21, and pK3 = 12.67, respectively. The hydrogen bonding network was significantly impacted by the PO_4^{3-} species, which was the main species in the pH range above 12.67. On flavonoids and triterpenoids, orthophosphate caused the emergence of novel connections between P-O groups and hydrogen groups. According to the findings, L-GO/MgMn-LDH-300 demonstrated high phosphate adsorption over a broad pH range, making it suitable for use in actual wastewater treatment.[328]

3.7 CONCLUSION

An overview of the development of graphene and its derivatives, as well as their preparation techniques, is provided in this chapter. The use of powdered, membrane-based, and porous graphene-based materials for the removal of wastewater contaminants, such as heavy metal ions, phosphates/nitrates, pharmaceuticals and so on, from wastewater is discussed. This chapter is an excellent piece that will serve as a guide to emerging researchers in graphene and graphene-based materials, especially in wastewater remediation.

REFERENCES

(1) Sattar, T. Current review on synthesis, composites and multifunctional properties of graphene. *Topics in Current Chemistry* 2019, *377*, 1–45.

(2) Shinohara, H.; Tiwari, A. *Graphene: An Introduction to the Fundamentals and Industrial Applications*; John Wiley & Sons, 2015.

(3) Boehm, H.-P.; Stumpp, E. Citation errors concerning the first report on exfoliated graphite. *Carbon* 2007, *7* (45), 1381–1383.

(4) Dreyer, D. R.; Ruoff, R. S.; Bielawski, C. W. From conception to realization: An historial account of graphene and some perspectives for its future. *Angewandte Chemie International Edition* 2010, *49* (49), 9336–9344.

(5) Hérold, C.; Hérold, A.; Lagrange, P. New synthesis routes for donor-type graphite intercalation compounds. *Journal of Physics and Chemistry of Solids* 1996, *57* (6–8), 655–662.

(6) Noel, M.; Santhanam, R. Electrochemistry of graphite intercalation compounds. *Journal of Power Sources* 1998, *72* (1), 53–65.

(7) Sorokina, N.; Nikol'skaya, I.; Ionov, S.; Avdeev, V. Acceptor-type graphite intercalation compounds and new carbon materials based on them. *Russian Chemical Bulletin* 2005, *54*, 1749–1767.

(8) Inagaki, M. Applications of graphite intercalation compounds. *Journal of Materials Research* 1989, *4* (6), 1560–1568.

(9) Liu, Z.-H.; Wang, Z.-M.; Yang, X.; Ooi, K. Intercalation of organic ammonium ions into layered graphite oxide. *Langmuir* 2002, *18* (12), 4926–4932.

(10) Belash, I.; Zharikov, O.; Palnichenko, A. Superconductivity of GIC with Li, Na and K. *Synthetic Metals* 1989, *34* (1–3), 455–460.

(11) Berger, C.; Song, Z.; Li, T.; Li, X.; Ogbazghi, A. Y.; Feng, R.; Dai, Z.; Marchenkov, A. N.; Conrad, E. H.; First, P. N. Ultrathin epitaxial graphite: 2D electron gas properties and a route toward graphene-based nanoelectronics. *The Journal of Physical Chemistry B* 2004, *108* (52), 19912–19916.

(12) Boehm, H. P.; Setton, R.; Stumpp, E. Nomenclature and terminology of graphite intercalation compounds (IUPAC Recommendations 1994). *Pure and Applied Chemistry* 1994, *66* (9), 1893–1901.

(13) Ang, P. K.; Wang, S.; Bao, Q.; Thong, J. T.; Loh, K. P. High-throughput synthesis of graphene by intercalation – Exfoliation of graphite oxide and study of ionic screening in graphene transistor. *ACS Nano* 2009, *3* (11), 3587–3594.

(14) Brodie XIII, B. C. On the atomic weight of graphite. *Philosophical Transactions of the Royal Society of London* 1859, (149), 249–259.

(15) Stankovich, S.; Dikin, D. A.; Piner, R. D.; Kohlhaas, K. A.; Kleinhammes, A.; Jia, Y.; Wu, Y.; Nguyen, S. T.; Ruoff, R. S. Synthesis of graphene-based nanosheets via chemical reduction of exfoliated graphite oxide. *Carbon* 2007, *45* (7), 1558–1565.

(16) Dreyer, D. R.; Jia, H. P.; Bielawski, C. W. Graphene oxide: A convenient carbocatalyst for facilitating oxidation and hydration reactions. *Angewandte Chemie International Edition* 2010, *38* (49), 6813–6816.

(17) Soldano, C.; Mahmood, A.; Dujardin, E. Production, properties and potential of graphene. *Carbon* 2010, *48* (8), 2127–2150.

(18) DiVincenzo, D.; Mele, E. Self-consistent effective-mass theory for intralayer screening in graphite intercalation compounds. *Physical Review B* 1984, *29* (4), 1685.

(19) Boehm, H. P. Chemical Identification of Surface Groups. In *Advances in Catalysis*, Eley, D. D., Pines, H., Weisz, P. B. Eds.; Vol. 16; Academic Press, 1966; pp. 179–274.

(20) Morgan, A.; Somorjai, G. Low energy electron diffraction studies of gas adsorption on the platinum (100) single crystal surface. *Surface Science* 1968, *12* (3), 405–425.

(21) May, J. W. Platinum surface LEED rings. *Surface Science* 1969, *17* (1), 267–270.

(22) Blakely, J.; Kim, J.; Potter, H. Segregation of carbon to the (100) surface of nickel. *Journal of Applied Physics* 1970, *41* (6), 2693–2697.

(23) Eizenberg, M.; Blakely, J. Carbon interaction with nickel surfaces: Monolayer formation and structural stability. *The Journal of Chemical Physics* 1979, *71* (8), 3467–3477.

(24) Eizenberg, M.; Blakely, J. Carbon monolayer phase condensation on Ni (111). *Surface Science* 1979, *82* (1), 228–236.

(25) Hamilton, J.; Blakely, J. Carbon layer formation on the Pt (111) surface as a function of temperature. *Journal of Vacuum Science and Technology* 1978, *15* (2), 559–562.

(26) Land, T.; Michely, T.; Behm, R.; Hemminger, J.; Comsa, G. STM investigation of single layer graphite structures produced on Pt (111) by hydrocarbon decomposition. *Surface Science* 1992, *264* (3), 261–270.

(27) Van Bommel, A.; Crombeen, J.; Van Tooren, A. LEED and Auger electron observations of the SiC (0001) surface. *Surface Science* 1975, *48* (2), 463–472.

(28) Badami, D. X-ray studies of graphite formed by decomposing silicon carbide. *Carbon* 1965, *3* (1), 53–57.

(29) Boehm, H.; Setton, R.; Stumpp, E. Nomenclature and terminology of graphite intercalation compounds. Carbon, 1986, 24(2), 241–245.

(30) Kaburagi, Y.; Asano, Y.; Hishiyama, Y. Kish graphite as a magnetic field sensor. *Carbon* 2004, *42* (15), 3266–3268.

(31) Charlier, J.-C.; Gonze, X.; Michenaud, J.-P. Graphite interplanar bonding: Electronic delocalization and van der Waals interaction. *Europhysics Letters* 1994, *28* (6), 403.

(32) Lu, X.; Huang, H.; Nemchuk, N.; Ruoff, R. S. Patterning of highly oriented pyrolytic graphite by oxygen plasma etching. *Applied Physics Letters* 1999, *75* (2), 193–195.

(33) Lu, X.; Yu, M.; Huang, H.; Ruoff, R. S. Tailoring graphite with the goal of achieving single sheets. *Nanotechnology* 1999, *10* (3), 269.

(34) Novoselov, K. S.; Geim, A. K.; Morozov, S. V.; Jiang, D.-e.; Zhang, Y.; Dubonos, S. V.; Grigorieva, I. V.; Firsov, A. A. Electric field effect in atomically thin carbon films. *Science* 2004, *306* (5696), 666–669.

(35) Huang, Y.; Pan, Y.-H.; Yang, R.; Bao, L.-H.; Meng, L.; Luo, H.-L.; Cai, Y.-Q.; Liu, G.-D.; Zhao, W.-J.; Zhou, Z. Universal mechanical exfoliation of large-area 2D crystals. *Nature Communications* 2020, *11* (1), 2453.

(36) Borand, G.; Akçamlı, N.; Uzunsoy, D. Structural characterization of graphene nanostructures produced via arc discharge method. *Ceramics International* 2021, *47* (6), 8044–8052.

(37) Kaur, A.; Kaur, J.; Singh, R. C. Tailor made exfoliated reduced graphene oxide nanosheets based on oxidative-exfoliation approach. *Fullerenes, Nanotubes and Carbon Nanostructures* 2018, *26* (1), 1–11.

(38) Xu, Y.; Cao, H.; Xue, Y.; Li, B.; Cai, W. Liquid-phase exfoliation of graphene: An overview on exfoliation media, techniques, and challenges. *Nanomaterials* 2018, *8* (11), 942.

(39) Silva, A. A.; Pinheiro, R. A.; Rodrigues, A. C.; Baldan, M. R.; Trava-Airoldi, V. J.; Corat, E. J. Graphene sheets produced by carbon nanotubes unzipping and their performance as supercapacitor. *Applied Surface Science* 2018, *446*, 201–208.

(40) Lee, X. J.; Hiew, B. Y. Z.; Lai, K. C.; Lee, L. Y.; Gan, S.; Thangalazhy-Gopakumar, S.; Rigby, S. Review on graphene and its derivatives: Synthesis methods and potential industrial implementation. *Journal of the Taiwan Institute of Chemical Engineers* 2019, *98*, 163–180.

(41) Saeed, M.; Alshammari, Y.; Majeed, S. A.; Al-Nasrallah, E. Chemical vapour deposition of graphene—Synthesis, characterisation, and applications: A review. *Molecules* 2020, *25* (17), 3856.

(42) Li, G.; Zhang, Y.-Y.; Guo, H.; Huang, L.; Lu, H.; Lin, X.; Wang, Y.-L.; Du, S.; Gao, H.-J. Epitaxial growth and physical properties of 2D materials beyond graphene: From monatomic materials to binary compounds. *Chemical Society Reviews* 2018, *47* (16), 6073–6100.

(43) Sun, Y.; Zhang, J. Strategies for scalable gas-phase preparation of free-standing graphene. *CCS Chemistry* 2021, *3* (4), 1058–1077.

(44) Yang, Y.; Liu, R.; Wu, J.; Jiang, X.; Cao, P.; Hu, X.; Pan, T.; Qiu, C.; Yang, J.; Song, Y. Bottom-up fabrication of graphene on silicon/silica substrate via a facile soft-hard template approach. *Scientific Reports* 2015, *5* (1), 1–7.

(45) Singh, V.; Joung, D.; Zhai, L.; Das, S.; Khondaker, S. I.; Seal, S. Graphene based materials: Past, present and future. *Progress in Materials Science* 2011, *56* (8), 1178–1271.

(46) Ciesielski, A.; Samorì, P. Graphene via sonication assisted liquid-phase exfoliation. *Chemical Society Reviews* 2014, *43* (1), 381–398.

(47) Kumar, N.; Salehiyan, R.; Chauke, V.; Botlhoko, O. J.; Setshedi, K.; Scriba, M.; Masukume, M.; Ray, S. S. Top-down synthesis of graphene: A comprehensive review. *FlatChem* 2021, *27*, 100224.

(48) Yi, M.; Shen, Z. A review on mechanical exfoliation for the scalable production of graphene. *Journal of Materials Chemistry A* 2015, *3* (22), 11700–11715.

(49) Chen, X.; Dobson, J. F.; Raston, C. L. Vortex fluidic exfoliation of graphite and boron nitride. *Chemical Communications* 2012, *48* (31), 3703–3705.

(50) Yasmin, L.; Chen, X.; Stubbs, K. A.; Raston, C. L. Optimising a vortex fluidic device for controlling chemical reactivity and selectivity. *Scientific Reports* 2013, *3* (1), 1–6.

(51) Chen, J.; Duan, M.; Chen, G. Continuous mechanical exfoliation of graphene sheets via three-roll mill. *Journal of Materials Chemistry* 2012, *22* (37), 19625–19628.

(52) Wahid, M. H.; Eroglu, E.; Chen, X.; Smith, S. M.; Raston, C. L. Functional multi-layer graphene–algae hybrid material formed using vortex fluidics. *Green Chemistry* 2013, *15* (3), 650–655.

(53) Tran, T. S.; Park, S. J.; Yoo, S. S.; Lee, T.-R.; Kim, T. High shear-induced exfoliation of graphite into high quality graphene by Taylor–Couette flow. *RSC Advances* 2016, *6* (15), 12003–12008.

(54) Sumdani, M.; Islam, M.; Yahaya, A.; Safie, S. Recent advances of the graphite exfoliation processes and structural modification of graphene: A review. *Journal of Nanoparticle Research* 2021, *23*, 1–35.

(55) Paton, K. R.; Varrla, E.; Backes, C.; Smith, R. J.; Khan, U.; O'Neill, A.; Boland, C.; Lotya, M.; Istrate, O. M.; King, P. Scalable production of large quantities of defect-free few-layer graphene by shear exfoliation in liquids. *Nature Materials* 2014, *13* (6), 624–630.

(56) Choi, W.; Lee, J.-W. *Graphene: Synthesis and Applications*; CRC Press, 2011.

(57) Viculis, L. M.; Mack, J. J.; Kaner, R. B. A chemical route to carbon nanoscrolls. *Science* 2003, *299* (5611), 1361–1361.

(58) Ereš, Z.; Hrabar, S. Low-cost synthesis of high-quality graphene in do-it-yourself CVD reactor. *Automatika* 2018, *59* (3–4), 254–260.

(59) Zhang, Z.; Fraser, A.; Ye, S.; Merle, G.; Barralet, J. Top-down bottom-up graphene synthesis. *Nano Futures* 2019, *3* (4), 042003.

(60) Lee, Y.; Bae, S.; Jang, H.; Jang, S.; Zhu, S.-E.; Sim, S. H.; Song, Y. I.; Hong, B. H.; Ahn, J.-H. Wafer-scale synthesis and transfer of graphene films. *Nano Letters* 2010, *10* (2), 490–493.

(61) Li, X.; Cai, W.; Colombo, L.; Ruoff, R. S. Evolution of graphene growth on Ni and Cu by carbon isotope labeling. *Nano Letters* 2009, *9* (12), 4268–4272.

(62) Wood, J. D.; Schmucker, S. W.; Lyons, A. S.; Pop, E.; Lyding, J. W. Effects of polycrystalline Cu substrate on graphene growth by chemical vapor deposition. *Nano Letters* 2011, *11* (11), 4547–4554.

(63) Usachov, D.; Fedorov, A.; Vilkov, O.; Adamchuk, V.; Yashina, L.; Bondarenko, L.; Saranin, A.; Grüneis, A.; Vyalikh, D. Experimental and computational insight into the properties of the lattice-mismatched structures: Monolayers of h-BN and graphene on Ir (111). *Physical Review B* 2012, *86* (15), 155151.

(64) Jean, F.; Zhou, T.; Blanc, N.; Felici, R.; Coraux, J.; Renaud, G. Effect of preparation on the commensurabilities and thermal expansion of graphene on Ir (111) between 10 and 1300 K. *Physical Review B* 2013, *88* (16), 165406.

(65) Jia-Tao, S.; Shi-Xuan, D.; Wen-De, X.; Hao, H.; Yu-Yang, Z.; Guo, L.; Hong-Jun, G. Effect of strain on geometric and electronic structures of graphene on a Ru (0001) surface. *Chinese Physics B* 2009, *18* (7), 3008.

(66) Gao, L.; Ren, W.; Xu, H.; Jin, L.; Wang, Z.; Ma, T.; Ma, L.-P.; Zhang, Z.; Fu, Q.; Peng, L.-M. Repeated growth and bubbling transfer of graphene with millimetre-size single-crystal grains using platinum. *Nature Communications* 2012, *3* (1), 699.

(67) Usachov, D. Y.; Fedorov, A. V.; Petukhov, A. E.; Vilkov, O. Y.; Rybkin, A. G.; Otrokov, M. M.; Arnau, A.; Chulkov, E. V.; Yashina, L. V.; Farjam, M. Epitaxial B-graphene: Large-scale growth and atomic structure. *ACS Nano* 2015, *9* (7), 7314–7322.

(68) Park, B.-J.; Choi, J.-S.; Eom, J.-H.; Ha, H.; Kim, H. Y.; Lee, S.; Shin, H.; Yoon, S.-G. Defect-free graphene synthesized directly at 150 C via chemical vapor deposition with no transfer. *ACS Nano* 2018, *12* (2), 2008–2016.

(69) Yao, Y.; Wong, C.-P. Monolayer graphene growth using additional etching process in atmospheric pressure chemical vapor deposition. *Carbon* 2012, *50* (14), 5203–5209.

(70) Seah, C.-M.; Chai, S.-P.; Mohamed, A. R. Mechanisms of graphene growth by chemical vapour deposition on transition metals. *Carbon* 2014, *70*, 1–21.

(71) Yan, K.; Fu, L.; Peng, H.; Liu, Z. Designed CVD growth of graphene via process engineering. *Accounts of Chemical Research* 2013, *46* (10), 2263–2274.

(72) Zhang, Y.; Zhang, L.; Zhou, C. Review of chemical vapor deposition of graphene and related applications. *Accounts of Chemical Research* 2013, *46* (10), 2329–2339.

(73) Greene, J. E. Tracing the 5000-year recorded history of inorganic thin films from ~3000 BC to the early 1900s AD. *Applied Physics Reviews* 2014, *1* (4), 041302.

(74) Choy, K.-L. *Chemical Vapour Deposition (CVD): Advances, Technology and Applications*; CRC Press, 2019.

(75) Shang, N. G.; Papakonstantinou, P.; McMullan, M.; Chu, M.; Stamboulis, A.; Potenza, A.; Dhesi, S. S.; Marchetto, H. Catalyst-free efficient growth, orientation and biosensing properties of multilayer graphene nanoflake films with sharp edge planes. *Advanced Functional Materials* 2008, *18* (21), 3506–3514.

(76) Obraztsov, A.; Zolotukhin, A.; Ustinov, A.; Volkov, A.; Svirko, Y.; Jefimovs, K. DC discharge plasma studies for nanostructured carbon CVD. *Diamond and Related Materials* 2003, *12* (3–7), 917–920.

(77) Li, M.; Liu, D.; Wei, D.; Song, X.; Wei, D.; Wee, A. T. S. Controllable synthesis of graphene by plasma-enhanced chemical vapor deposition and its related applications. *Advanced Science* 2016, *3* (11), 1600003.

(78) Rhallabi, A.; Catherine, Y. Computer simulation of a carbon-deposition plasma in CH_4. *IEEE Transactions on Plasma Science* 1991, *19* (2), 270–277.

(79) Mutsukura, N.; Inoue, S. I.; Machi, Y. Deposition mechanism of hydrogenated hard-carbon films in a CH4 rf discharge plasma. *Journal of Applied Physics* 1992, *72* (1), 43–53.

(80) Pack, J.; Voshall, R.; Phelps, A.; Kline, L. Longitudinal electron diffusion coefficients in gases: Noble gases. *Journal of Applied Physics* 1992, *71* (11), 5363–5371.

(81) Herrebout, D.; Bogaerts, A.; Yan, M.; Gijbels, R.; Goedheer, W.; Dekempeneer, E. One-dimensional fluid model for an rf methane plasma of interest in deposition of diamond-like carbon layers. *Journal of Applied Physics* 2001, *90* (2), 570–579.

(82) Hinkov, I.; Pashova, K.; Farhat, S. Modeling of plasma-enhanced chemical vapor deposition growth of graphene on cobalt substrates. *Diamond and Related Materials* 2019, *93*, 84–95.

(83) Sun, J.; Rattanasawatesun, T.; Tang, P.; Bi, Z.; Pandit, S.; Lam, L.; Wasén, C.; Erlandsson, M.; Bokarewa, M.; Dong, J. Insights into the mechanism for vertical graphene growth by plasma-enhanced chemical vapor deposition. *ACS Applied Materials & Interfaces* 2022, *14* (5), 7152–7160.

(84) Choucair, M.; Thordarson, P.; Stride, J. A. Gram-scale production of graphene based on solvothermal synthesis and sonication. *Nature Nanotechnology* 2009, *4* (1), 30–33.

(85) Hibino, H.; Kageshima, H.; Nagase, M. Graphene growth on silicon carbide. *NTT Technical Review (Web)* 2010, *8* (8), 1-6.

(86) Juang, Z.-Y.; Wu, C.-Y.; Lo, C.-W.; Chen, W.-Y.; Huang, C.-F.; Hwang, J.-C.; Chen, F.-R.; Leou, K.-C.; Tsai, C.-H. Synthesis of graphene on silicon carbide substrates at low temperature. *Carbon* 2009, *47* (8), 2026–2031.

(87) Pan, Y.; Zhang, H.; Shi, D.; Sun, J.; Du, S.; Liu, F.; Gao, H. J. Highly ordered, millimeter-scale, continuous, single-crystalline graphene monolayer formed on Ru (0001). *Advanced Materials* 2009, *21* (27), 2777–2780.

(88) Chen, Z.; Lin, Y.-M.; Rooks, M. J.; Avouris, P. Graphene nano-ribbon electronics. *Physica E: Low-Dimensional Systems and Nanostructures* 2007, *40* (2), 228–232.

(89) Kosynkin, D. V.; Higginbotham, A. L.; Sinitskii, A.; Lomeda, J. R.; Dimiev, A.; Price, B. K.; Tour, J. M. Longitudinal unzipping of carbon nanotubes to form graphene nanoribbons. *Nature* 2009, *458* (7240), 872–876.

(90) Jiao, L.; Zhang, L.; Wang, X.; Diankov, G.; Dai, H. Narrow graphene nanoribbons from carbon nanotubes. *Nature* 2009, *458* (7240), 877–880.

(91) Cano-Márquez, A. G.; Rodríguez-Macías, F. J.; Campos-Delgado, J.; Espinosa-González, C. G.; Tristán-López, F.; Ramírez-González, D.; Cullen, D. A.; Smith, D. J.; Terrones, M.; Vega-Cantú, Y. I. Ex-MWNTs: Graphene sheets and ribbons produced by lithium intercalation and exfoliation of carbon nanotubes. *Nano Letters* 2009, *9* (4), 1527–1533.

(92) Thess, A.; Lee, R.; Nikolaev, P.; Dai, H.; Petit, P.; Robert, J.; Xu, C.; Lee, Y. H.; Kim, S. G.; Rinzler, A. G. Crystalline ropes of metallic carbon nanotubes. *Science* 1996, *273* (5274), 483–487.

(93) Liu, J.; Kim, G. H.; Xue, Y.; Kim, J. Y.; Baek, J. B.; Durstock, M.; Dai, L. Graphene oxide nanoribbon as hole extraction layer to enhance efficiency and stability of polymer solar cells. *Advanced Materials* 2014, *26* (5), 786–790.

(94) Mahmoud, W. E.; Al-Hazmi, F. S.; Al-Harbi, G. H. Wall by wall controllable unzipping of MWCNTs via intercalation with oxalic acid to produce multilayers graphene oxide ribbon. *Chemical Engineering Journal* 2015, *281*, 192–198.

(95) Shinde, D. B.; Debgupta, J.; Kushwaha, A.; Aslam, M.; Pillai, V. K. Electrochemical unzipping of multi-walled carbon nanotubes for facile synthesis of high-quality graphene nanoribbons. *Journal of the American Chemical Society* 2011, *133* (12), 4168–4171. DOI: 10.1021/ja1101739

(96) Guo, S.; Dong, S. Graphene nanosheet: Synthesis, molecular engineering, thin film, hybrids, and energy and analytical applications. *Chemical Society Reviews* 2011, *40* (5), 2644–2672.

(97) Kim, S.; Song, Y.; Takahashi, T.; Oh, T.; Heller, M. J. An aqueous single reactor arc discharge process for the synthesis of graphene nanospheres. *Small* 2015, *11* (38), 5041–5046.

(98) Subrahmanyam, K.; Panchakarla, L.; Govindaraj, A.; Rao, C. Simple method of preparing graphene flakes by an arc-discharge method. *The Journal of Physical Chemistry C* 2009, *113* (11), 4257–4259.

(99) Rao, C. E. E.; Sood, A. E.; Subrahmanyam, K. E.; Govindaraj, A. Graphene: The new two-dimensional nanomaterial. *Angewandte Chemie International Edition* 2009, *48* (42), 7752–7777.

(100) Wang, Z.; Li, N.; Shi, Z.; Gu, Z. Low-cost and large-scale synthesis of graphene nanosheets by arc discharge in air. *Nanotechnology* 2010, *21* (17), 175602.

(101) Kim, S.; Song, Y.; Ibsen, S.; Ko, S.-Y.; Heller, M. J. Controlled degrees of oxidation of nanoporous graphene filters for water purification using an aqueous arc discharge. *Carbon* 2016, *109*, 624–631.

(102) Hossain, M. F.; Akther, N.; Zhou, Y. Recent advancements in graphene adsorbents for wastewater treatment: Current status and challenges. *Chinese Chemical Letters* 2020, *31* (10), 2525–2538.

 103) Su, C.-Y.; Xu, Y.; Zhang, W.; Zhao, J.; Tang, X.; Tsai, C.-H.; Li, L.-J. Electrical and spectroscopic characterizations of ultra-large reduced graphene oxide monolayers. *Chemistry of Materials* 2009, *21* (23), 5674–5680.

(104) Perrozzi, F.; Prezioso, S.; Ottaviano, L. Graphene oxide: From fundamentals to applications. *Journal of Physics: Condensed Matter* 2014, *27* (1), 013002.

(105) Lerf, A.; He, H.; Forster, M.; Klinowski, J. Structure of graphite oxide revisited. *The Journal of Physical Chemistry B* 1998, *102* (23), 4477–4482.

(106) He, H.; Riedl, T.; Lerf, A.; Klinowski, J. Solid-state NMR studies of the structure of graphite oxide. *The Journal of Physical Chemistry* 1996, *100* (51), 19954–19958.

(107) Erickson, K.; Erni, R.; Lee, Z.; Alem, N.; Gannett, W.; Zettl, A. Determination of the local chemical structure of graphene oxide and reduced graphene oxide. *Advanced Materials* 2010, *22* (40), 4467–4472.

(108) Rourke, J. P.; Pandey, P. A.; Moore, J. J.; Bates, M.; Kinloch, I. A.; Young, R. J.; Wilson, N. R. The real graphene oxide revealed: Stripping the oxidative debris from the graphene-like sheets. *Angewandte Chemie International Edition* 2011, *50* (14), 3173–3177.

(109) Eigler, S.; Hirsch, A. Chemistry with graphene and graphene oxide—Challenges for synthetic chemists. *Angewandte Chemie International Edition* 2014, *53* (30), 7720–7738.

(110) Dimiev, A.; Kosynkin, D. V.; Alemany, L. B.; Chaguine, P.; Tour, J. M. Pristine graphite oxide. *Journal of the American Chemical Society* 2012, *134* (5), 2815–2822.

(111) Chen, W.; Yan, L.; Bangal, P. R. Preparation of graphene by the rapid and mild thermal reduction of graphene oxide induced by microwaves. *Carbon* 2010, *48* (4), 1146–1152.

(112) Marcano, D. C.; Kosynkin, D. V.; Berlin, J. M.; Sinitskii, A.; Sun, Z.; Slesarev, A.; Alemany, L. B.; Lu, W.; Tour, J. M. Improved synthesis of graphene oxide. *ACS Nano* 2010, *4* (8), 4806–4814.

(113) Singh, R. K.; Kumar, R.; Singh, D. P. Graphene oxide: Strategies for synthesis, reduction and frontier applications. *RSC Advances* 2016, *6* (69), 64993–65011.

(114) Dideikin, A. T.; Vul', A. Y. Graphene oxide and derivatives: The place in graphene family. *Frontiers in Physics* 2019, *6*, 149.

(115) Dong, L.; Yang, J.; Chhowalla, M.; Loh, K. P. Synthesis and reduction of large sized graphene oxide sheets. *Chemical Society Reviews* 2017, *46* (23), 7306–7316.

(116) Lin, S.; Dong, L.; Zhang, J.; Lu, H. Room-temperature intercalation and ~1000-fold chemical expansion for scalable preparation of high-quality graphene. *Chemistry of Materials* 2016, *28* (7), 2138–2146.

(117) Sun, L. Structure and synthesis of graphene oxide. *Chinese Journal of Chemical Engineering* 2019, *27* (10), 2251–2260.

(118) Staudenmaier, L. Verfahren zur darstellung der graphitsäure. *Berichte der Deutschen Chemischen Gesellschaft* 1898, *31* (2), 1481–1487.

(119) Hummers Jr, W. S.; Offeman, R. E. Preparation of graphitic oxide. *Journal of the American Chemical Society* 1958, *80* (6), 1339–1339.

(120) Fu, L.; Liu, H.; Zou, Y.-H.; Li, B. Technology research on oxidative degree of graphite oxide prepared by Hummers method. *Carbon* 2005, *4*, 10–14.

(121) Shen, J.; Hu, Y.; Shi, M.; Lu, X.; Qin, C.; Li, C.; Ye, M. Fast and facile preparation of graphene oxide and reduced graphene oxide nanoplatelets. *Chemistry of Materials* 2009, *21* (15), 3514–3520.

(122) Sun, L.; Fugetsu, B. Mass production of graphene oxide from expanded graphite. *Materials Letters* 2013, *109*, 207–210.

(123) Eigler, S.; Enzelberger-Heim, M.; Grimm, S.; Hofmann, P.; Kroener, W.; Geworski, A.; Dotzer, C.; Röckert, M.; Xiao, J.; Papp, C. Wet chemical synthesis of graphene. *Advanced Materials* 2013, *25* (26), 3583–3587.

(124) Chen, J.; Li, Y.; Huang, L.; Li, C.; Shi, G. High-yield preparation of graphene oxide from small graphite flakes via an improved Hummers method with a simple purification process. *Carbon* 2015, *81*, 826–834.

(125) Panwar, V.; Chattree, A.; Pal, K. A new facile route for synthesizing of graphene oxide using mixture of sulfuric–nitric–phosphoric acids as intercalating agent. *Physica E: Low-Dimensional Systems and Nanostructures* 2015, *73*, 235–241.

(126) Peng, L.; Xu, Z.; Liu, Z.; Wei, Y.; Sun, H.; Li, Z.; Zhao, X.; Gao, C. An iron-based green approach to 1-h production of single-layer graphene oxide. *Nature Communications* 2015, *6* (1), 5716.

(127) Rosillo-Lopez, M.; Salzmann, C. G. A simple and mild chemical oxidation route to high-purity nano-graphene oxide. *Carbon* 2016, *106*, 56–63.

(128) Yu, H.; Zhang, B.; Bulin, C.; Li, R.; Xing, R. High-efficient synthesis of graphene oxide based on improved hummers method. *Scientific Reports* 2016, *6* (1), 36143.

(129) Dimiev, A. M.; Ceriotti, G.; Metzger, A.; Kim, N. D.; Tour, J. M. Chemical mass production of graphene nanoplatelets in ~100% yield. *ACS Nano* 2016, *10* (1), 274–279.

(130) Pei, S.; Wei, Q.; Huang, K.; Cheng, H.-M.; Ren, W. Green synthesis of graphene oxide by seconds timescale water electrolytic oxidation. *Nature Communications* 2018, *9* (1), 145.

(131) Ranjan, P.; Agrawal, S.; Sinha, A.; Rao, T. R.; Balakrishnan, J.; Thakur, A. D. A low-cost non-explosive synthesis of graphene oxide for scalable applications. *Scientific Reports* 2018, *8* (1), 12007.

(132) Vats, N.; Rauschenbach, S.; Sigle, W.; Sen, S.; Abb, S.; Portz, A.; Dürr, M.; Burghard, M.; Van Aken, P.; Kern, K. Electron microscopy of polyoxometalate ions on graphene by electrospray ion beam deposition. *Nanoscale* 2018, *10* (10), 4952–4961.

(133) Kornilov, D. Y.; Gubin, S. Graphene oxide: Structure, properties, synthesis, and reduction (a review). *Russian Journal of Inorganic Chemistry* 2020, *65*, 1965–1976.

(134) Gómez-Navarro, C.; Meyer, J. C.; Sundaram, R. S.; Chuvilin, A.; Kurasch, S.; Burghard, M.; Kern, K.; Kaiser, U. Atomic structure of reduced graphene oxide. *Nano Letters* 2010, *10* (4), 1144–1148.

(135) Gao, Y.; Qin, Y.; Zhang, M.; Xu, L.; Yang, Z.; Xu, Z.; Wang, Y.; Men, M. Revealing the role of oxygen-containing functional groups on graphene oxide for the highly efficient adsorption of thorium ions. *Journal of Hazardous Materials* 2022, *436*, 129148.

(136) Dave, S. H.; Gong, C.; Robertson, A. W.; Warner, J. H.; Grossman, J. C. Chemistry and structure of graphene oxide via direct imaging. *ACS Nano* 2016, *10* (8), 7515–7522.

(137) Sadhukhan, S.; Bhattacharyya, A.; Rana, D.; Ghosh, T. K.; Orasugh, J. T.; Khatua, S.; Acharya, K.; Chattopadhyay, D. Synthesis of RGO/NiO nanocomposites adopting a green approach and its photocatalytic and antibacterial properties. *Materials Chemistry and Physics* 2020, *247*, 122906.

(138) Pantelic, R. S.; Meyer, J. C.; Kaiser, U.; Baumeister, W.; Plitzko, J. M. Graphene oxide: A substrate for optimizing preparations of frozen-hydrated samples. *Journal of Structural Biology* 2010, *170* (1), 152–156.

(139) Meyer, J. C.; Geim, A. K.; Katsnelson, M. I.; Novoselov, K. S.; Booth, T. J.; Roth, S. The structure of suspended graphene sheets. *Nature* 2007, *446* (7131), 60–63.

(140) Meyer, J. C.; Geim, A.; Katsnelson, M.; Novoselov, K.; Obergfell, D.; Roth, S.; Girit, C.; Zettl, A. On the roughness of single-and bi-layer graphene membranes. *Solid State Communications* 2007, *143* (1–2), 101–109.

(141) Aliyev, E.; Filiz, V.; Khan, M. M.; Lee, Y. J.; Abetz, C.; Abetz, V. Structural characterization of graphene oxide: Surface functional groups and fractionated oxidative debris. *Nanomaterials* 2019, *9* (8), 1180.

(142) Lian, Q.; Islam, F.; Ahmad, Z. U.; Lei, X.; Depan, D.; Zappi, M.; Gang, D. D.; Holmes, W.; Yan, H. Enhanced adsorption of resorcinol onto phosphate functionalized graphene oxide synthesized via Arbuzov reaction: A proposed mechanism of hydrogen bonding and π-π interactions. *Chemosphere* 2021, *280*, 130730.

(143) Gao, W.; Alemany, L. B.; Ci, L.; Ajayan, P. M. New insights into the structure and reduction of graphite oxide. *Nature Chemistry* 2009, *1* (5), 403–408.

(144) Orasugh, J. T.; Ray, S. S. Nanocellulose-graphene oxide-based nanocomposite for adsorptive water treatment. In Hato, M. J.; Ray, S. S., *Functional Polymer Nanocomposites for Wastewater Treatment*; Springer, 2022; pp. 1–53.

(145) Pei, S.; Cheng, H.-M. The reduction of graphene oxide. *Carbon* 2012, *50* (9), 3210–3228.

(146) Wang, H.; Robinson, J. T.; Li, X.; Dai, H. Solvothermal reduction of chemically exfoliated graphene sheets. *Journal of the American Chemical Society* 2009, *131* (29), 9910–9911.

(147) Rabchinskii, M. K.; Shnitov, V. V.; Dideikin, A. T.; Aleksenskii, A. E.; Vul', S. P.; Baidakova, M. V.; Pronin, I. I.; Kirilenko, D. A.; Brunkov, P. N.; Weise, J. Nanoscale perforation of graphene oxide during photoreduction process in the argon atmosphere. *The Journal of Physical Chemistry C* 2016, *120* (49), 28261–28269.

(148) Khalil, A. M. E.; Memon, F. A.; Tabish, T. A.; Salmon, D.; Zhang, S.; Butler, D. Nanostructured porous graphene for efficient removal of emerging contaminants (pharmaceuticals) from water. *Chemical Engineering Journal* 2020, *398*, 125440. DOI: 10.1016/j.cej.2020.125440

(149) Li, M.; Wang, C.; O'Connell, M. J.; Chan, C. K. Carbon nanosphere adsorbents for removal of arsenate and selenate from water. *Environmental Science: Nano* 2015, *2* (3), 245–250.

(150) Wang, Y.; Liang, S.; Chen, B.; Guo, F.; Yu, S.; Tang, Y. Synergistic removal of Pb (II), Cd (II) and humic acid by Fe_3O_4@ mesoporous silica-graphene oxide composites. *PloS One* 2013, *8* (6), e65634.

(151) Liu, J.; Tang, J.; Gooding, J. J. Strategies for chemical modification of graphene and applications of chemically modified graphene. *Journal of Materials Chemistry* 2012, *22* (25), 12435–12452.

(152) Kumarathilaka, P.; Jayaweera, V.; Wijesekara, H.; Kottegoda, I.; Rosa, S.; Vithanage, M. Insights into starch coated nanozero valent iron-graphene composite for Cr (VI) removal from aqueous medium. *Journal of Nanomaterials* 2016, *2016*.

(153) Babel, S.; Kurniawan, T. A. Low-cost adsorbents for heavy metals uptake from contaminated water: A review. *Journal of Hazardous Materials* 2003, *97* (1), 219–243.

(154) Lee, Y.-C.; Yang, J.-W. Self-assembled flower-like TiO_2 on exfoliated graphite oxide for heavy metal removal. *Journal of Industrial and Engineering Chemistry* 2012, *18* (3), 1178–1185.

(155) Zare-Dorabei, R.; Ferdowsi, S. M.; Barzin, A.; Tadjarodi, A. Highly efficient simultaneous ultrasonic-assisted adsorption of Pb(II), Cd(II), Ni(II) and Cu (II) ions from aqueous solutions by graphene oxide modified with 2,2'-dipyridylamine: Central composite design optimization. *Ultrasonics Sonochemistry* 2016, *32*, 265–276.

(156) Zhang, Y.; Yan, L.; Xu, W.; Guo, X.; Cui, L.; Gao, L.; Wei, Q.; Du, B. Adsorption of Pb(II) and Hg(II) from aqueous solution using magnetic $CoFe_2O_4$-reduced graphene oxide. *Journal of Molecular Liquids* 2014, *191*, 177–182.

(157) Rajivgandhi, G.; Rtv, V.; Nandhakumar, R.; Murugan, S.; Alharbi, N. S.; Kadaikunnan, S.; Khaled, J. M.; Alanzi, K. F.; Li, W.-J. Adsorption of nickel ions from electroplating effluent by graphene oxide and reduced graphene oxide. *Environmental Research* 2021, *199*, 111322.

(158) Chen, R.; Cheng, Y.; Wang, P.; Wang, Q.; Wan, S.; Huang, S.; Su, R.; Song, Y.; Wang, Y. Enhanced removal of Co(II) and Ni(II) from high-salinity aqueous solution using reductive self-assembly of three-dimensional magnetic fungal hyphal/graphene oxide nanofibers. *Science of The Total Environment* 2021, *756*, 143871.

(159) Zhang, C.-Z.; Yuan, Y.; Li, T. Adsorption and desorption of heavy metals from water using aminoethyl reduced graphene oxide. *Environmental Engineering Science* 2018, *35* (9), 978–987.

(160) Chaabane, L.; Beyou, E.; El Ghali, A.; Baouab, M. H. V. Comparative studies on the adsorption of metal ions from aqueous solutions using various functionalized graphene oxide sheets as supported adsorbents. *Journal of Hazardous Materials* 2020, *389*, 121839.

(161) Yuan, Y.; Zhang, G.; Li, Y.; Zhang, G.; Zhang, F.; Fan, X. Poly (amidoamine) modified graphene oxide as an efficient adsorbent for heavy metal ions. *Polymer Chemistry* 2013, *4* (6), 2164–2167.

(162) Guo, F.-Y.; Liu, Y.-G.; Wang, H.; Zeng, G.-M.; Hu, X.-J.; Zheng, B.-H.; Li, T.-T.; Tan, X.-F.; Wang, S.-F.; Zhang, M.-M. Adsorption behavior of Cr (VI) from aqueous solution onto magnetic graphene oxide functionalized with 1,2-diaminocyclohexanetetraac etic acid. *RSC Advances* 2015, *5* (56), 45384–45392.

(163) Zhang, K.; Li, H.; Xu, X.; Yu, H. Synthesis of reduced graphene oxide/NiO nanocomposites for the removal of Cr (VI) from aqueous water by adsorption. *Microporous and Mesoporous Materials* 2018, *255*, 7–14.

(164) Liu, M.; Wen, T.; Wu, X.; Chen, C.; Hu, J.; Li, J.; Wang, X. Synthesis of porous Fe$_3$O$_4$ hollow microspheres/graphene oxide composite for Cr (VI) removal. *Dalton Transactions* 2013, *42* (41), 14710–14717.

(165) Luo, X.; Wang, C.; Luo, S.; Dong, R.; Tu, X.; Zeng, G. Adsorption of As (III) and As (V) from water using magnetite Fe3O4-reduced graphite oxide–MnO$_2$ nanocomposites. *Chemical Engineering Journal* 2012, *187*, 45–52.

(166) Chandra, V.; Park, J.; Chun, Y.; Lee, J. W.; Hwang, I.-C.; Kim, K. S. Water-dispersible magnetite-reduced graphene oxide composites for arsenic removal. *ACS Nano* 2010, *4* (7), 3979–3986.

(167) Das, T. K.; Sakthivel, T. S.; Jeyaranjan, A.; Seal, S.; Bezbaruah, A. N. Ultra-high arsenic adsorption by graphene oxide iron nanohybrid: Removal mechanisms and potential applications. *Chemosphere* 2020, *253*, 126702.

(168) Thakur, S.; Das, G.; Raul, P. K.; Karak, N. Green one-step approach to prepare sulfur/reduced graphene oxide nanohybrid for effective mercury ions removal. *The Journal of Physical Chemistry C* 2013, *117* (15), 7636–7642.

(169) Yap, P. L.; Tung, T. T.; Kabiri, S.; Matulick, N.; Tran, D. N. H.; Losic, D. Polyamine-modified reduced graphene oxide: A new and cost-effective adsorbent for efficient removal of mercury in waters. *Separation and Purification Technology* 2020, *238*, 116441.

(170) Saxena, R.; Saxena, M.; Lochab, A. Recent progress in nanomaterials for adsorptive removal of organic contaminants from wastewater. *ChemistrySelect* 2020, *5* (1), 335–353.

(171) Benotti, M. J.; Trenholm, R. A.; Vanderford, B. J.; Holady, J. C.; Stanford, B. D.; Snyder, S. A. Pharmaceuticals and endocrine disrupting compounds in US drinking water. *Environmental Science & Technology* 2009, *43* (3), 597–603.

(172) Baccar, R.; Sarrà, M.; Bouzid, J.; Feki, M.; Blánquez, P. Removal of pharmaceutical compounds by activated carbon prepared from agricultural by-product. *Chemical Engineering Journal* 2012, *211*, 310–317.

(173) Lima, E. C. Removal of emerging contaminants from the environment by adsorption. *Ecotoxicology and Environmental Safety* 2018, *150*, 1–17.

(174) Lambropoulou, D. A.; Nollet, L. M. *Transformation Products of Emerging Contaminants in the Environment: Analysis, Processes, Occurrence, Effects and Risks*; John Wiley & Sons, 2014.

(175) Organization, W. H. Pharmaceuticals in Drinking-Water. 1. Water Pollutants, Chemical. 2. Pharmaceutical Preparations. 3. Water Purification. 4. Potable Water. I. World Health Organization 2012. ISBN 978 92 4 150208 5.

(176) Fekadu, S.; Alemayehu, E.; Dewil, R.; Van der Bruggen, B. Pharmaceuticals in freshwater aquatic environments: A comparison of the African and European challenge. *Science of the Total Environment* 2019, *654*, 324–337.

(177) Rizzo, L.; Manaia, C.; Merlin, C.; Schwartz, T.; Dagot, C.; Ploy, M.; Michael, I.; Fatta-Kassinos, D. Urban wastewater treatment plants as hotspots for antibiotic resistant bacteria and genes spread into the environment: A review. *Science of the Total Environment* 2013, *447*, 345–360.

(178) Al-Khateeb, L. A.; Almotiry, S.; Salam, M. A. Adsorption of pharmaceutical pollutants onto graphene nanoplatelets. *Chemical Engineering Journal* 2014, *248*, 191–199.

(179) Rostamian, R.; Behnejad, H. A comparative adsorption study of sulfamethoxazole onto graphene and graphene oxide nanosheets through equilibrium, kinetic and thermodynamic modeling. *Process Safety and Environmental Protection* 2016, *102*, 20–29.

(180) Brito, C. H. V.; Gloria, D. C. S.; de Barros Santos, E.; Domingues, R. A.; Valente, G. T.; Vieira, N. C. S.; Gonçalves, M. Porous activated carbon/graphene oxide composite for efficient adsorption of pharmaceutical contaminants. *Chemical Engineering Research and Design* 2023, *191*, 387–400.

(181) Yagub, M. T.; Sen, T. K.; Afroze, S.; Ang, H. M. Dye and its removal from aqueous solution by adsorption: A review. *Advances in Colloid and Interface Science* 2014, *209*, 172–184.

(182) Robati, D.; Rajabi, M.; Moradi, O.; Najafi, F.; Tyagi, I.; Agarwal, S.; Gupta, V. K. Kinetics and thermodynamics of malachite green dye adsorption from aqueous solutions on graphene oxide and reduced graphene oxide. *Journal of Molecular Liquids* 2016, *214*, 259–263.

(183) Soudagar, S.; Akash, S.; Sree Venkat, M.; Rao Poiba, V.; Vangalapati, M. Adsorption of methylene blue dye on nano graphene oxide-thermodynamics and kinetic studies. *Materials Today: Proceedings* 2022, *59*, 667–672.

(184) Dan, S.; Bagheri, H.; Shahidizadeh, A.; Hashemipour, H. Performance of graphene Oxide/SiO2 Nanocomposite-based: Antibacterial Activity, dye and heavy metal removal. *Arabian Journal of Chemistry* 2023, *16* (2), 104450.

(185) Yao, L. W.; Ahmed Khan, F. S.; Mubarak, N. M.; Karri, R. R.; Khalid, M.; Walvekar, R.; Abdullah, E. C.; Mazari, S. A.; Ahmad, A.; Dehghani, M. H. Insight into immobilization efficiency of Lipase enzyme as a biocatalyst on the graphene oxide for adsorption of Azo dyes from industrial wastewater effluent. *Journal of Molecular Liquids* 2022, *354*, 118849.

(186) Paton-Carrero, A.; Sanchez, P.; Sánchez-Silva, L.; Romero, A. Graphene-based materials behaviour for dyes adsorption. *Materials Today Communications* 2022, *30*, 103033.

(187) Jun, T.-S.; Park, N.-H.; So, D.-S.; Lee, J.-W.; Shim, K. B.; Ham, H. Phosphate removing by graphene oxide in aqueous solution. *Journal of the Korean Crystal Growth and Crystal Technology* 2013, *23* (6), 325–328.

(188) Sereshti, H.; Zamiri Afsharian, E.; Esmaeili Bidhendi, M.; Rashidi Nodeh, H.; Afzal Kamboh, M.; Yilmaz, M. Removal of phosphate and nitrate ions aqueous using strontium magnetic graphene oxide nanocomposite: Isotherms, kinetics, and thermodynamics studies. *Environmental Progress & Sustainable Energy* 2020, *39* (2), e13332.

(189) Gao, Y.; Li, Y.; Zhang, L.; Huang, H.; Hu, J.; Shah, S. M.; Su, X. Adsorption and removal of tetracycline antibiotics from aqueous solution by graphene oxide. *Journal of Colloid and Interface Science* 2012, *368* (1), 540–546.

(190) Wu, S.; Zhao, X.; Li, Y.; Du, Q.; Sun, J.; Wang, Y.; Wang, X.; Xia, Y.; Wang, Z.; Xia, L. Adsorption properties of doxorubicin hydrochloride onto graphene oxide: Equilibrium, kinetic and thermodynamic studies. *Materials* 2013, *6* (5), 2026–2042.

(191) Song, W.; Yang, T.; Wang, X.; Sun, Y.; Ai, Y.; Sheng, G.; Hayat, T.; Wang, X. Experimental and theoretical evidence for competitive interactions of tetracycline and sulfamethazine with reduced graphene oxides. *Environmental Science: Nano* 2016, *3* (6), 1318–1326.

(192) Yang, J.; Shojaei, S.; Shojaei, S. Removal of drug and dye from aqueous solutions by graphene oxide: Adsorption studies and chemometrics methods. *NPJ Clean Water* 2022, *5* (1), 5.

(193) Peng, G.; Zhang, M.; Deng, S.; Shan, D.; He, Q.; Yu, G. Adsorption and catalytic oxidation of pharmaceuticals by nitrogen-doped reduced graphene oxide/Fe_3O_4 nanocomposite. *Chemical Engineering Journal* 2018, *341*, 361–370.

(194) Moradi, O.; Alizadeh, H.; Sedaghat, S. Removal of pharmaceuticals (diclofenac and amoxicillin) by maltodextrin/reduced graphene and maltodextrin/reduced graphene/copper oxide nanocomposites. *Chemosphere* 2022, *299*, 134435.

(195) Ganesan, P.; Kamaraj, R.; Vasudevan, S. Application of isotherm, kinetic and thermodynamic models for the adsorption of nitrate ions on graphene from aqueous solution. *Journal of the Taiwan Institute of Chemical Engineers* 2013, *44* (5), 808–814.

(196) Bakry, A. M.; Alamier, W. M.; Salama, R. S.; Samy El-Shall, M.; Awad, F. S. Remediation of water containing phosphate using ceria nanoparticles decorated partially reduced graphene oxide (CeO_2-PRGO) composite. *Surfaces and Interfaces* 2022, *31*, 102006.

(197) Luo, W.; Cao, X.; Xu, X.; Huang, S.; Liu, C.; Tomljanovic, T. Developmental transcriptome analysis and identification of genes involved in formation of intestinal air-breathing function of Dojo loach, Misgurnus anguillicaudatus. *Scientific Reports* 2016, *6* (1), 31845.

(198) Sakulpaisan, S.; Vongsetskul, T.; Reamouppaturm, S.; Luangkachao, J.; Tantirungrotechai, J.; Tangboriboonrat, P. Titania-functionalized graphene oxide for an efficient adsorptive removal of phosphate ions. *Journal of Environmental Management* 2016, *167*, 99–104.

(199) Vicente-Martínez, Y.; Caravaca, M.; Soto-Meca, A.; De Francisco-Ortiz, O.; Gimeno, F. Graphene oxide and graphene oxide functionalized with silver nanoparticles as adsorbents of phosphates in waters. A comparative study. *Science of the Total Environment* 2020, *709*, 136111.

(200) Li, Y.; Du, Q.; Liu, T.; Peng, X.; Wang, J.; Sun, J.; Wang, Y.; Wu, S.; Wang, Z.; Xia, Y.; Xia, L. Comparative study of methylene blue dye adsorption onto activated carbon, graphene oxide, and carbon nanotubes. *Chemical Engineering Research and Design* 2013, *91* (2), 361–368.

(201) Fan, L.; Luo, C.; Sun, M.; Qiu, H.; Li, X. Synthesis of magnetic β-cyclodextrin–chitosan/graphene oxide as nanoadsorbent and its application in dye adsorption and removal. *Colloids and Surfaces B: Biointerfaces* 2013, *103*, 601–607.

(202) Ramesha, G. K.; Vijaya Kumara, A.; Muralidhara, H. B.; Sampath, S. Graphene and graphene oxide as effective adsorbents toward anionic and cationic dyes. *Journal of Colloid and Interface Science* 2011, *361* (1), 270–277.

(203) Gupta, K.; Khatri, O. P. Reduced graphene oxide as an effective adsorbent for removal of malachite green dye: Plausible adsorption pathways. *Journal of Colloid and Interface Science* 2017, *501*, 11–21.

(204) Othman, N. H.; Alias, N. H.; Shahruddin, M. Z.; Abu Bakar, N. F.; Nik Him, N. R.; Lau, W. J. Adsorption kinetics of methylene blue dyes onto magnetic graphene oxide. *Journal of Environmental Chemical Engineering* 2018, *6* (2), 2803–2811.

(205) Konicki, W.; Aleksandrzak, M.; Mijowska, E. Equilibrium, kinetic and thermodynamic studies on adsorption of cationic dyes from aqueous solutions using graphene oxide. *Chemical Engineering Research and Design* 2017, *123*, 35–49.

(206) Diraki, A.; Mackey, H.; McKay, G.; Abdala, A. A. Removal of oil from oil–water emulsions using thermally reduced graphene and graphene nanoplatelets. *Chemical Engineering Research and Design* 2018, *137*, 47–59.

(207) Diraki, A.; Mackey, H. R.; McKay, G.; Abdala, A. Removal of emulsified and dissolved diesel oil from high salinity wastewater by adsorption onto graphene oxide. *Journal of Environmental Chemical Engineering* 2019, *7* (3), 103106.

(208) Hafiz, M. A.; Hawari, A. H.; Altaee, A. A hybrid forward osmosis/reverse osmosis process for the supply of fertilizing solution from treated wastewater. *Journal of Water Process Engineering* 2019, *32*, 100975.

(209) Speed, D. E. 11 - Environmental aspects of planarization processes. In *Advances in Chemical Mechanical Planarization (CMP)*, Babu, S. Ed.; Woodhead Publishing, 2016; pp 229–269.

(210) Nair, R. R.; Wu, H. A.; Jayaram, P. N.; Grigorieva, I. V.; Geim, A. K. Unimpeded permeation of water through helium-leak – Tight graphene-based membranes. *Science* 2012, *335* (6067), 442–444.

(211) Liu, G.; Jin, W.; Xu, N. Graphene-based membranes. *Chemical Society Reviews* 2015, *44* (15), 5016–5030.

(212) Dreyer, D. R.; Park, S.; Bielawski, C. W.; Ruoff, R. S. The chemistry of graphene oxide. *Chemical Society Reviews* 2010, *39* (1), 228–240.

(213) Xu, Q.; Xu, H.; Chen, J.; Lv, Y.; Dong, C.; Sreeprasad, T. S. Graphene and graphene oxide: Advanced membranes for gas separation and water purification. *Inorganic Chemistry Frontiers* 2015, *2* (5), 417–424.

(214) Alkhouzaam, A.; Qiblawey, H. Functional GO-based membranes for water treatment and desalination: Fabrication methods, performance and advantages. A review. *Chemosphere* 2021, *274*, 129853.

(215) Sun, M.; Li, J. Graphene oxide membranes: Functional structures, preparation and environmental applications. *Nano Today* 2018, *20*, 121–137.

(216) Mazinani, S.; Darvishmanesh, S.; Ehsanzadeh, A.; Van der Bruggen, B. Phase separation analysis of extem/solvent/non-solvent systems and relation with membrane morphology. *Journal of Membrane Science* 2017, *526*, 301–314.

(217) Jung, J. T.; Kim, J. F.; Wang, H. H.; Di Nicolo, E.; Drioli, E.; Lee, Y. M. Understanding the non-solvent induced phase separation (NIPS) effect during the fabrication of microporous PVDF membranes via thermally induced phase separation (TIPS). *Journal of Membrane Science* 2016, *514*, 250–263.

(218) Sadrzadeh, M.; Bhattacharjee, S. Rational design of phase inversion membranes by tailoring thermodynamics and kinetics of casting solution using polymer additives. *Journal of Membrane Science* 2013, *441*, 31–44.

(219) Qian, X.; Li, N.; Wang, Q.; Ji, S. Chitosan/graphene oxide mixed matrix membrane with enhanced water permeability for high-salinity water desalination by pervaporation. *Desalination* 2018, *438*, 83–96.

(220) Nigiz, F. U. Graphene oxide-sodium alginate membrane for seawater desalination through pervaporation. *Desalination* 2020, *485*, 114465.

(221) Chen, J.; Guo, Y.; Huang, L.; Xue, Y.; Geng, D.; Liu, H.; Wu, B.; Yu, G.; Hu, W.; Liu, Y. Controllable fabrication of ultrathin free-standing graphene films. *Philosophical Transactions of the Royal Society A: Mathematical, Physical and Engineering Sciences* 2014, *372* (2013), 20130017.

(222) Yang, Z.; Ma, X.-H.; Tang, C. Y. Recent development of novel membranes for desalination. *Desalination* 2018, *434*, 37–59.

(223) Liu, H.; Wang, H.; Zhang, X. Facile fabrication of freestanding ultrathin reduced graphene oxide membranes for water purification. *Advanced Materials* 2015, *27* (2), 249–254.

(224) Chen, H.; Wu, H.; Wang, Q.; Ji, L.; Zhang, T.; Yang, H. Separation performance of Hg^{2+} in desulfurization wastewater by the graphene oxide polyethersulfone membrane. *Energy & Fuels* 2019, *33* (9), 9241–9248.

(225) Tsou, C.-H.; An, Q.-F.; Lo, S.-C.; De Guzman, M.; Hung, W.-S.; Hu, C.-C.; Lee, K.-R.; Lai, J.-Y. Effect of microstructure of graphene oxide fabricated through different self-assembly techniques on 1-butanol dehydration. *Journal of Membrane Science* 2015, *477*, 93–100.

(226) Xu, W. L.; Fang, C.; Zhou, F.; Song, Z.; Liu, Q.; Qiao, R.; Yu, M. Self-assembly: A facile way of forming ultrathin, high-performance graphene oxide membranes for water purification. *Nano Letters* 2017, *17* (5), 2928–2933.

(227) Chen, X.; Qiu, M.; Ding, H.; Fu, K.; Fan, Y. A reduced graphene oxide nanofiltration membrane intercalated by well-dispersed carbon nanotubes for drinking water purification. *Nanoscale* 2016, *8* (10), 5696–5705.

(228) Li, F.; Yu, Z.; Shi, H.; Yang, Q.; Chen, Q.; Pan, Y.; Zeng, G.; Yan, L. A Mussel-inspired method to fabricate reduced graphene oxide/g-C_3N_4 composites membranes for catalytic decomposition and oil-in-water emulsion separation. *Chemical Engineering Journal* 2017, *322*, 33–45.

(229) Kang, X.; Cheng, Y.; Wen, Y.; Qi, J.; Li, X. Bio-inspired co-deposited preparation of GO composite loose nanofiltration membrane for dye contaminated wastewater sustainable treatment. *Journal of Hazardous Materials* 2020, *400*, 123121.

(230) Sun, J.; Qian, X.; Wang, Z.; Zeng, F.; Bai, H.; Li, N. Tailoring the microstructure of poly (vinyl alcohol)-intercalated graphene oxide membranes for enhanced desalination performance of high-salinity water by pervaporation. *Journal of Membrane Science* 2020, *599*, 117838.

(231) Qian, Y.; Zhang, X.; Liu, C.; Zhou, C.; Huang, A. Tuning interlayer spacing of graphene oxide membranes with enhanced desalination performance. *Desalination* 2019, *460*, 56–63.

(232) Lian, B.; Deng, J.; Leslie, G.; Bustamante, H.; Sahajwalla, V.; Nishina, Y.; Joshi, R. Surfactant modified graphene oxide laminates for filtration. *Carbon* 2017, *116*, 240–245.

(233) Zhao, X.; Su, Y.; Liu, Y.; Li, Y.; Jiang, Z. Free-standing graphene oxide-palygorskite nanohybrid membrane for oil/water separation. *ACS Applied Materials & Interfaces* 2016, *8* (12), 8247–8256.

(234) Jia, F.; Xiao, X.; Nashalian, A.; Shen, S.; Yang, L.; Han, Z.; Qu, H.; Wang, T.; Ye, Z.; Zhu, Z. Advances in graphene oxide membranes for water treatment. *Nano Research* 2022, *15* (7), 6636–6654.

(235) Yang, S.-T.; Chang, Y.; Wang, H.; Liu, G.; Chen, S.; Wang, Y.; Liu, Y.; Cao, A. Folding/aggregation of graphene oxide and its application in Cu^{2+} removal. *Journal of Colloid and Interface Science* 2010, *351* (1), 122–127.

(236) Pei, J.; Huang, L.; Jiang, H.; Liu, H.; Liu, X.; Hu, X. Inhibitory effect of hydrogen ion on the copper ions separation from acid solution across graphene oxide membranes. *Separation and Purification Technology* 2019, *210*, 651–658.

(237) Yu, X.; Liu, W.; Deng, X.; Yan, S.; Su, Z. Gold nanocluster embedded bovine serum albumin nanofibers-graphene hybrid membranes for the efficient detection and separation of mercury ion. *Chemical Engineering Journal* 2018, *335*, 176–184.

(238) Zhang, Y.; Chung, T.-S. Graphene oxide membranes for nanofiltration. *Current Opinion in Chemical Engineering* 2017, *16*, 9–15.

(239) Zunita, M.; Irawanti, R.; Koesmawati, T. A.; Lugito, G.; Wentena, I. G. Graphene oxide (Go) membrane in removing heavy metals from wastewater: A review. *Chemical Engineering Transactions* 2020, *82*, 415–420.

(240) Sun, P.; Zhu, M.; Wang, K.; Zhong, M.; Wei, J.; Wu, D.; Xu, Z.; Zhu, H. Selective ion penetration of graphene oxide membranes. *ACS Nano* 2013, *7* (1), 428–437.

(241) Zhang, Y.; Zhang, S.; Gao, J.; Chung, T.-S. Layer-by-layer construction of graphene oxide (GO) framework composite membranes for highly efficient heavy metal removal. *Journal of Membrane Science* 2016, *515*, 230–237.

(242) Mi, B. Graphene oxide membranes for ionic and molecular sieving. *Science* 2014, *343* (6172), 740–742.

(243) Zhang, P.; Gong, J.-L.; Zeng, G.-M.; Deng, C.-H.; Yang, H.-C.; Liu, H.-Y.; Huan, S.-Y. Cross-linking to prepare composite graphene oxide-framework membranes with high-flux for dyes and heavy metal ions removal. *Chemical Engineering Journal* 2017, *322*, 657–666.

(244) Rao, Z.; Feng, K.; Tang, B.; Wu, P. Surface decoration of amino-functionalized metal–organic framework/graphene oxide composite onto polydopamine-coated membrane substrate for highly efficient heavy metal removal. *ACS Applied Materials & Interfaces* 2017, *9* (3), 2594–2605.

(245) Ahmad, H.; Huang, Z.; Kanagaraj, P.; Liu, C. Separation and preconcentration of arsenite and other heavy metal ions using graphene oxide laminated with protein molecules. *Journal of Hazardous Materials* 2020, *384*, 121479.

(246) Musielak, M.; Gagor, A.; Zawisza, B.; Talik, E.; Sitko, R. Graphene Oxide/Carbon Nanotube Membranes for Highly Efficient Removal of Metal Ions from Water. *ACS Applied Materials & Interfaces* 2019, *11* (31), 28582–28590.

(247) Fang, F.; Kong, L.; Huang, J.; Wu, S.; Zhang, K.; Wang, X.; Sun, B.; Jin, Z.; Wang, J.; Huang, X.-J. Removal of cobalt ions from aqueous solution by an amination graphene oxide nanocomposite. *Journal of Hazardous Materials* 2014, *270*, 1–10.

(248) Guo, Y.; Jia, Z.; Wang, S.; Su, Y.; Ma, H.; Wang, L.; Meng, W. Sandwich membranes based on PVDF-g-4VP and surface modified graphene oxide for Cu (II) adsorption. *Journal of Hazardous Materials* 2019, *377*, 17–23.

(249) Zhao, G.; Li, J.; Ren, X.; Chen, C.; Wang, X. Few-layered graphene oxide nanosheets as superior sorbents for heavy metal ion pollution management. *Environmental Science & Technology* 2011, *45* (24), 10454–10462.

(250) Han, S.; Li, W.; Xi, H.; Yuan, R.; Long, J.; Xu, C. Plasma-assisted in-situ preparation of graphene-Ag nanofiltration membranes for efficient removal of heavy metal ions. *Journal of Hazardous Materials* 2022, *423*, 127012.

(251) Ibrahim, Y.; Wadi, V. S.; Ouda, M.; Naddeo, V.; Banat, F.; Hasan, S. W. Highly selective heavy metal ions membranes combining sulfonated polyethersulfone and self-assembled manganese oxide nanosheets on positively functionalized graphene oxide nanosheets. *Chemical Engineering Journal* 2022, *428*, 131267.

(252) Wu, T.; Wang, Z.; Lu, Y.; Liu, S.; Li, H.; Ye, G.; Chen, J. Graphene oxide membranes for tunable ion sieving in acidic radioactive waste. *Advanced Science* 2021, *8* (7), 2002717.

(253) He, M.; Wang, L.; Zhang, Z.; Zhang, Y.; Zhu, J.; Wang, X.; Lv, Y.; Miao, R. Stable forward osmosis nanocomposite membrane doped with sulfonated graphene oxide@ metal–organic frameworks for heavy metal removal. *ACS Applied Materials & Interfaces* 2020, *12* (51), 57102–57116.

(254) Hadi Najafabadi, H.; Irani, M.; Roshanfekr Rad, L.; Heydari Haratameh, A.; Haririan, I. Removal of Cu^{2+}, Pb^{2+} and Cr^{6+} from aqueous solutions using a chitosan/graphene oxide composite nanofibrous adsorbent. *RSC Advances* 2015, *5* (21), 16532–16539.

(255) Koushkbaghi, S.; Jafari, P.; Rabiei, J.; Irani, M.; Aliabadi, M. Fabrication of PET/PAN/GO/Fe_3O_4 nanofibrous membrane for the removal of Pb (II) and Cr (VI) ions. *Chemical Engineering Journal* 2016, *301*, 42–50.

(256) Zhang, Y.; Zhang, S.; Chung, T.-S. Nanometric graphene oxide framework membranes with enhanced heavy metal removal via nanofiltration. *Environmental Science & Technology* 2015, *49* (16), 10235–10242.

(257) Hu, M.; Mi, B. Layer-by-layer assembly of graphene oxide membranes via electrostatic interaction. *Journal of Membrane Science* 2014, *469*, 80–87.

(258) Rezaee, R.; Nasseri, S.; Mahvi, A. H.; Nabizadeh, R.; Mousavi, S. A.; Rashidi, A.; Jafari, A.; Nazmara, S. Fabrication and characterization of a polysulfone-graphene oxide nanocomposite membrane for arsenate rejection from water. *Journal of Environmental Health Science and Engineering* 2015, *13* (1), 1–11.

(259) Mukherjee, R.; Bhunia, P.; De, S. Impact of graphene oxide on removal of heavy metals using mixed matrix membrane. *Chemical Engineering Journal* 2016, *292*, 284–297.

(260) Yang, R.; Fan, Y.; Yu, R.; Dai, F.; Lan, J.; Wang, Z.; Chen, J.; Chen, L. Robust reduced graphene oxide membranes with high water permeance enhanced by K$^+$ modification. *Journal of Membrane Science* 2021, *635*, 119437.

(261) Chandio, I.; Janjhi, F. A.; Memon, A. A.; Memon, S.; Ali, Z.; Thebo, K. H.; Pirzado, A. A. A.; Hakro, A. A.; Khan, W. S. Ultrafast ionic and molecular sieving through graphene oxide based composite membranes. *Desalination* 2021, *500*, 114848.

(262) Dai, F.; Zhou, F.; Chen, J.; Liang, S.; Chen, L.; Fang, H. Ultrahigh water permeation with a high multivalent metal ion rejection rate through graphene oxide membranes. *Journal of Materials Chemistry A* 2021, *9* (17), 10672–10677.

(263) Humplik, T.; Lee, J.; O'Hern, S.; Fellman, B.; Baig, M.; Hassan, S.; Atieh, M.; Rahman, F.; Laoui, T.; Karnik, R. Nanostructured materials for water desalination. *Nanotechnology* 2011, *22* (29), 292001.

(264) Li, S.; Chen, Y.; He, X.; Mao, X.; Zhou, Y.; Xu, J.; Yang, Y. Modifying reduced graphene oxide by conducting polymer through a hydrothermal polymerization method and its application as energy storage electrodes. *Nanoscale Research Letters* 2019, *14* (1), 1–12.

(265) Nguyen, M. N.; Weidler, P. G.; Schwaiger, R.; Schäfer, A. I. Interactions between carbon-based nanoparticles and steroid hormone micropollutants in water. *Journal of Hazardous Materials* 2021, *402*, 122929.

(266) Yu, L.; Zhang, Y.; Zhang, B.; Liu, J.; Zhang, H.; Song, C. Preparation and characterization of HPEI-GO/PES ultrafiltration membrane with antifouling and antibacterial properties. *Journal of Membrane Science* 2013, *447*, 452–462.

(267) Wang, J.; Li, N.; Zhao, Y.; Xia, S. Graphene oxide modified semi-aromatic polyamide thin film composite membranes for PPCPs removal. *Desalination and Water Treatment* 2017, *66*, 166–175.

(268) Wang, X.; Guo, Y.; Jia, Z.; Ma, H.; Liu, C.; Liu, Z.; Shi, Q.; Ren, B.; Li, L.; Zhang, X.; Hu, Y. Fabrication of graphene oxide/polydopamine adsorptive membrane by stepwise in-situ growth for removal of rhodamine B from water. *Desalination* 2021, *516*, 115220.

(269) Hu, M.; Mi, B. Enabling graphene oxide nanosheets as water separation membranes. *Environmental Science & Technology* 2013, *47* (8), 3715–3723.

(270) Qiu, Z.; Ji, X.; He, C. Fabrication of a loose nanofiltration candidate from polyacrylonitrile/graphene oxide hybrid membrane via thermally induced phase separation. *Journal of Hazardous Materials* 2018, *360*, 122–131.

(271) Nicolopoulou-Stamati, P.; Maipas, S.; Kotampasi, C.; Stamatis, P.; Hens, L. Chemical pesticides and human health: The urgent need for a new concept in agriculture. *Frontiers in Public Health* 2016, *4*, 148.

(272) Silvestro, I.; Ciarlantini, C.; Francolini, I.; Tomai, P.; Gentili, A.; Dal Bosco, C.; Piozzi, A. Chitosan–graphene oxide composite membranes for solid-phase extraction of pesticides. *International Journal of Molecular Sciences* 2021, *22* (16), 8374.

(273) Salehi, S.; Hosseinifard, M. Optimized removal of phosphate and nitrate from aqueous media using zirconium functionalized nanochitosan-graphene oxide composite. *Cellulose* 2020, *27*, 8859–8883.

(274) Janjhi, F. A.; Chandio, I.; Memon, A. A.; Ahmed, Z.; Thebo, K. H.; Pirzado, A. A. A.; Hakro, A. A.; Iqbal, M. Functionalized graphene oxide based membranes for ultrafast molecular separation. *Separation and Purification Technology* 2021, *274*, 117969.

(275) Amiri, S.; Asghari, A.; Vatanpour, V.; Rajabi, M. Fabrication and characterization of a novel polyvinyl alcohol-graphene oxide-sodium alginate nanocomposite hydrogel blended PES nanofiltration membrane for improved water purification. *Separation and Purification Technology* 2020, *250*, 117216.

(276) Fang, S.-Y.; Zhang, P.; Gong, J.-L.; Tang, L.; Zeng, G.-M.; Song, B.; Cao, W.-C.; Li, J.; Ye, J. Construction of highly water-stable metal-organic framework UiO-66 thin-film composite membrane for dyes and antibiotics separation. *Chemical Engineering Journal* 2020, *385*, 123400.

(277) Luque-Alled, J. M.; Abdel-Karim, A.; Alberto, M.; Leaper, S.; Perez-Page, M.; Huang, K.; Vijayaraghavan, A.; El-Kalliny, A. S.; Holmes, S. M.; Gorgojo, P. Polyethersulfone membranes: From ultrafiltration to nanofiltration via the incorporation of APTS functionalized-graphene oxide. *Separation and Purification Technology* 2020, *230*, 115836.

(278) Thebo, K. H.; Qian, X.; Wei, Q.; Zhang, Q.; Cheng, H.-M.; Ren, W. Reduced graphene oxide/metal oxide nanoparticles composite membranes for highly efficient molecular separation. *Journal of Materials Science & Technology* 2018, *34* (9), 1481–1486.

(279) Lu, J.-J.; Gu, Y.-H.; Chen, Y.; Yan, X.; Guo, Y.-J.; Lang, W.-Z. Ultrahigh permeability of graphene-based membranes by adjusting D-spacing with poly (ethylene imine) for the separation of dye wastewater. *Separation and Purification Technology* 2019, *210*, 737–745.

(280) Wu, H.; Tang, B.; Wu, P. Development of novel SiO_2–GO nanohybrid/polysulfone membrane with enhanced performance. *Journal of Membrane Science* 2014, *451*, 94–102.

(281) Wu, L.; Liu, Y.; Hu, J.; Feng, X.; Ma, C.; Wen, C. Preparation of polyvinylidene fluoride composite ultrafiltration membrane for micro-polluted surface water treatment. *Chemosphere* 2021, *284*, 131294.

(282) Mustafa, B.; Mehmood, T.; Wang, Z.; Chofreh, A. G.; Shen, A.; Yang, B.; Yuan, J.; Wu, C.; Liu, Y.; Lu, W.; Hu, W. Next-generation graphene oxide additives composite membranes for emerging organic micropollutants removal: Separation, adsorption and degradation. *Chemosphere* 2022, *308*, 136333.

(283) Bahamon, D.; Vega, L. F. Molecular simulations of phenol and ibuprofen removal from water using multilayered graphene oxide membranes. *Molecular Physics* 2019, *117* (23–24), 3703–3714.

(284) Sheng, J.; Yin, H.; Qian, F.; Huang, H.; Gao, S.; Wang, J. Reduced graphene oxide-based composite membranes for in-situ catalytic oxidation of sulfamethoxazole operated in membrane filtration. *Separation and Purification Technology* 2020, *236*, 116275.

(285) Yuan, W.; Chen, J.; Shi, G. Nanoporous graphene materials. *Materials Today* 2014, *17* (2), 77–85.

(286) Yasaei, P.; Kumar, B.; Hantehzadeh, R.; Kayyalha, M.; Baskin, A.; Repnin, N.; Wang, C.; Klie, R. F.; Chen, Y. P.; Král, P. Chemical sensing with switchable transport channels in graphene grain boundaries. *Nature Communications* 2014, *5* (1), 1–8.

(287) Tabish, T. A.; Memon, F. A.; Gomez, D. E.; Horsell, D. W.; Zhang, S. A facile synthesis of porous graphene for efficient water and wastewater treatment. *Scientific Reports* 2018, *8* (1), 1–14.

(288) Alazmi, A.; El Tall, O.; Rasul, S.; Hedhili, M. N.; Patole, S. P.; Costa, P. M. A process to enhance the specific surface area and capacitance of hydrothermally reduced graphene oxide. *Nanoscale* 2016, *8* (41), 17782–17787.

(289) Venkatesan, B. M.; Bashir, R. Nanopore sensors for nucleic acid analysis. *Nature Nanotechnology* 2011, *6* (10), 615–624.

(290) Chang, Z.; Xu, J.; Zhang, X. Recent progress in electrocatalyst for Li-O_2 batteries. *Advanced Energy Materials* 2017, *7* (23), 1700875.

(291) Song, N.; Gao, X.; Ma, Z.; Wang, X.; Wei, Y.; Gao, C. A review of graphene-based separation membrane: Materials, characteristics, preparation and applications. *Desalination* 2018, *437*, 59–72.

(292) Liang, X.; Jung, Y.-S.; Wu, S.; Ismach, A.; Olynick, D. L.; Cabrini, S.; Bokor, J. Formation of bandgap and subbands in graphene nanomeshes with sub-10 nm ribbon width fabricated via nanoimprint lithography. *Nano Letters* 2010, *10* (7), 2454–2460.

(293) Kim, M.; Safron, N. S.; Han, E.; Arnold, M. S.; Gopalan, P. Fabrication and characterization of large-area, semiconducting nanoperforated graphene materials. *Nano Letters* 2010, *10* (4), 1125–1131.

(294) Sinitskii, A.; Tour, J. M. Patterning graphene through the self-assembled templates: Toward periodic two-dimensional graphene nanostructures with semiconductor properties. *Journal of the American Chemical Society* 2010, *132* (42), 14730–14732.

(295) Akhavan, O. Graphene nanomesh by ZnO nanorod photocatalysts. *ACS Nano* 2010, *4* (7), 4174–4180.

(296) Fischbein, M. D.; Drndić, M. Electron beam nanosculpting of suspended graphene sheets. *Applied Physics Letters* 2008, *93* (11), 113107.

(297) Lemme, M. C.; Bell, D. C.; Williams, J. R.; Stern, L. A.; Baugher, B. W.; Jarillo-Herrero, P.; Marcus, C. M. Etching of graphene devices with a helium ion beam. *ACS Nano* 2009, *3* (9), 2674–2676.

(298) Zeng, Z.; Huang, X.; Yin, Z.; Li, H.; Chen, Y.; Li, H.; Zhang, Q.; Ma, J.; Boey, F.; Zhang, H. Fabrication of graphene nanomesh by using an anodic aluminum oxide membrane as a template. *Advanced Materials* 2012, *24* (30), 4138–4142.

(299) Yang, S.; Feng, X.; Wang, L.; Tang, K.; Maier, J.; Müllen, K. Graphene-based nanosheets with a sandwich structure. *Angewandte Chemie International Edition* 2010, *49* (28), 4795–4799.

(300) Diankov, G.; Neumann, M.; Goldhaber-Gordon, D. Extreme monolayer-selectivity of hydrogen-plasma reactions with graphene. *ACS Nano* 2013, *7* (2), 1324–1332.

(301) Wasalathilake, K. C.; Ayoko, G.; Yan, C. Porous graphene materials for energy storage and conversion applications. *Recent Advances in Graphene Research* 2016, *2*, 196.

(302) Huang, X.; Qian, K.; Yang, J.; Zhang, J.; Li, L.; Yu, C.; Zhao, D. Functional nanoporous graphene foams with controlled pore sizes. *Advanced Materials* 2012, *24* (32), 4419–4423.

(303) Su, D. Macroporous 'bubble' graphene film via template-directed ordered-assembly for high rate supercapacitors. *Chemical Communications* 2012, *48* (57), 7149–7151.

(304) Zhang, Y.; Wan, Q.; Yang, N. Recent advances of porous graphene: Synthesis, functionalization, and electrochemical applications. *Small* 2019, *15* (48), 1903780.

(305) Xu, Y.; Lin, Z.; Zhong, X.; Huang, X.; Weiss, N. O.; Huang, Y.; Duan, X. Holey graphene frameworks for highly efficient capacitive energy storage. *Nature Communications* 2014, *5* (1), 1–8.

(306) Liu, L.; Ryu, S.; Tomasik, M. R.; Stolyarova, E.; Jung, N.; Hybertsen, M. S.; Steigerwald, M. L.; Brus, L. E.; Flynn, G. W. Graphene oxidation: Thickness-dependent etching and strong chemical doping. *Nano Letters* 2008, *8* (7), 1965–1970.

(307) Hooch Antink, W.; Choi, Y.; Seong, K. D.; Kim, J. M.; Piao, Y. Recent progress in porous graphene and reduced graphene oxide-based nanomaterials for electrochemical energy storage devices. *Advanced Materials Interfaces* 2018, *5* (5), 1701212.

(308) Fan, Z.; Zhao, Q.; Li, T.; Yan, J.; Ren, Y.; Feng, J.; Wei, T. Easy synthesis of porous graphene nanosheets and their use in supercapacitors. *Carbon* 2012, *50* (4), 1699–1703.

(309) Romanos, J.; Beckner, M.; Rash, T.; Firlej, L.; Kuchta, B.; Yu, P.; Suppes, G.; Wexler, C.; Pfeifer, P. Nanospace engineering of KOH activated carbon. *Nanotechnology* 2011, *23* (1), 015401.

(310) Lin, Y.; Watson, K. A.; Kim, J.-W.; Baggett, D. W.; Working, D. C.; Connell, J. W. Bulk preparation of holey graphene via controlled catalytic oxidation. *Nanoscale* 2013, *5* (17), 7814–7824.

(311) Ning, G.; Fan, Z.; Wang, G.; Gao, J.; Qian, W.; Wei, F. Gram-scale synthesis of nano-mesh graphene with high surface area and its application in supercapacitor electrodes. *Chemical Communications* 2011, *47* (21), 5976–5978.

(312) O'Hern, S. C.; Boutilier, M. S.; Idrobo, J.-C.; Song, Y.; Kong, J.; Laoui, T.; Atieh, M.; Karnik, R. Selective ionic transport through tunable subnanometer pores in single-layer graphene membranes. *Nano Letters* 2014, *14* (3), 1234–1241.

(313) Celebi, K.; Buchheim, J.; Wyss, R. M.; Droudian, A.; Gasser, P.; Shorubalko, I.; Kye, J.-I.; Lee, C.; Park, H. G. Ultimate permeation across atomically thin porous graphene. *Science* 2014, *344* (6181), 289–292.

(314) Bell, D. C.; Lemme, M. C.; Stern, L. A.; Williams, J. R.; Marcus, C. M. Precision cutting and patterning of graphene with helium ions. *Nanotechnology* 2009, *20* (45), 455301.

(315) Wei, Y.; Pastuovic, Z.; Shen, C.; Murphy, T.; Gore, D. B. Ion beam engineered graphene oxide membranes for mono-/di-valent metal ions separation. *Carbon* 2020, *158*, 598–606.

(316) Zhou, F.; Yu, J.; Jiang, X. 3D porous graphene synthesised using different hydrothermal treatment times for the removal of lead ions from an aqueous solution. *Micro & Nano Letters* 2017, *12* (5), 308–311.

(317) Tian, H.; Guo, J.; Pang, Z.; Hu, M.; He, J. A sulfur, nitrogen dual-doped porous graphene nanohybrid for ultraselective Hg (ii) separation over Pb (ii) and Cu (ii). *Nanoscale* 2020, *12* (31), 16543–16555.

(318) Zhao, X.-R.; Xu, X.; Teng, J.; Zhou, N.; Zhou, Z.; Jiang, X.-Y.; Jiao, F.-P.; Yu, J.-G. Three-dimensional porous graphene oxide-maize amylopectin composites with controllable pore-sizes and good adsorption-desorption properties: Facile fabrication and reutilization, and the adsorption mechanism. *Ecotoxicology and Environmental Safety* 2019, *176*, 11–19.

(319) Zhang, Y.; Liu, Y.; Wang, X.; Sun, Z.; Ma, J.; Wu, T.; Xing, F.; Gao, J. Porous graphene oxide/carboxymethyl cellulose monoliths, with high metal ion adsorption. *Carbohydrate Polymers* 2014, *101*, 392–400.

(320) Tabasi, E.; Vafa, N.; Firoozabadi, B.; Salmankhani, A.; Nouranian, S.; Habibzadeh, S.; Mashhadzadeh, A. H.; Spitas, C.; Saeb, M. R. Ion rejection performances of functionalized porous graphene nanomembranes for wastewater purification: A molecular dynamics simulation study. *Colloids and Surfaces A: Physicochemical and Engineering Aspects* 2023, *656*, 130492.

(321) Yang, K.; Li, X.; Cui, J.; Zhang, M.; Wang, Y.; Lou, Z.; Shan, W.; Xiong, Y. Facile synthesis of novel porous graphene-like carbon hydrogel for highly efficient recovery of precious metal and removal of organic dye. *Applied Surface Science* 2020, *528*, 146928.

(322) Lingamdinne, L. P.; Choi, Y.-L.; Kim, I.-S.; Chang, Y.-Y.; Koduru, J. R.; Yang, J.-K. Porous graphene oxide based inverse spinel nickel ferrite nanocomposites for the enhanced adsorption removal of arsenic. *RSC Advances* 2016, *6* (77), 73776–73789.

(323) Niu, Z.; Liu, L.; Zhang, L.; Chen, X. Porous graphene materials for water remediation. *Small* 2014, *10* (17), 3434–3441.

(324) Khalil, A. M.; Han, L.; Maamoun, I.; Tabish, T. A.; Chen, Y.; Eljamal, O.; Zhang, S.; Butler, D.; Memon, F. A. Novel graphene-based foam composite as a highly reactive filter medium for the efficient removal of gemfibrozil from (waste) water. *Advanced Sustainable Systems* 2022, *6* (8), 2200016.

(325) Shan, D.; Deng, S.; Li, J.; Wang, H.; He, C.; Cagnetta, G.; Wang, B.; Wang, Y.; Huang, J.; Yu, G. Preparation of porous graphene oxide by chemically intercalating a rigid molecule for enhanced removal of typical pharmaceuticals. *Carbon* 2017, *119*, 101–109.

(326) Li, D.; Zhang, H.; Zhang, L.; Wang, P.; Xu, H.; Xuan, J. Rapid synthesis of porous graphene microspheres through a three-dimensionally printed inkjet nozzle for selective pollutant removal from water. *ACS Omega* 2019, *4* (24), 20509–20518.

(327) Yu, B.; Zhang, X.; Xie, J.; Wu, R.; Liu, X.; Li, H.; Chen, F.; Yang, H.; Ming, Z.; Yang, S.-T. Magnetic graphene sponge for the removal of methylene blue. *Applied Surface Science* 2015, *351*, 765–771.

(328) Lai, Y.-T.; Huang, Y.-S.; Chen, C.-H.; Lin, Y.-C.; Jeng, H.-T.; Chang, M.-C.; Chen, L.-J.; Lee, C.-Y.; Hsu, P.-C.; Tai, N.-H. Green treatment of phosphate from wastewater using a porous bio-templated graphene oxide/mgmn-layered double hydroxide composite. *iScience* 2020, *23* (5), 101065.

(329) Kong, D.; Zheng, X.; Tao, Y.; Lv, W.; Gao, Y.; Zhi, L.; Yang, Q.-H. Porous graphene oxide-based carbon artefact with high capacity for methylene blue adsorption. *Adsorption* 2016, *22*, 1043–1050.

(330) Yao, T.; Qiao, L.; Du, K. High tough and highly porous graphene/carbon nanotubes hybrid beads enhanced by carbonized polyacrylonitrile for efficient dyes adsorption. *Microporous and Mesoporous Materials* 2020, *292*, 109716.

(331) Deb, H.; Islam, M. Z.; Ahmed, A.; Hasan, M. K.; Hossain, M. K.; Hu, H.; Chen, C.; Yang, S.; Zhang, Y.; Yao, J. Kinetics & dynamic studies of dye adsorption by porous graphene nano-adsorbent for facile toxic wastewater remediation. *Journal of Water Process Engineering* 2022, *47*, 102818.

(332) Ma, J.; Yang, M.; Yu, F.; Zheng, J. Water-enhanced removal of ciprofloxacin from water by porous graphene hydrogel. *Scientific Reports* 2015, *5* (1), 1–10.

4 Adsorption and Its Kinetics, Isotherms, as Well as Thermodynamics

4.1 INTRODUCTION

This chapter presents the core of the present book in that it discusses adsorption, its kinetics, as well as the isotherms which are of great importance to researchers within the niche of adsorption.

4.1.1 ADSORPTION

Adsorption is a surface phenomenon that takes place when adsorbates transfer to adsorbents. Adsorption techniques have been extensively used for water as well as wastewater treatment for decades since they are affordable, efficient, straightforward, and ecologically beneficial. Chemical adsorption involves the creation of chemical bonds between the absorbent and adsorbate, while physical adsorption involves the creation of van der Waals force(s) and ion exchange; these are the most common adsorption mechanisms (Figure 4.1). Among all these approaches, isotherm modeling of adsorption data is the most convenient and commonly used. Furthermore, adsorption isotherm models could provide details on the maximum adsorption capacity, which is useful in evaluating the effectiveness of adsorbent materials.

Surface characteristics like surface area to volume ratio size as well as shape have a substantial influence on graphene adsorption. Adsorption of said adsorbate here on adsorbent is typically affected by surface area, as the adsorption capacity of an adsorbent is mostly determined by the entire surface area per unit volume of adsorbed species. This comprises the explicit surface area, which is the entire internal surface area ascribed to the adsorbent's pore size distribution. Whenever the surface gets higher, the area of binding sites exposed to adsorbate during the adsorption increases. To optimize total surface area, the adsorbent should have a high porosity and a narrow particle size dispersion to boost adsorption capacity. Graphene possesses a very high specific surface area and no porosity. The insertion of porosity is an efficient and feasible way for improving graphene's adsorption capability. Porosity can be introduced by combining graphene with other porous materials of inorganic or organic origin. Such porosity could be introduced to graphene-based adsorbent materials using a special class of polymeric materials referred to as PIM (Polymer of Intrinsic Microporosity). In assorted applications, such as adsorption, gas separation, as well as catalysis, polymers of PIM have demonstrated behavior that is equivalent to that of crystalline and amorphous microporous but also mesoporous materials.[1,2] PIMs possess a backbone made of fused rings and a site of contortion, which are

DOI: 10.1201/9781032621302-4

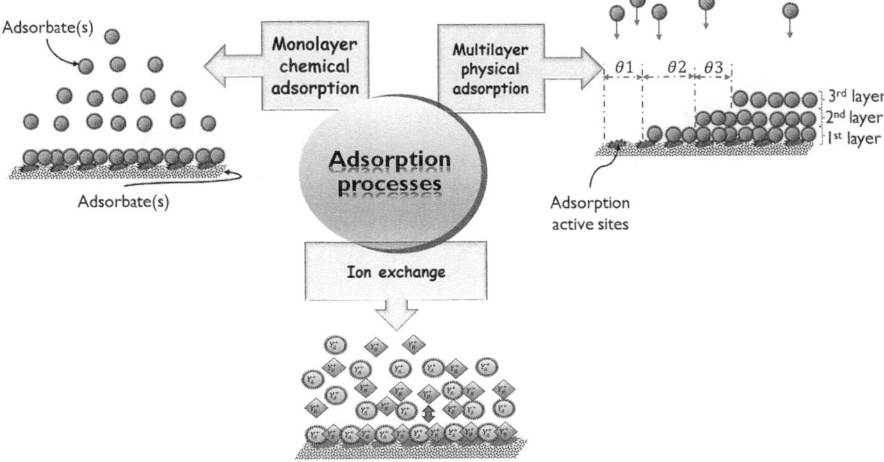

FIGURE 4.1 Typical scheme illustrating potential adsorption processes.

uncommon structural properties that cause them to have a high free volume because they can't pack space effectively.[2] Studies on this material revealed its high surface area, high thermal as well as chemical stability, along with potential use in adsorption.[1] It is also easily processed into a variety of application-driven forms, such as membranes, powders, as well as fibers, and they are soluble in common organic solvents.[1,3] The first generation of these polymeric materials known as PIM-1 (poly(1-trimethylsilyl-1-propyne) (PTMSP) as well as poly(4-methyl-2-pentyne) (PMP)) received a lot of attention in the adsorption of material systems like the removal of a variety of dyes[4] as well as phenol[5] from aqueous solutions and indeed the exclusion of aniline from air as well as aqueous phases,[6] although many PIMs have been successfully synthesized. Alnajrani et al. have researched and shown the use of PIM-1 in the elimination of emerging organic contaminants like antibiotics.[1] A group of authors have recently reported the modification of graphene oxide by covalently grafting PIM-1 via the facile dip coating approach.[7] The structural architecture of the system they synthesized is depicted in Figure 4.2 showing the covalently attached PIM-1 on the edges of graphene oxide.

4.1.2 Adsorption Kinetics

The primary goal of an adsorption kinetic study is to give an overall understanding of the reaction path followed and the time required to achieve equilibrium. Adsorption is a complex issue that typically involves both surface adsorption and diffusion further into pores. Based on previous research, interpreting adsorption kinetics aids in determining the extent of adsorbate uptake by an adsorbent and resolving the adsorption procedure mechanism. In general, two kinetic models are used: the Lagergren model, also known as the 'pseudo-first-order kinetic model (PFOM),' and the 'pseudo-second-order kinetic model (PSOM). However, this section will attempt to describe most of the models, as per literature.

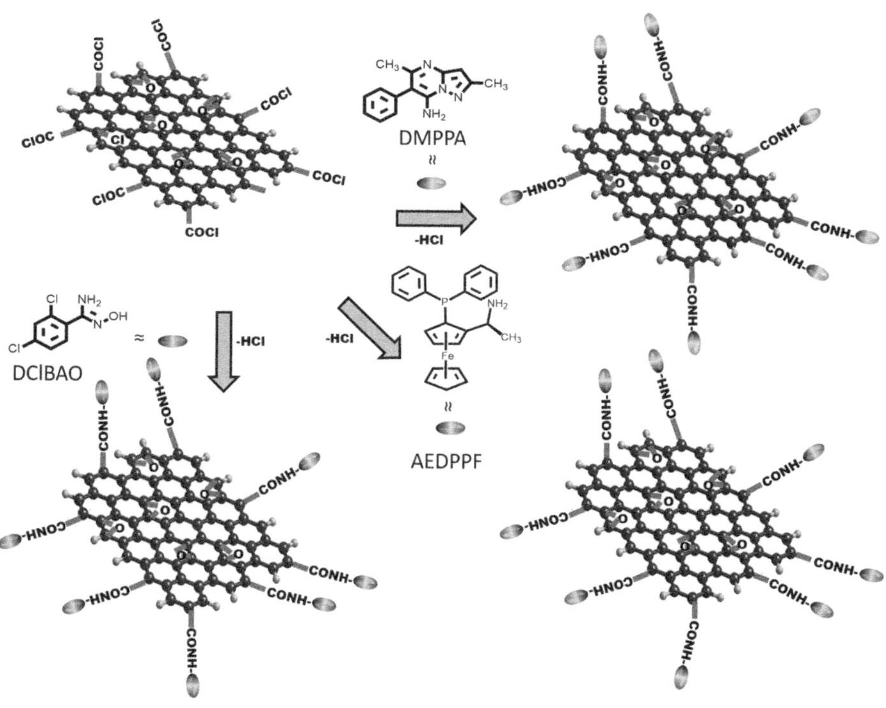

FIGURE 4.2 Covalently modified graphene oxide and polymer of intrinsic microporosity (PIM-1) in mixed matrix thin-film composite membranes. (Reproduced with permission from Elvin M. Aliyev et al.[7] Copyright 2018, Springer Nature.)

4.1.2.1 Lagergren's Pseudo-First-Order Kinetic Model

Depending upon the PFOM equation, the liquid-solid adsorption system most frequently employs the Lagergren kinetic.[8] The PFOM, as per Rodrigues and Silva,[9] the following ordinary first-order differential equation (4.1) represents the adsorption kinetics of a species in an adsorbent particle[9]:

$$\frac{dQ}{dt} = K_1\left(Q_e - Q_t\right) \tag{4.1}$$

Q_e and Q_t whose unit is in mg/g denotes the adsorbates amounts adsorbed at equilibrium but also at time t while K_1 represents the rate constant equilibrium within the PSOM in L/m.

The "expanse towards equilibrium," which is defined as the difference between the ultimate concentration of the phase adsorbed in equilibrium with the fluid phase and the average concentration of the species in the adsorbed phase in differential equation (4.1) above, is a function of the adsorption capacity: The distance at equilibrium for a clean particle is equal to the ultimate concentration of the adsorbed phase in equilibrium, with the fluid phase at time $t = 0$, and as time passes, the distance at equilibrium shrinks until it is null when equilibrium is reached: Average adsorbent dose of species in the adsorbed phase and the ultimate adsorbed phase dosage in equilibrium with the fluid phase differ by zero.[9]

The following PFO Lagergren equation is derived by integrating the differential equation (4.1) above for the boundary conditions: $qt = 0$ to $t = 0$ and $qt = qt$ to $t = t$[10,11]:

$$\ln\left(Q_e - Q_t\right) = \ln\left(Q_e\right) - K_1 t \tag{4.2}$$

where Q_e represents the adsorbate amount in the adsorbent at equilibrium (mg/g); Q_t refers to the quantity of adsorbate there in adsorbent at time t (mg/g); K_1 represents the Lagergren's first order rate constant, and t represents the time of contact (min).

To verify the said isotherm, it is useful to graph $\ln\left(Q_e - Q_t\right)$ as a function of contact time t and note the coefficient of determination R^2.

Though the kinetics of dye adsorption onto graphene, its derivatives, as well as composites/hybrids, mostly follow second-order kinetics.[12,13,14,15] However, some reports have been shown to follow both first-order and second-order kinetic models.[16,17] For instance, Verma et al. reported R^2 value of 0.999 for PSOM kinetic with

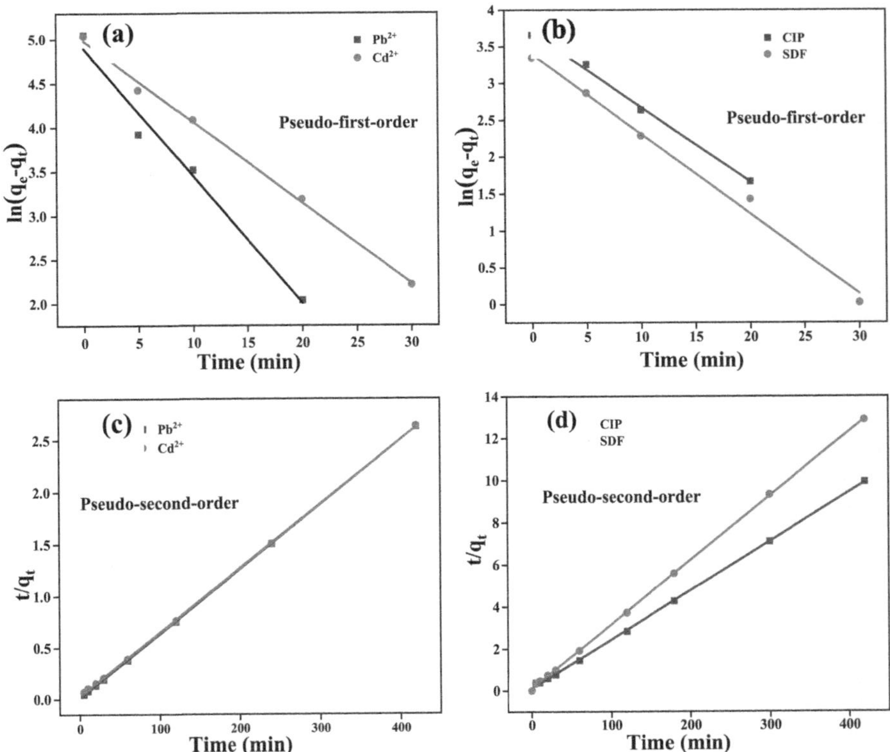

FIGURE 4.3 PFO linear plots for (a) Pb^{2+} and Cd^{2+} ions, (b) CIP and SDF, and PSO linear plots for (c) Pb^{2+} and Cd^{2+} ions, and (d) CIP and SDF onto GO–CS–EDTA polymer. (Reproduced with permission from Verma et al.[18] Copyright 2022, Elsevier Science Ltd.)

TABLE 4.1

Model Parameters of PFO and PSO Kinetics for Adsorption of Metal Ions and Organic Contaminants onto GO−CS−EDTA Adsorbent

System	$q_{e,exp}$ (mg g^{-1})	PFO $q_{e,cal}$ (mg g^{-1})	k_1 (min^{-1})	R^2	PSO $q_{e,cal}$ (mg g^{-1})	k_2 (g mg^{-1} min^{-1})	R^2
Pb^{2+}	156.49	159.98	0.144	0.979	160.77	0.0086	0.999
Cd^{2+}	152.78	159.62	0.091	0.988	160.25	0.0041	0.999
CIP	42.33	40.04	0.108	0.986	43.12	0.0038	0.999
SDF	32.60	29.66	0.101	0.988	32.89	0.0028	0.999

Source: Reproduced with copyright permission from Verma et al.,[18] Elsevier, 2022.

the use of GO-CS-EDTA engineered adsorbent and postulated that the adsorption process was basically chemosorption for both inorganic as well as organic contaminants, as in Figure 4.3 and Table 4.1.[18] They observed that the inorganic metallic ions were adsorbed more quicker than their organic counterpart, as presented in Table 4.1.

Several other studies on wastewater contaminants have likewise shown to follow PSOM as revealed by Verma et al.[18]

With regards to reports showing that both adsorption kinetic processes playing part in the adsorption mechanism may involve two or more kinetics, the Al-Shemy et al.[16] report has revealed that their synthesized hybrid graphene-based adsorbent utilized for the adsorptive removal of methylene blue from water displayed a PSOM kinetics though the looking at the R^2 value, one could only envisage that the adsorption kinetics followed both PSOM as well as PFOM kinetics as presented in Table 4.2 as well as Figure 4.4.

TABLE 4.2

Kinetic Parameters for the Adsorption of MB for Dye Concentration of 100 mg/L with Dosage of Adsorbents 2.5 g/L at 298 K

Scaffold	$Q_{e(exp.)}$, mg/g	Pseudo-First-Order $Q_{e(calc.)}$, mg/g	k_1, 1/min	R^2	Pseudo-Second-Order $Q_{e(calc.)}$, mg/g	k_2, g/mg.min	h, mg/g.min	R^2
Control	36.24	43.00	0.90×10^{-2}	0.95	51.55	11.02×10^{-5}	0.26	0.98
10NC	36.51	41.77	0.90×10^{-2}	0.96	49.75	13.41×10^{-5}	0.29	0.98
15NC	29.40	32.24	1.04×10^{-2}	0.97	35.21	35.03×10^{-5}	0.37	0.99
20NC	20.63	23.31	1.24×10^{-2}	0.98	24.57	49.76×10^{-5}	0.28	0.99

Source: Reproduced with permission from Al-Shemy et al.[16] Copyright 2022, Elsevier Science Ltd.

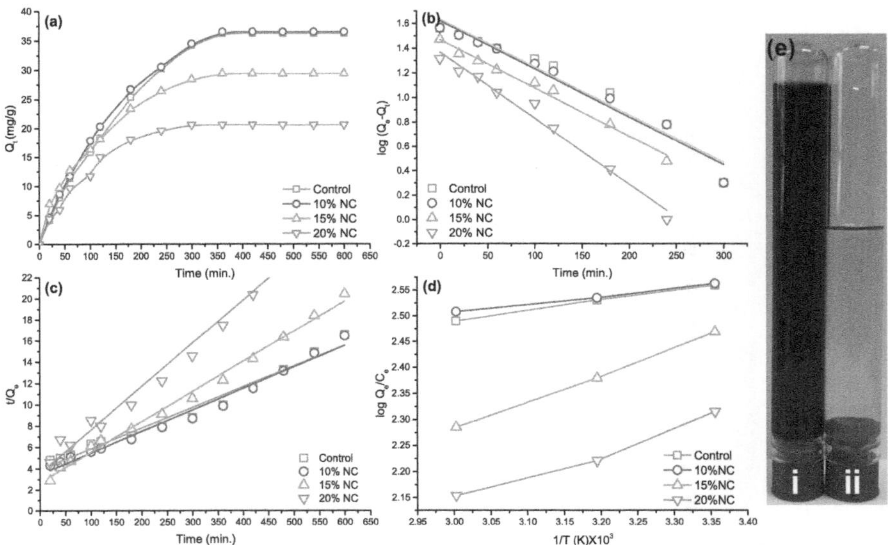

FIGURE 4.4 (a) Kinetics, (b) Pseudo first-order plots, and (c) Pseudo second-order plots of MB adsorptions by the prepared scaffolds at 298 K and pH 6. (d) Thermodynamic plots of MB adsorptions by the prepared scaffolds for dye concentrations of 100.0 mg/L and scaffolds dosage of 2.5 g /L, for 200 min at pH 6.0. (e) Photograph to compare MB dye before (i) and after (ii) adsorption by bio-nanocomposites scaffolds of SA/GO with 10% NC at equilibrium. (Reproduced with copyright permission from Al-Shemy et al.,[16] Elsevier 2022.)

4.1.2.2 Kinetic Model of Pseudo-Second-Order

Adsorption occurs on two surface sites in pseudo-second-order kinetics (also known as "Blanchard's model" by some authors) and can thus be expressed by the second-order differential equation (4.3)[19,20]:

The mathematics equation for the following pseudo-second-order kinetic model, equation (4.3) below, was generated by integrating this differential equation (4.3) below for the boundary conditions: $qt = 0$ to $t = 0$ and $q_t = qt$ to $t = t$ [10,11]:

$$t/Q_t = \frac{1}{K_2 Q_e^2} + \frac{1}{Q_e} t \tag{4.3}$$

where K_2 is a constant rate of pseudo-second order, t is the time of contact (min), Q_e is the amount of adsorbate in the adsorbent at equilibrium (mg/g), Q_t is the quantity of adsorbate in the adsorbent at time t (mg/g).

It is convenient to graphically display t/Q_t according to the contact time t and record the coefficient of determination R^2 in order to test this isotherm.

As earlier stated, this kinetic model is the most followed by most graphene-based adsorbents adsorption process, especially for dyes as well as antibiotic adsorbates, as shown in Tables 4.3 and 4.4.[16,17,21–23]

TABLE 4.3

GBAs and Their Application for Pharmaceutical Traces Removal

Formulation	Pharmaceutical Traces Removed	Adsorbent Dosage	pH for Max. Adsorption	Temp. (K)	Maximum Adsorption Capacity	Adsorption Isotherm & Kinetics		
						Isotherm	Kinetics	Ref.
GO	Carbamazepine (CBZ)	1 g/L	2.0	308.15	9.2 mg/g	LM	PSOM	25
Graphene	Sulfamethoxazole (SMX)	60 mg/L	8.0	—	210.08 mg/g	LM	PSOM	26
	Acetaminophen (ACM)	60 mg/L	6.0	—	56.21 mg/g	LM	PSOM	26
GO	Metformin	40 mg/L	6.0	288–303	50.47 mg/g	Freundlich isotherm (FRIM)	PSOM	27
GO	Acetaminophen	200 mg	8.0	296–3323	18.06 mg/g	Type IV isotherm with an H3 type hysteresis loop	PSOM	28
	Caffeine	200 mg	8.0	77	19.72 mg/g	Type IV isotherm with an H3 type hysteresis loop	PSOM	28
	Aspirin	200 mg	12.0	77	13.02 mg/g	Type IV isotherm with an H3 type hysteresis loop	PSOM	28
rGO@magnetite (RGO-M)	Ciprofloxacin (CIP)	0.2 g/L	6.2	298	18.22 mg/g	LM	PSOM	29
	Norfloxacin (NOR)	0.2 g/L	6.2	298	22.20 mg/g	LM	PSOM	29
GO	Tetracycline	0.181 mg/mL	3.6	298.15	313 mg/g	LM	PSOM	30
GO	Trimethoprim (TMP)	15 mg/L			204.08	LM	PSOM	31
	Isoniazid (INH)	15 mg/L			13.89	LM	PSOM	31

Source: Reproduced with copyright permission from Temane, Orasugh, and Ray.[24] Copyright 2022, Springer Nature.

TABLE 4.4

GBAs and Their Application for Dyes/Pigments Removal

Formulation	Dyes/Pigment Removed	Adsorbent Dosage	pH for Max. Adsorption	Temp. (K)	Maximum Adsorption Capacity	Adsorption Isotherm & Kinetics		Ref.
						Isotherm	Kinetics	
Polypyrrole functionalized cobalt oxide graphene (COPYGO)	MB	0.01g/L	7.2	308.5	663.018 mg/g	LM	PSOM	32
GO@chitosan	MB	50 mg/L	7.1	RT	141 ± 6.60 mg/g	LM	PSOM	33
	Crystal violet (CV)	50 mg/L	7.1	RT	121 ± 3.50 mg/g	LM	PSOM	33
P(AA-MMA)/MGO/CA-CD/NH$_2$	MB	0.5 g/L	12.0	318	3185 mg/g	LM	PSOM	34
	Malachite green (MG)	0.5 g/L	8.0	298	3315 mg/g	LM	PSOM	34
	CR	0.5 g/L	6.0	298	1058 mg/g	LM	PSOM	34
Montmorillonite-rGO composite aerogel (M@rGO)	MB	—	6.5		450.90 mg/g	LM	PSOM	35
Cellulose@GO (CGA)	MB	50 mg/100 mL	—	RT	490 mg/g	LM	PSOM	36
Polyaniline/dicarboxyl acid cellulose @GO (PANI@DCC@GO)	Reactive brilliant red K-2G	40 mg/150 mL	2.0	308.15	733.3 mg/g	FM	PSOM	37
Montmorillonite/GO/CoFe$_2$O$_4$ (MMT/GO/CoFe$_2$O$_4$)	Methyl violet (MV)	1 g/L	9.0	298.15	97.26 mg/g	LM	PSOM	38
Polysaccharide@GO (PS-GO)	MB	5 mg/L	10.0	298	789 mg/g	LM	PSOM	39
CNC: GO	Basic blue 7 (BB7)	0.1 g/L	6.0	RT	891	Redlich-Peterson (RPM)	PSOM	40
	Reactive orange 122 (RO)	0.1 g/L	6.0	RT	711	RPM	PSOM	40
rGO	MG	10 mg/L	8.0	303	279.85	LM	PSOM	41
	Rhodamine B (RhB)	0.1 g/L	6.0	RT	560	RPM	PSOM	40

4.1.2.3 Kinetic Model of Avrami

The Avrami model is reliant upon the use of thermal breakdown for adsorption. Along with determining fractional kinetic orders, it calculates the adsorption rates with regard to the initial concentration, as well as the adsorption time.[42] According to this model, the reaction occurs on the solid support's active surface sites, as well as its equation has the following two forms, double logarithmic and exponential, as shown by equations (4.4) and (4.5), respectively[43]:

$$Q_t = Q_{av}\left(1 - e^{(k_{av}t)^{n_{av}}}\right) \tag{4.4}$$

$$\ln\left(-\ln\left(1 - Q_t\right)\right) = \ln\left(K_{av}\right) + n_{av}\ln\left(t\right) \tag{4.5}$$

where Q_{av} is the amount of adsorption predicted by Avrami (in mg/g); K_{av} signifies rate of adsorption; t is the period of contact (in min); n_{av} is the adsorption order model; and Q_t is the amount of adsorbate in the adsorbent at time t (mg/g).

It is convenient to graph Q_t as a function of contact time t for the exponential model or $\ln\left(-\ln\left(1 - Q_t\right)\right)$ as a function of $\ln\left(t\right)$ for the double logarithmic model and record the coefficient of determination R^2 in order to test this isotherm.

The Avrami model is dependent on two variables derived from the Kolmogorov-Erofeev-Kazeeva-Avrami-Mampel equation, according to equations (4.4) and (4.5):

- The total adsorption speed (k)
- The adsorption mechanism-specific parameter (n), the values of which can be used to test for probable changes in the adsorption processes as a function of contact time and temperature. In general, n is the criterion that determines the domain in which these heterogeneous reactions occur. The value indicates whether or not the adsorption process can be limited by a surface reaction $(n > 1)$. If $n = 1$, the interaction is immediate.[43]

The main disadvantage of this model is that it occasionally exhibits the presence of two and/or three linearized areas as a function of time and adsorption temperatures.[43] As a result, it may be essential to consider two or three different sets of values for $n\{n_1, n_2, n_3\}$ as well as $k\{k_1, k_2, k_3\}$.[43] This model was used by George and Sugunan[43] to characterize the kinetics of adsorption of the lipase enzyme of Candida rugosa on pure and mesoporous silica materials produced hydrothermally and modified with glutaraldehyde. The enzyme-support interaction was well adjusted to the Avrami model (the coefficient of determination R^2 for the various mesoporous materials is between 0.90 and 0.98), adsorption is controlled by surface kinetics $(n > 1)$, and the enzyme had a relatively high affinity for the active sites of mesoporous silica based on the values obtained from the speed constant k and the activation energy, and it was around 49 kJ/mol.[43] The Avrami kinetic model, as per Ahmad et al.,[44] could've been utilized if adsorption is sluggish and/or there is far more than one process of adsorption.[44] A few reports have reported the obedience of graphene-based materials with regard to this model.[42] In most cases, the obedience of graphene-based adsorbents to this model is always in conjunction with other models simultaneously. For instance, Muthusaravanan et al. reveal in their study where they utilized GO for the adsorptive removal of atrazine that R^2 values gotten from the kinetics studies pointed Avrami

TABLE 4.5

Adsorption Kinetics Parameters and Statistical Values for Batch Adsorption of Atrazine onto GO Nanosheets

Model	Parameters	Concentration (mg/L)				
		10	20	30	40	50
Ho $q_t = \dfrac{q_e^2 k_{Ho} t}{1 + k_{Ho} q_e t}$	Q_e (mg/g)	32.20	84.74	107.78	163.20	190.12
	k_{Ho} (g mol^{-1} min^{-1})	4.45×10^{-05}	5.5×10^{-06}	6.47×10^{-06}	6.67×10^{-06}	6.92×10^{-06}
	R^2	0.9917	0.9920	0.9907	0.9789	0.9478
	RMSE	1.1410	2.2483	3.7017	6.8647	12.7745
	χ^2	0.1679	0.3971	0.5699	2.1793	8.7473
	Error	ERRSQ	HYBRID	MPSD	ERRSQ	EABS
	SNE	4.21	4.98	4.79	4.99	4.98
Avrami $q_t = q_e\left(1 - e^{(-k_{av}t)^{n_{av}}}\right)$	Q_e (mg/g)	21.83	42.35	78.48	120.06	152.66
	k_{av} (min^{-1})	0.0192	0.0519	0.0353	0.0139	0.0106
	n_{av}	0.7688	0.2996	0.2082	0.1955	0.1031
	R^2	0.9920	0.9823	0.9938	0.9987	0.9841
	RMSE	1.1376	3.355	3.0245	5.7022	7.037
	χ^2	0.2047	0.7793	0.4579	0.1559	1.5891
	Error	ERRSQ	EABS	ERRSQ	ARE	HYBRID
	SNE	3.93	4.80	4.99	4.92	4.98
Sobkowsk-Czerwi $q_t = \dfrac{q_e K_{sc} t}{K_{sc} t + 1}$	Q_e (mg/g)	29.18	61.04	111.64	134.14	180.54
	K_{sc} (min^{-1})	0.0130	0.0108	0.0069	0.0016	0.0012
	R^2	0.9919	0.9712	0.9616	0.9904	0.9839
	RMSE	1.1419	2.3614	3.5265	3.9035	7.0852
	χ^2	0.1681	0.3955	0.5706	0.9285	1.6504
	Error	ERRSQ	ERRSQ	MPSD	ARE	EABS
	SNE	4.33	4.68	4.97	4.98	4.99

Source: Reproduced with permission from Muthusaravanan et al.[42] Copyright 2021, Elsevier Science Ltd.

($R^2 = 0.9920$) and Sobkowsk-Czerwi ($R^2 = 0.9919$) models to be favored in comparison to the other models used.[42] Though, as per their report in Table 4.5 below, one could see that Ho kinetic model was also fairly obeyed in this regard.

Al-Qadri et al. have revealed in their study on the use of GO, rGO, as well as their composites aimed at the adsorptive removal of bisphenol A (BPA) that the kinetic model obeyed by the adsorption process was Avrami, though closely followed by PSOM kinetic model and then PFOM kinetic model.[45] The results they obtained are as presented in Table 4.6 below.

This model has been mostly followed where the adsorbate is an organic contaminant, especially the emerging organic contaminants, though in few instances inorganic,[46] and has been studied with respect to graphene-based adsorbents by limited researchers.[45,47,48] Though in a recent report, a group of researchers have shown that the adsorption kinetics of Acid black 1 dye (AB 1) as well as Cr(VI) followed PSOM but was also closely followed by the Avrami kinetic model where the R^2 for PFOM,

TABLE 4.6

The Kinetic Parameters for BPA Adsorption on GO-diethylenetriamine (DETA) and GO-diethylamine (DEA) Predicted Using the First Order, Second Order, and Avrami Models

Model	Parameter	GO-DEA	GO-DETA
First order	q_e (mg/g)	139.0 ± 8.6	151.8 ± 5.8
	k_1 (min^{-1})	0.0506 ± 0.0108	0.0958 ± 0.0151
	R^2	0.8749	0.9367
Second order	q_e (mg/g)	156.2 ± 9.8	163.0 ± 4.8
	k_2 (g mg^{-1} min^{-1})	0.0004 ± 0.0001	0.0009 ± 0.0002
	R^2	0.9250	0.9742
Avrami	q_e (mg/g)	187.9 ± 36.2	167.7 ± 4.2
	k_a (min^{-n})	0.0033 ± 0.0045	0.0149 ± 0.0050
	n	0.4684 ± 0.0770	0.4737 ± 0.0307
	R^2	0.9853	0.9966

Source: Reproduced with permission from Al-Qadri et al.,[45] Copyright 2022, Elsevier Science Ltd.

PSOM, and Avrami models were found as presented in Table 4.7.[47] These authors proposed that the diffusion of adsorbate molecules into the interior pores of the adsorbent, which would be likely to be a slow process and thus is considered to be the rate-determining step, occurs after the transport of adsorbate molecules from the aqueous solution to the surface of the adsorbent particles.[47]

Another report by Balasubramani et al. also revealed that in their study on the adsorptive removal of Flupentixol (FPL) as modelled by PFOM, Sobkowsk-Czerwi, PSOM, and vrami kinetic models showed that though the most prevalent model was PFOM but both Sobkowsk-Czerwi and Avrami kinetic models were also closely followed.[48] As a result, they proposed that the interaction between the functional groups on the surface of GO and GOC and the FPL molecules was what caused the FPL to adsorb onto GO and GOC. Additionally, the mechanism of FPL adsorption onto GO and GOC undoubtedly followed a number of kinetic orders that changed as the FPL came into contact with GO and GOC. The high initial FPL adsorption (30 min) was attributed to the reactive active sites that were present on the surfaces of GO and GOC, and the adsorption changed into diffusive as time went on, according to a reasonable agreement with the Sobkowsk-Czerwi kinetic equation.

So, it can be deduced that the obedience of this model by graphene-based adsorbents depends greatly on the kind of adsorbent as well as the adsorbate.

4.1.2.4   Bangham Kinetic Model

The Bangham kinetic model investigates the slow diffusion phase of the adsorption process.[49] The empirical mathematical equation (4.6) can be used to illustrate the model:

$$\log\left(\log\left(\frac{C_t}{C_i} \right) - Q_t m \right) = \log K_0 + Q_t \log\left(t \right) \tag{4.6}$$

TABLE 4.7

Kinetic Parameters for Acid Black 1 and Cr(VI) Ions onto the Amino-Functionalized SiO_2@$CoFe_2O_4$-GO Nanocomposites

Adsorbate	Pseudo-First-Order Model					
	$q_{e(exp)}$ (mg g^{-1})	k_1 (min^{-1})	$q_{e(cal)}$ (mg g^{-1})	RMSE	R^2	Empty Cell
AB 1	7.85	0.03	7.49	0.34	0.974	
Cr(VI)	21.01	0.05	19.79	5.51	0.813	

Adsorbate	Pseudo-Second-Order Model				
	$q_{e(exp)}$ (mg g^{-1})	k_2 (min^{-1})	$q_{e(cal)}$ (mg g^{-1})	RMSE	R^2
AB 1	7.85	0.005	8.80	0.24	0.998
Cr(VI)	21.01	0.003	22.39	4.72	0.992

Adsorbate	Avrami Model					
	$q_{e(exp)}$ (mg g^{-1})	K_{av}	n_{av}	$q_{e(cal)}$ (mg g^{-1})	RMSE	R^2
AB 1	7.85	0.005	6.73	7.49	0.34	0.974
Cr(VI)	21.01	0.011	4.55	19.79	5.51	0.974

Adsorbate	Intra-Particle Diffusion Model				
	C_i (mg L^{-1})	I (mg g^{-1})	K_p (mg g^{-1} min$^{-0.5}$)	RMSE	R^2
AB 1	10	2.15	0.45	1.33	0.964
Cr(VI)	10	0	2.13	8.17	0.944

Source: Reproduced with permission from Santhosh et al.,[47] Copyright 2017, Elsevier Science Ltd.

where C_t is the solution concentration at the fixed bed outlet at time t; C_i is the initial concentration of the adsorbate (mg/L); Q_t is the amount of adsorbate in the adsorbent at time t (mg/g); m is the mass of the adsorbent in 1 L of the adsorbate (g/L); K_0 is the constant rate of Bangham's model; and t is the time of contact (min).

To test this isotherm, plot $\log\left(\log\left(\dfrac{C_t}{C_i}\right) - Q_t m\right)$ as a function of $\log(t)$ and record the coefficient of determination R^2.

This model has been adopted by only a few researchers in the kinetic studies with respect to graphene-based adsorbents toward adsorptive removal of diverse contaminants from wastewater,[50–53] though the results are, in most instances, inferior to PFOM and PSOM kinetics.[52] In a particular study, it has been shown that the kinetics of HPA-GO (hyperbranched polyamine modified graphene oxide) for adsorptive removal of b(II) partly followed Bangham kinetic model with a chi square value of 0.9460 in comparison to the PSOM chi square value of 0.9984.[54] This proved that some of the adsorbates were adsorbed by the adsorbent via the pores within the adsorbent. In another study, it has been shown that the removal of MB CSC-5GO composite though followed PSOM kinetics at large but partly also obeyed Elovich ($R^2 = 0.96912$) followed by Bangham ($R^2 = 0.95762$) kinetic models.

As per available literature, this model has been followed by only a handful of graphene-based adsorbents.

4.1.2.5 Boyd Kinetic Model

Boyd's film model examines whether the main locations of mass transfer resistance are in the thin film (also known as the boundary layer) surrounding adsorbents or during diffusion inside pores.

Diffusion via pores is taken into account using the Bangham model. It is a model that distinguishes between extraparticle and intraparticle diffusion: If the plot of B_t as a function of time is a straight line through the origin, the sorption is controlled by intraparticle diffusion; otherwise, the sorption is governed by diffusion in the film (it is limited by extraparticular transport). This model is represented by the equation (4.7)[55]:

$$B_t = -0.4977 - \ln\left(1 - \frac{Q_t}{Q_e}\right) \tag{4.7}$$

where B_t represents the Boyd constant, Q_t represents the amount of adsorbate in the adsorbent at time t (mg/g), and Q_e represents the amount of adsorbate in the adsorbent at equilibrium (mg/g).

To test this isotherm, it is useful to plot B_t as a function of contact time t and note the coefficient of determination R^2.

The Boyds' kinetic model utilization in the study of graphene-based materials has been limited, seeing it has been adopted by only a few researchers,[56–58] especially where the adsorbent materials perform as an adsorbent and a filter material. A report on the adsorptive removal of As(III) and MB on rGO-MnO$_2$-BC has been shown to follow the Weber Morris model as well as the Boyd model, as shown in Table 4.8 below, where they concluded that the physical sorption of MB and As(III) onto the rGO-MnO$_2$/BC surface, was regulated by an intraparticle diffusion step, and may be accounted for by the kinetic investigations.[59]

TABLE 4.8

Weber Morris and Boyd Parameters for Adsorption of As(III) and MB onto rGO@MnO$_2$-BC

| | Weber Morris Model | | | | | | |
| | First Straight (Blue) Line | | | Second Straight (Red) Line | | | Boyd Model |
Pollutant	K_{d1}	C_1	R^2	K_{d2}	C_2	R^2	R^2
MB	0.29	7.45	0.983	0.016	9.80	0.880	0.960
As(III)	0.46	0.60	0.988	0.001	0.98	0.767	0.951

Source: Reproduced with permission from Tara et al.[59] Copyright 2020, Elsevier Science Ltd.

It will be better if researchers working within the niche of water purification and using the adsorption techniques to consider at least 80–100% of the adsorption kinetic theories in their investigations: this will help to holistically ascertain the actual mechanisms involved.

4.1.2.6 Elovich Kinetic Model

In recent years, the Elovich equation has been widely applied to characterize not only the kinetics of gas adsorption on solids but also the adsorption of contaminants from aqueous solutions.[60] This model is represented by the equation (4.8)[61]:

$$Q_t = \beta \ln(\alpha\beta) - \ln(t) \tag{4.8}$$

where Q_t is the amount of adsorbate in the adsorbent at time t (mg/g); β denotes the number of adsorption sites accessible; α represents initial adsorption rate (mg g^{-1} min); and t is the period of contact (min). To test this isotherm, plot Q_t as a function of $\ln(t)$ and note the coefficient of determination R^2.

Liu et al. have shown in their studies (as presented in Table 4.9) for the adsorptive removal of methylene blue dye using a composite "CSC-5GO" that although the R^2 value proved that the adsorption process followed PSOM kinetics, having an R^2 value of 0.99898 but closely followed by PFOM and Elvoich kinetic models, proving

---

TABLE 4.9

Kinetic Parameters of Various Models Fitted to Experimental Data of CSC-5GO

Kinetic Model	Parameter	MB C_0 (mg L^{-1})				
		375	500	625	750	875
Pseudo-first-order	k_1 (mg min g^{-1})	0.16377	0.13476	0.11513	0.10776	0.10804
	q_e (mg g^{-1})	98.62	48.63	55.00	64.84	77.80
	R^2	0.97881	0.93482	0.9752	0.95798	0.87662
Pseudo-second-order	k_2 (mg min g^{-1})	0.00350	0.00520	0.00569	0.00390	0.00283
	q_e (mg g^{-1})	156.25	188.68	227.27	263.16	322.58
	R^2	0.99898	0.99923	0.99941	0.99857	0.99804
Elovich	α (mmol g^{-1} min^{-1})	1819.30	381,045	321,699	2,714,704	2,003,102
	β (g mmol^{-1})	0.05641	0.07527	0.05973	0.06228	0.04902
	R^2	0.96912	0.90599	0.90832	0.99591	0.97759
Intra-particle diffusion	k_{dif} (mg g^{-1} min$^{-1/2}$)	9.49	6.84	8.67	8.73	10.89
	C	95.78	143.86	174.36	200.34	244.32
	R^2	0.87473	0.75215	0.76316	0.92854	0.87701
Bangham	k_b (mg g^{-1})	91.88	137.81	167.02	195.63	237.27
	m	7.12	12.26	11.86	14.32	13.69
	R^2	0.95762	0.88482	0.88508	0.99183	0.96802

Source: Reproduced with permission from Liu et al.,[62] Copyright 2018, Elsevier Science Ltd.

that the adsorption process is not just based on ion transfer as well as its sharing but also electrostatic interactions as well as chelation.[62]

In order to more thoroughly determine the real mechanisms at play, it will be preferable for researchers working in the field of water purification and employing adsorption techniques to take at least 80–100% of the adsorption kinetic theories into account in their studies.

4.1.2.7 Intraparticle Diffusion Kinetic Model

The adsorption process by porous substances is divided into four stages:

i. Solute transfer from solution to the boundary layer surrounding the particle
ii. Solute transfer from the border layer to the adsorbent surface
iii. Solute transfer to the adsorbent sites: Diffusion in micro- and macropores
iv. Interactions amongst solute molecules plus active areas on the surface: Complexation, adsorption, as well as precipitation

The work by Weber and Morris[63] on the adsorption of simple aromatic compounds on activated carbon shows that the concentration of adsorbate (C_t) is a linear function of the square root of the contact time $t^{\frac{1}{2}}$, whose slope is assayed to the speed constant according to the following equation (4.9)[17,63]:

$$Q_t = K_{ID}\sqrt{t} + I \tag{4.9}$$

where in Q_t is the amount of adsorbed species in the adsorbent at time t (mg/g), K_{ID} denotes the constant rate of intraparticle diffusion, t denotes the period of contact (min), and I denotes the intercept of the kinetic model for intraparticle diffusion. It is appropriate to graph Q_t as a function of the square root of the contact time and to record the coefficient of predictability R^2 in order to test this isotherm.

4.1.3 ADSORPTION ISOTHERMS

An adsorption isotherm is a crucial curve that describes the phenomena that govern the retention (or release) and mobility of a chemical from aqueous porous media or aquatic habitats to a solid-phase at constant temperature and pH.[17,64–68] When an adsorbate-containing phase has been in contact with the adsorbent for a sufficient amount of time, its adsorbate concentration in the bulk solution is in dynamic balance with the interface concentration,[69] adsorption equilibrium (the ratio between the adsorbed amount and the remaining in the solution) is established. The mathematical correlation, which plays a key role in the modeling analysis, operational design, and applicable practice of adsorption systems, is typically illustrated graphically by expressing the solid-phase versus its residual concentration.[70] Their physicochemical parameters, along with the underlying thermodynamic hypotheses, offer clarity into the adsorption process, surface qualities, as well as adsorbent affinity.[17,70] The first strategy to be mentioned is kinetic analysis. Adsorption equilibrium is described as a state of dynamic equilibrium in which the rates of adsorption as well as desorption are identical.[17] Whereas thermodynamics, as the second approach's foundation, can give a framework for generating various forms of adsorption isotherm models,[71] potential theory, as the third approach, often conveys the key notion in the development of characteristic curves. An intriguing tendency in isotherm modeling is the derivation in more than one approach, pointing to differences in the physical understanding of the model parameters.[72]

The Langmuir model (LM),[73] linear model, Freundlich model (FM),[74] Sips model (SM),[75] Temkin model (TM),[76] and Brunauer, Emmett, and Teller (BET) model have all been used in adsorption systems.[77] Among these models are the linear, Freundlich, Sips, Temkin, and others, which are empirical and lack effective theoretical foundation. These models cannot provide the adsorption mechanisms. As a result, the origins and physical implications of these models should be examined. As a result, the origins and physical implications of these models should be examined. Furthermore, the isotherms have been categorized based on the number of model parameters[69,78,79] or their shapes (S-, L-, H-, as well as C-shaped).[80] The classification based on the number of parameters, on the other hand, lacks a theoretical foundation. The Langmuir as well as the BET isotherms, for example, are two parameter models that reflect chemical and physical adsorption, respectively. The amount of model parameters cannot give mechanism information. The categorization based on model shapes does have a drawback as well because the majority of adsorption equilibrium data of liquid-solid systems is L-shaped. As a result, isotherms should be grouped more logically based on their physical definitions.

The physical meaning and classification of these isotherms have not been adequately researched, to the best of our knowledge. Furthermore, some isotherms are sometimes utilized in wrong or inappropriate forms. The adsorption potential of the Dubinin-Radushkevich (D-R) model, for example, has been estimated incorrectly as $\varepsilon = RT \ln\left(1 + \dfrac{1}{C_e}\right)$, $C_e \left(mgL^{-1}\right)$ is the equilibrium concentration, $R\left(8.314\dfrac{J}{molK}\right)$ is the universal gas constant, and $T\left(K\right)$ is the temperature.[69]

The most commonly used form of the BET model, as reported by Foo and Hameed,[78] Staudt et al.,[68] Petkovska,[81] and others, has been revealed to have inadequate estimations of model parameters in the liquid-solid system.[82]

The linear regression method was employed to estimate the model parameters in the majority of literature.[78] The linear regression approach is easy to use. The linearization of adsorption models, on the other hand, can modify the independent and dependent variables and produce propagating errors. In this sense, the model parameter estimate is erroneous and biased.[69] Nonlinear regression can provide accurate model parameter estimations, although it is more involved than the linear regression method. As a result, practical techniques for solving nonlinear isotherms should be devised. Figure 4.5 illustrates the various adsorption isotherms as per literature.

4.1.3.1 Two-Parameter Isotherms

4.1.3.1.1 *Langmuir Isotherm Model*

The Langmuir adsorption isotherm, which was initially designed to explain gas-solid-phase adsorption onto activated carbon, has long been used to quantify and compare the performance of various bio-sorbents.[73] This empirical model's formulation

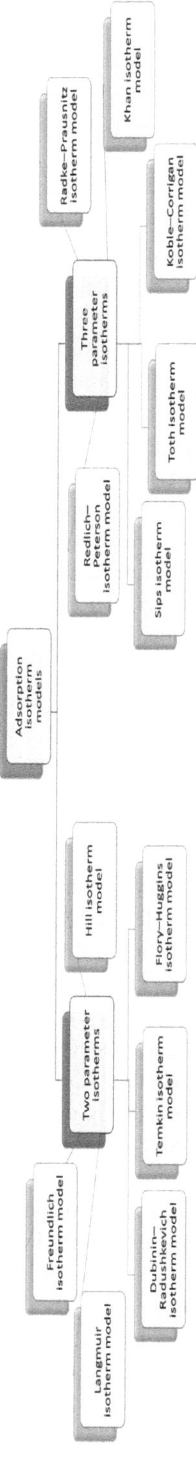

FIGURE 4.5   Adsorption isotherm models scheme.

assumes monolayer adsorption (the adsorbed layer is one molecule thick), with adsorption occurring only at a finite (fixed) number of definite localized sites that are identical and equivalent, with no lateral interaction and steric hindrance between the adsorbed molecules, even on adjacent sites. Langmuir isotherm refers to homogenous adsorption in which each molecule has constant enthalpies and sorption activation energy (all sites have identical affinity for the adsorbate),[73] with no adsorbate transmigration along the surface plane.[73]

It is represented graphically by a plateau, an equilibrium saturation points where once a molecule occupies a site, no further adsorption can occur.[14,83] Furthermore, Langmuir's theory has linked the quick decrease of intermolecular attractive forces to the increase in the distance.

As a result, a dimensionless constant known as the separation factor (RL) can be represented as follows[84]:

$$R_L = 1 \Big/ \left(1 + K_L C_0\right) \tag{4.10}$$

where K_L (L mg^{-1}) denotes the Langmuir constant and C_0 denotes the initial adsorbate concentration (mg L^{-1}).

A lower R_L value indicates that adsorption is more favorable in this circumstance. In more detail, the R_L value indicates whether the adsorption is unfavorable ($R_L > 1$), linear ($R_L = 1$), favorable ($0 < R_L < 1$), or irreversible ($R_L = 0$).

This models' nonlinear form is written thus.

$$q_e = \left(Q_0 b C_e\right) \Big/ \left(1 + b C_e\right) \tag{4.11}$$

Its linear form is.

$$\frac{C_e}{q_e} = \frac{1}{bQ_0} + \frac{C_e}{Q_0} \tag{4.12}$$

$$1/q_e = \frac{1}{Q_0} + \frac{1}{bQ_0 C_e} \tag{4.13}$$

$$q_e = Q_0 - \frac{q_e}{bC_e} \tag{4.14}$$

$$\frac{q_e}{C_e} = bQ_0 - bq_e \tag{4.15}$$

Their corresponding plots are.

$$\frac{C_e}{q_e} \text{ vs } C_e \tag{4.16}$$

$$\frac{1}{q_e} \text{ vs } \frac{1}{C_e} \tag{4.17}$$

$$q_e \text{ vs } \frac{q_e}{bC_e} \tag{4.18}$$

and

$$\frac{C_e}{C_e} \text{ vs } q_e \tag{4.19}$$

4.1.3.1.2 Freundlich Isotherm Model

Freundlich isotherm[85] is the first known connection indicating non-ideal and reversible adsorption that is not limited to monolayer formation. This empirical model is applicable to multilayer adsorption with non-uniform distribution of adsorption heat and affinities throughout the heterogeneous surface.[86] Historically, it was created for the adsorption of animal charcoal, indicating that the adsorbate-to-adsorbent ratio was not constant at varying solution concentrations.[69] According to this viewpoint, the amount adsorbed is the sum of adsorption on all sites (each with its own bond energy), with the stronger binding sites occupied first, until the adsorption energy is exponentially diminished upon completion of the adsorption process.[87]

Freundlich isotherms are currently commonly used in heterogeneous systems, particularly for organic chemicals or highly interacting species on activated carbon and molecular sieves. The slope, which varies from 0 to 1, is a measure of adsorption intensity or surface heterogeneity, becoming more heterogeneous as it approaches zero. A value less than one indicates chemisorption, whereas $1/n$ greater than one indicates cooperative adsorption.[88] Table 4.1 lists its linearized and non-linearized equations. Recently, the Freundlich isotherm has been questioned for lacking a fundamental thermodynamic basis and failing to approach Henry's rule at vanishing concentrations.[89]

The nonlinear equation for this isotherm is.

$$q_e = K_F C_e^{\frac{1}{n}} \tag{4.20}$$

Its linear expression is.

$$\log(q_e) = \log(K_F) + \frac{1}{n}\left(\log C_e\right) \tag{4.21}$$

While it is plotted as $\log q_e$ vs $\log C_e$. K_F is the Freundlich constant and $1/n$ is the heterogeneity factor.

4.1.3.1.3 Dubinin–Radushkevich Isotherm Model

The Dubinin-Radushkevich isotherm[89] was developed as an empirical model for the adsorption of subcritical vapors onto micropore materials via a pore filling mechanism. It is commonly used to express the adsorption mechanism[90] onto a heterogeneous surface with a Gaussian energy distribution.[91] The model has frequently successfully fitted high solute activity and intermediate concentration range data, but it has unsatisfactory asymptotic features and does not predict Henry's law at low pressure.[92] The approach was commonly used to discriminate between physical and chemical adsorption of metal ions,[69] and its mean free energy, E per molecule of adsorbate (for removing a molecule from its location in the sorption space to infinity), can be calculated using the ratio[93]:

$$E = \left(\frac{1}{\sqrt{2B_{DR}}} \right) \tag{4.22}$$

where B_{DR} is the isotherm constant.

Meanwhile, the parameter ε can be interpreted as follows:

$$\varepsilon = RT \ln\left(1 + \frac{1}{C_e}\right) \tag{4.23}$$

wherein $R, T, \& C_e$ denote the gas constant (8.314 J/mol K), absolute temperature (K), & adsorbate equilibrium concentration (mg L^{-1}).

The Dubinin-Radushkevich isotherm model is temperature-dependent, which means that when adsorption data at different temperatures are plotted as a function of the logarithm of quantity adsorbed vs square of potential energy, all relevant data will lie on the same curve, known as the response curve.

The nonlinear equation for this isotherm is.

$$q_e = \left(q_s\right)\exp\left(-k_{ad}\varepsilon^2\right) \tag{4.24}$$

Its linear expression is.

$$\ln\left(q_e\right) = \ln\left(q_s\right) - k_{ad}\varepsilon^2 \tag{4.25}$$

While it is plotted as $\ln q_e$ vs ε^2.

4.1.3.1.4 Temkin Isotherm Model

The Temkin isotherm was an early model used to describe the adsorption of hydrogen onto platinum electrodes in acidic liquids. The isotherm[77] includes a component that explicitly accounts for adsorbent-adsorbate interactions. The model assumes that the heat of adsorption (function of temperature) of all molecules in the layer decreases linearly rather than logarithmically with coverage by neglecting extremely low and high concentrations.[77] The derivation, as specified by the equation, is characterized by a homogeneous distribution of binding energies (up to some maximum binding energy). The Temkin equation is excellent for forecasting gas phase equilibrium (where the organization in a tightly packed structure with similar orientation is not required), but complex adsorption systems, including liquid-phase adsorption isotherms, are usually not suitable for representation.[94]

This isotherm is represented by a nonlinear equation thus:

$$q_e = \frac{RT}{b_T} \ln A_T C_e \tag{4.26}$$

While its linear form is written as

$$q_e = \frac{RT}{b_T} \ln A_T + \left(RT \big/ b_T \right) \ln C_e \tag{4.27}$$

This model is plotted using q_e vs $\ln C_e$.

4.1.3.1.5   Flory-Huggin's Isotherm Model

The Flory-Huggins isotherm model,[69] which is used to describe the feasibility and spontaneous nature of an adsorption process by occasionally determining the degree of surface coverage features of adsorbate onto adsorbent. In this context, is the degree of surface covering, while K_{FH} and n_{FH} are the equilibrium constant and model exponent, respectively. Its equilibrium constant, K_{FH}, is related to the equation,[95] which is used to calculate spontaneous free Gibbs energy.

$$\Delta G^o = -RT \ln\left(K_{FH} \right) \tag{4.28}$$

The nonlinear FGM is written as.

$$\frac{\theta}{C_0} = K_{FH} \left(1 - \theta \right)^{n_{FH}} \tag{4.29}$$

While its linear form is represented in the form.

$$\log\left(\theta / C_0 \right) = \log\left(K_{FH} \right) + n_{FH} \log\left(1 - \theta \right) \tag{4.30}$$

It is plotted as $\log\left(\theta \big/ C_0 \right)$ vs $\log\left(1 - \theta \right)$ $\tag{4.31}$

where ΔG^o is the standard Gibbs free energy change, R is the gas constant, T is the temperature, K_{FH} is the Flory–Huggins equilibrium constant. θ is the surface coverage (the fraction of the adsorption sites occupied by the adsorbate), C_0 is the initial concentration of the adsorbate, and n_{FH} is the Flory–Huggins exponent, which indicates the degree of interaction between the adsorbate molecules on the surface.

4.1.3.1.6   Hill Isotherm Model

The Hill equation,[96] which evolved from the NICA[97] model, was proposed to describe the binding of various species onto homogeneous substrates. The model posits that adsorption is a cooperative phenomenon, with the ability of a ligand to bind at one site on a macromolecule influencing other binding sites on the very same macromolecule.[98]

His nonlinear equation is written in the form

$$q_e = \frac{q_{sH} C_e^{n_H}}{K_D + C_e^{n_H}}$$
(4.32)

Its linear form is written as

$$\log\left(\frac{q_e}{q_{sH} - q_e}\right) = n_H \log(C_e) - \log(K_D)$$
(4.33)

And it is plotted as $\log\left(\dfrac{q_e}{q_{sH} - q_e}\right)$ vs $\log(C_e)$

q_e is the amount of adsorbate adsorbed per unit mass of adsorbent at equilibrium, q_{sH} is the maximum adsorption capacity, C_e is the equilibrium concentration of the adsorbate, K_D is the dissociation constant (concentration at which half of the adsorption sites are occupied), and n_H is the Hill coefficient, indicating the degree of cooperativity.

4.1.3.2 Three Parameter Isotherms

4.1.3.2.1 Redlich–Peterson Isotherm Model

The Redlich-Peterson isotherm[99] is a hybrid isotherm that incorporates both the Langmuir and Freundlich isotherms into an empirical equation.[100] Because of its versatility, the model has a linear dependence on concentration in the numerator and an exponential function in the denominator[101] to simulate adsorption equilibria over a wide concentration range and may be utilized in either homogeneous or heterogeneous systems.[102] In most cases, a minimization approach is used to solve the equations by maximizing the correlation coefficient between the experimental data points and theoretical model predictions using Microsoft Excel's solver add-in function.[103] In the limit, it approaches the Freundlich isotherm model at high concentration (as the exponent β tends to zero) and is consistent with the ideal Langmuir condition at low concentration (as the β values are all close to one).[104]

The nonlinear plot for this model is

$$q_e = \frac{K_R C_e}{1 + a_R C_e^{\beta}}$$
(4.34)

Its linear form is written as

$$\ln\left(K_R \frac{C_e}{q_e} - 1\right) = g \ln(C_e) + \ln(a_R)$$
(4.35)

While it is plotted as $\ln\left(K_R \dfrac{C_e}{q_e} - 1\right)$ vs $\ln(C_e)$

q_e is the amount of adsorbate adsorbed per unit mass of adsorbent at equilibrium, C_e is the equilibrium concentration of the adsorbate, K_R is the Redlich–Peterson isotherm constant, a_R is the Redlich–Peterson isotherm parameter, and β is an exponent between 0 and 1.

4.1.3.2.2   Sips Isotherm Model

The Sips isotherm[105] is a hybrid form of the Langmuir and Freundlich expressions developed for predicting heterogeneous adsorption systems and avoiding the limitation of growing adsorbate concentration associated with the Freundlich isotherm model. It reduces to the Freundlich isotherm at low adsorbate concentrations but predicts a monolayer adsorption capacity characteristic of the Langmuir isotherm at high concentrations.[90] In general, the operating circumstances, such as pH, temperature, as well as concentration, have the greatest influence on the model parameters.[106]

Sips isotherm model in its nonlinear form is expressed as

$$q_e = \left(Q_s K_s C_e^{\beta_s}\right) \Big/ \left(1 + K_s C_e^{\beta_s}\right) \tag{4.36}$$

Its linear form is represented by

$$\beta_s \ln(C) = -\ln\left(K_s \Big/ q_e\right) + \ln(a_S) \tag{4.37}$$

While it is plotted as $\ln\left(K_s \big/ q_e\right)$ vs $\ln(C_e)$

K_s stands for the equilibrium constant. A Langmuir equation will be created if the value of β_S is equal to 1. Or maybe this isotherm reduces to the Freundlich isotherm as either C_e or K_s gets closer to 0.

4.1.3.2.3   Toth Isotherm Model

Toth isotherm model[107] is another empirical equation designed to improve Langmuir isotherm fits (experimental data), and it is beneficial in characterizing heterogeneous adsorption systems that fulfil both the low and high-end concentration boundaries.[95] Its correlation is based on asymmetrical quasi-Gaussian energy distribution, with most of its sites having adsorption energy that is less than the peak (maximum) and otherwise mean value.[89]

This model, in its nonlinear form, is expressed as

$$q_e = \left(K_T C_e\right) \Big/ \left(a_T + C_e\right)^{1/t} \tag{4.38}$$

Its linear form is represented by

$$\ln\left(q_e\big/K_T\right) = \ln\left(C_e\right) - \frac{1}{t}\ln\left(a_T + C_e\right)$$

(4.39)

While it is plotted as $\ln\left(q_e\big/K_T\right)$ vs $\ln\left(C_e\right)$.

q_e is the amount of adsorbate adsorbed per unit mass of adsorbent at equilibrium, C_e is the equilibrium concentration of the adsorbate, K_T, a_T, and t are model constants.

4.1.3.2.4   Koble–Corrigan Isotherm Model

The Koble-Corrigan isotherm,[108] like the Sips isotherm model, is a three-parameter equation that incorporates both the Langmuir and Freundlich isotherm models for describing equilibrium adsorption data. Using trial and error optimization, the isotherm constants A, B, and n are calculated from the linear plot.

This model, in its nonlinear form, is expressed as

$$q_e = \frac{AC_e^m}{1 + BC_e^m}$$

(4.40)

Its linear form is *represented* by

$$\frac{1}{q_e} = \frac{1}{AC_e^m} + \frac{B}{A}$$

(4.41)

q_e is the amount of adsorbate adsorbed per unit mass of adsorbent at equilibrium, C_e is the equilibrium concentration of the adsorbate, and A, B, and m are model constants.

4.1.3.2.5   Khan Isotherm Model

Khan isotherm is a generalized model proposed for pure solutions,[109] where b_k and a_k are the model constant and model exponent, respectively. Its maximal uptake levels can be well determined at reasonably high correlation coefficients and low ERRSQ or chi-square values.[110]

4.1.3.2.6   Radke–Prausnitz Isotherm Model

The correlation of the Radke-Prausnitz isotherm is typically well predicted by strong RMSE and chi-square values. Its model exponent is denoted by β_R, where a_R and r_R are the constants of the model.[95]

$$q_e = a_{RP}r_R C_e^{\beta_R}\big/a_{RP} + r_R C_e^{\beta_R - 1}$$

(4.42)

This model, in its nonlinear form, is expressed as

$$q_e = \frac{q_s b_K C_e}{\left(1 + b_K C_e\right)^{\beta_R}} \tag{4.43}$$

a_{RP} and r_R are Radke–Prausnitz isotherm constants, and β_R is an exponent.

4.1.3.3 Multilayer Physisorption Isotherms

Brunauer-Emmett-Teller (BET) Isotherm is an equation to describe what is most commonly used in gas-solid equilibrium systems.[111] This was created to create multilayer adsorption systems with relative pressure ranges ranging from 0.05 to 0.30, corresponding to monolayer cover ranging from 0.50 to 1.50. Its attrition model for the liquid-solid interface is as follows:

$$q_e = \left. \left(q_s C_{BET} C_e\right) \middle/ \left(\left(C_s - C_e\right)\left(1 + \left(\left(C_{BET} - 1\right)\left(C_e/C_s\right)\right)\right)\right) \right. \tag{4.44}$$

where C_{BET}, C_s, q_s, as well as q_e denote the BET adsorption isotherm (L mg^{-1}), adsorbate monolayer saturation concentration (L mg^{-1}), theoretical isotherm saturation capacity (mg/g), and equilibrium adsorption capacity (mg g^{-1}), respectively. Because C_{BET} and $C_{BET}\left(C_e/C_s\right)$ are substantially more than one, the expression is streamlined as follows:

$$q_e = \left. q_s \middle/ \left(1 - \left(C_e/C_s\right)\right) \right. \tag{4.45}$$

Similarly, another multilayer adsorption derivation from potential theory, Frenkel-Halsey-Hill (FHH) isotherm,[112] can be stated as:

$$\ln\left(C_e/C_s\right) = -\left(\alpha/RT\right)\left(q_s/q_e d\right)^r \tag{4.46}$$

d, and r denote the interlayer spacing sign (m), the isotherm constant (J mr/mole), and the inverse power of distance from the surface. Similarly, the MacMillan-Teller (MET) isotherm[113] is an adsorption model derived from the incorporation of surface tension effects in the BET isotherm.

$$q_e = q_s \left(\left. k \middle/ \ln\left(C_s/C_e\right) \right. \right)^{1/3} \tag{4.47}$$

where k denotes an isothermal constant as C_s/C_e approaches unity, the logarithmic term is estimated as follows:

$$q_e = q_s \left(\frac{kC_e}{(C_s - C_e)} \right)^{1/3} \tag{4.48}$$

Considering relative pressures greater than 0.8, the empirical isotherm is an acceptable fit to the Frenkel-Halsey-Hill (FHH) or MacMillan-Teller (MET) isotherms, but it approximates the BET isotherm.

The nonlinear expression of this model assumes the form

$$q_e = \frac{q_s C_{BET} C_e}{\left((C_s - C_e) + (C_{BET} - 1)\left(\frac{C_e}{C_s} \right) \right)} \tag{4.49}$$

Its linear expression assumes the form

$$\frac{C_e}{q_e [C_s - C_e]} = \frac{1}{q_e C_{BET}} + \frac{(C_{BET} - 1)}{q_s C_{BET}} \frac{C_e}{C_s} \tag{4.50}$$

While it is plotted using $\dfrac{C_e}{q_e [C_s - C_e]}$ vs $\dfrac{C_e}{C_s}$.

4.1.4 ADSORPTION THERMODYNAMICS

The computation of phase equilibrium between a gaseous mixture and a solid adsorbent is the most important application of thermodynamics to adsorption. The adsorption isotherm, which gives the amount of adsorbate adsorbed in the nanopores as a function of external pressure, serves as the foundation for thermodynamic calculations.

The thermodynamic quantities Gibbs free energy (G^0), enthalpy (H^0), as well as entropy change (S^0) are frequently taken into account in adsorption studies to evaluate the process and are calculated by Equations (4.51, 4.52, and 4.53).[114]

$$\Delta G^0 = -RT \ln K_0 \tag{4.51}$$

$$\Delta G^0 = \Delta H^0 - T \Delta S^0 \tag{4.52}$$

$$\ln K_0 = \left(\frac{\Delta S^0}{R} \right) - \left(\frac{\Delta H^0}{RT} \right) \tag{4.53}$$

from which T is the absolute temperature, R denotes the universal gas constant, but also K_0 is the kinetic equilibrium constant.

In a broad sense, graphene nanocomposites positive S^0 values, indicating augmented arbitrariness at the solid-liquid interface, but also negative G^0 values, indicating that the process of adsorption is spontaneous as well as endothermic, respectively, for pollutants like pharmaceuticals, dyes, as well as heavy metals.[114] Additionally, the values of H^0 below 50 kJ mol^{-1} typically indicates physisorption, whereas chemisorption is denoted by higher values.[17]

4.2 CONCLUSION

This chapter has presented, in general, the adsorption process, kinetic, isotherm, as well as thermodynamics of adsorption with special emphasis on graphene-based adsorbents with reference to the available literature. This information provided herewith is expected to be of immense importance to researchers and industrialists working within the niche of graphene and its composites or hybrids towards application as adsorbents for wastewater remediation. Academicians will also utilize the information available in this chapter to educate their students on the adsorption process, isotherm, kinetics, and thermodynamics with the aim of preparing them for real-life application of this knowledge.

REFERENCES

(1) Alnajrani, M. N.; Alsager, O. A. Removal of antibiotics from water by polymer of intrinsic microporosity: Isotherms, kinetics, thermodynamics, and adsorption mechanism. *Scientific Reports* 2020, *10* (1), 794.

(2) Heuchel, M.; Fritsch, D.; Budd, P. M.; McKeown, N. B.; Hofmann, D. Atomistic packing model and free volume distribution of a polymer with intrinsic microporosity (PIM-1). *Journal of Membrane Science* 2008, *318* (1–2), 84–99.

(3) Satilmis, B.; Budd, P. M.; Uyar, T. Systematic hydrolysis of PIM-1 and electrospinning of hydrolyzed PIM-1 ultrafine fibers for an efficient removal of dye from water. *Reactive and Functional Polymers* 2017, *121*, 67–75.

(4) Satilmis, B.; Alnajrani, M. N.; Budd, P. M. Hydroxyalkylaminoalkylamide PIMs: Selective adsorption by ethanolamine-and diethanolamine-modified PIM-1. *Macromolecules* 2015, *48* (16), 5663–5669.

(5) Zhang, C.; Li, P.; Cao, B. Electrospun polymer of intrinsic microporosity fibers and their use in the adsorption of contaminants from a nonaqueous system. *Journal of Applied Polymer Science* 2016, *133* (22), 43457.

(6) Satilmis, B.; Uyar, T. Removal of aniline from air and water by polymers of intrinsic microporosity (PIM-1) electrospun ultrafine fibers. *Journal of Colloid and Interface Science* 2018, *516*, 317–324.

(7) Aliyev, E. M.; Khan, M. M.; Nabiyev, A. M.; Alosmanov, R. M.; Bunyad-zadeh, I. A.; Shishatskiy, S.; Filiz, V. Covalently modified graphene oxide and polymer of intrinsic microporosity (PIM-1) in mixed matrix thin-film composite membranes. *Nanoscale Research Letters* 2018, *13* (1), 359.

(8) Lagergren, S. Zur theorie der sogenannten adsorption geloster stoffe. *Kungliga svenska vetenskapsakademiens. Handlingar* 1898, *24*, 1–39.

(9) Rodrigues, A. E.; Silva, C. M. What's wrong with Lagergreen pseudo first order model for adsorption kinetics? *Chemical Engineering Journal* 2016, *306*, 1138–1142.

(10) Çiftçi, T. D.; Henden, E. Nickel/nickel boride nanoparticles coated resin: A novel adsorbent for arsenic (III) and arsenic (V) removal. *Powder Technology* 2015, *269*, 470–480.

(11) Moussout, H.; Ahlafi, H.; Aazza, M.; Maghat, H. Critical of linear and nonlinear equations of pseudo-first order and pseudo-second order kinetic models. *Karbala International Journal of Modern Science* 2018, *4* (2), 244–254.

(12) Zhang, X.; Yu, H.; Yang, H.; Wan, Y.; Hu, H.; Zhai, Z.; Qin, J. Graphene oxide caged in cellulose microbeads for removal of malachite green dye from aqueous solution. *Journal of Colloid and Interface Science* 2015, *437*, 277–282.

(13) Li, L.; Fan, L.; Duan, H.; Wang, X.; Luo, C. Magnetically separable functionalized graphene oxide decorated with magnetic cyclodextrin as an excellent adsorbent for dye removal. *RSC Advances* 2014, *4* (70), 37114–37121.

(14) Bhattacharyya, A.; Ghorai, S.; Rana, D.; Roy, I.; Sarkar, G.; Saha, N. R.; Orasugh, J. T.; De, S.; Sadhukhan, S.; Chattopadhyay, D. Design of an efficient and selective adsorbent of cationic dye through activated carbon - graphene oxide nanocomposite: Study on mechanism and synergy. *Materials Chemistry and Physics* 2021, *260*, 124090.

(15) Rotte, N. K.; Yerramala, S.; Boniface, J.; Srikanth, V. V. S. S. Equilibrium and kinetics of Safranin O dye adsorption on MgO decked multi-layered graphene. *Chemical Engineering Journal* 2014, *258*, 412–419.

(16) Al-Shemy, M. T.; Al-Sayed, A.; Dacrory, S. Fabrication of sodium alginate/graphene oxide/nanocrystalline cellulose scaffold for methylene blue adsorption: Kinetics and thermodynamics study. *Separation and Purification Technology* 2022, *290*, 120825.

(17) Zaman, A.; Orasugh, J. T.; Banerjee, P.; Dutta, S.; Ali, M. S.; Das, D.; Bhattacharya, A.; Chattopadhyay, D. Facile one-pot in-situ synthesis of novel graphene oxide-cellulose nanocomposite for enhanced azo dye adsorption at optimized conditions. *Carbohydrate Polymers* 2020, *246*, 116661.

(18) Verma, M.; Kumar, A.; Lee, I.; Kumar, V.; Park, J.-H.; Kim, H. Simultaneous capturing of mixed contaminants from wastewater using novel one-pot chitosan functionalized with EDTA and graphene oxide adsorbent. *Environmental Pollution* 2022, *304*, 119130.

(19) Naderi, P.; Shirani, M.; Semnani, A.; Goli, A. Efficient removal of crystal violet from aqueous solutions with Centaurea stem as a novel biodegradable bioadsorbent using response surface methodology and simulated annealing: Kinetic, isotherm and thermodynamic studies. *Ecotoxicology and Environmental Safety* 2018, *163*, 372–381.

(20) Rout, S.; Kumar, A.; Ravi, P. M.; Tripathi, R. M. Pseudo second order kinetic model for the sorption of U (VI) onto soil: A comparison of linear and non-linear methods. *International Journal of Environmental Sciences* 2015, *6* (1), 145–154.

(21) Bai, S.; Shen, X.; Zhong, X.; Liu, Y.; Zhu, G.; Xu, X.; Chen, K. One-pot solvothermal preparation of magnetic reduced graphene oxide-ferrite hybrids for organic dye removal. *Carbon* 2012, *50* (6), 2337–2346.

(22) Bhattacharyya, A.; Banerjee, B.; Ghorai, S.; Rana, D.; Roy, I.; Sarkar, G.; Saha, N. R.; De, S.; Ghosh, T. K.; Sadhukhan, S.; Chattopadhyay, D. Development of an auto-phase separable and reusable graphene oxide-potato starch based cross-linked bio-composite adsorbent for removal of methylene blue dye. *International Journal of Biological Macromolecules* 2018, *116*, 1037–1048.

(23) Bhattacharyya, A.; Ghorai, S.; Rana, D.; Roy, I.; Sarkar, G.; Saha, N. R.; Orasugh, J. T.; De, S.; Sadhukhan, S.; Chattopadhyay, D. Design of an efficient and selective adsorbent of cationic dye through activated carbon – graphene oxide nanocomposite: Study on mechanism and synergy. *Materials Chemistry and Physics* 2021, 260.

(24) Temane, L. T.; Orasugh, J. T.; Ray, S. S. Adsorptive removal of pollutants using graphene-based materials for water purification. In *Two-Dimensional Materials for Environmental Applications*, Kumar, N., Gusain, R., Sinha Ray, S. Eds.; Springer International Publishing, 2023; pp. 179–244.

(25) Bhattacharya, S.; Banerjee, P.; Das, P.; Bhowal, A.; Majumder, S. K.; Ghosh, P. Removal of aqueous carbamazepine using graphene oxide nanoplatelets: Process modelling and optimization. *Sustainable Environment Research* 2020, *30* (1), 1–12.

(26) Rosli, F. A.; Ahmad, H.; Jumbri, K.; Abdullah, A. H.; Kamaruzaman, S.; Fathihah Abdullah, N. A. Efficient removal of pharmaceuticals from water using graphene nanoplatelets as adsorbent. *Royal Society Open Science* 2021, *8* (1), 201076.

(27) Zhu, S.; Liu, Y.-G.; Liu, S.-B.; Zeng, G.-M.; Jiang, L.-H.; Tan, X.-F.; Zhou, L.; Zeng, W.; Li, T.-T.; Yang, C.-P. Adsorption of emerging contaminant metformin using graphene oxide. *Chemosphere* 2017, *179*, 20–28.

(28) Al-Khateeb, L. A.; Almotiry, S.; Salam, M. A. Adsorption of pharmaceutical pollutants onto graphene nanoplatelets. *Chemical Engineering Journal* 2014, *248*, 191–199.

(29) Tang, Y.; Guo, H.; Xiao, L.; Yu, S.; Gao, N.; Wang, Y. Synthesis of reduced graphene oxide/magnetite composites and investigation of their adsorption performance of fluoroquinolone antibiotics. *Colloids and Surfaces A: Physicochemical and Engineering Aspects* 2013, *424*, 74–80.

(30) Gao, Y.; Li, Y.; Zhang, L.; Huang, H.; Hu, J.; Shah, S. M.; Su, X. Adsorption and removal of tetracycline antibiotics from aqueous solution by graphene oxide. *Journal of Colloid Interface Science* 2012, *368* (1), 540–546.

(31) Çalışkan Salihi, E.; Wang, J.; Kabacaoğlu, G.; Kırkulak, S.; Šiller, L. Graphene oxide as a new generation adsorbent for the removal of antibiotics from waters. *Separation Science and Technology* 2021, *56* (3), 453–461.

(32) Anuma, S.; Mishra, P.; Bhat, B. R. Polypyrrole functionalized Cobalt oxide Graphene (COPYGO) nanocomposite for the efficient removal of dyes and heavy metal pollutants from aqueous effluents. *Journal of Hazardraous Materials* 2021, *416*, 125929.

(33) Verma, M.; Lee, I.; Oh, J.; Kumar, V.; Kim, H. Synthesis of EDTA-functionalized graphene oxide-chitosan nanocomposite for simultaneous removal of inorganic and organic pollutants from complex wastewater. *Chemosphere* 2022, *287*, 132385.

(34) Yan, J.; Li, K. A magnetically recyclable polyampholyte hydrogel adsorbent functionalized with β-cyclodextrin and graphene oxide for cationic/anionic dyes and heavy metal ion wastewater remediation. *Separation and Purification Technology* 2021, *277*, 119469.

(35) Zhou, S.; Yin, J.; Ma, Q.; Baihetiyaer, B.; Sun, J.; Zhang, Y.; Jiang, Y.; Wang, J.; Yin, X. Montmorillonite-reduced graphene oxide composite aerogel (M−rGO): A green adsorbent for the dynamic removal of cadmium and methylene blue from wastewater. *Sepration and Purifiction Technology* 2022, *296*, 121416.

(36) Joshi, P.; Sharma, O. P.; Ganguly, S. K.; Srivastava, M.; Khatri, O. P. Fruit waste-derived cellulose and graphene-based aerogels: Plausible adsorption pathways for fast and efficient removal of organic dyes. *Journal of Colloid Interface Science* 2022, *608*, 2870–2883.

(37) Liu, T.; Wang, Z.; Wang, X.; Yang, G.; Liu, Y. Adsorption-photocatalysis performance of polyaniline/dicarboxyl acid cellulose@graphene oxide for dye removal. *International Journal of Biological Macromolecules* 2021, *182*, 492–501.

(38) Foroutan, R.; Mohammadi, R.; MousaKhanloo, F.; Sahebi, S.; Ramavandi, B.; Kumar, P. S.; Vardhan, K. H. Performance of montmorillonite/graphene oxide/CoFe$_2$O$_4$ as a magnetic and recyclable nanocomposite for cleaning methyl violet dye-laden wastewater. *Advanced Powder Technology* 2020, *31* (9), 3993–4004.

(39) Qi, Y.; Yang, M.; Xu, W.; He, S.; Men, Y. Natural polysaccharides-modified graphene oxide for adsorption of organic dyes from aqueous solutions. *Journal of Colloid Interface Science* 2017, *486*, 84–96.

(40) da Silva, P. M. M.; Camparotto, N. G.; Figueiredo Neves, T.; Mastelaro, V. R.; Nunes, B.; Siqueira Franco Picone, C.; Prediger, P. Instantaneous adsorption and synergic effect in simultaneous removal of complex dyes through nanocellulose/graphene oxide nanocomposites: Batch, fixed-bed experiments and mechanism. *Environmental Nanotechnology, Monitoring & Management* 2021, *16*, 100584.

(41) Rout, D. R.; Jena, H. M. Removal of malachite green dye from aqueous solution using reduced graphene oxide as an adsorbent. *Materials Today: Proceedings* 2021, *47*, 1173–1182.

(42) Muthusaravanan, S.; Balasubramani, K.; Suresh, R.; Ganesh, R. S.; Sivarajasekar, N.; Arul, H.; Rambabu, K.; Bharath, G.; Sathishkumar, V. E.; Murthy, A. P.; Banat, F. Adsorptive removal of noxious atrazine using graphene oxide nanosheets: Insights to process optimization, equilibrium, kinetics, and density functional theory calculations. *Environmental Research* 2021, *200*, 111428.

(43) George, R.; Sugunan, S. Kinetics of adsorption of lipase onto different mesoporous materials: Evaluation of Avrami model and leaching studies. *Journal of Molecular Catalysis B: Enzyme* 2014, *105*, 26–32.

(44) Ahmad, A. A.; Din, A. T. M.; Yahaya, N. K. E.; Khasri, A.; Ahmad, M. A. Adsorption of basic green 4 onto gasified Glyricidia sepium woodchip based activated carbon: Optimization, characterization, batch and column study. *Arabian Journal of Chemistry* 2020, *13* (8), 6887–6903.

(45) Al-Qadri, A. A. Q.; Drmosh, Q. A.; Onaizi, S. A. Enhancement of bisphenol a removal from wastewater via the covalent functionalization of graphene oxide with short amine molecules. *Case Studies in Chemical and Environmental Engineering* 2022, *6*, 100233.

(46) Lopes, E. C. N.; dos Anjos, F. S. C.; Vieira, E. F. S.; Cestari, A. R. An alternative Avrami equation to evaluate kinetic parameters of the interaction of Hg(II) with thin chitosan membranes. *Journal of Colloid and Interface Science* 2003, *263* (2), 542–547.

(47) Santhosh, C.; Daneshvar, E.; Kollu, P.; Peräniemi, S.; Grace, A. N.; Bhatnagar, A. Magnetic SiO_2@$CoFe_2O_4$ nanoparticles decorated on graphene oxide as efficient adsorbents for the removal of anionic pollutants from water. *Chemical Engineering Journal* 2017, *322*, 472–487.

(48) Balasubramani, K.; Sivarajasekar, N.; Muthusaravanan, S.; Ram, K.; Naushad, M.; Ahamad, T.; Sharma, G. Efficient removal of antidepressant Flupentixol using graphene oxide/cellulose nanogel composite: Particle swarm algorithm based artificial neural network modelling and optimization. *Journal of Molecular Liquids* 2020, *319*, 114371.

(49) Malana, M. A.; Qureshi, R. B.; Ashiq, M. N. Adsorption studies of arsenic on nano aluminium doped manganese copper ferrite polymer (MA, VA, AA) composite: Kinetics and mechanism. *Chemical Engineering Journal* 2011, *172* (2–3), 721–727.

(50) Sun, Y.; Liu, X.; Lv, X.; Wang, T.; Xue, B. Synthesis of novel lignosulfonate-modified graphene hydrogel for ultrahigh adsorption capacity of Cr(VI) from wastewater. *Journal of Cleaner Production* 2021, *295*, 126406.

(51) Chakrapani, C.; Babu, C.; Vani, K.; Rao, K. S. Adsorption kinetics for the removal of fluoride from aqueous solution by activated carbon adsorbents derived from the peels of selected citrus fruits. *E-Journal of Chemistry* 2010, *7* (S1), S419–S427.

(52) Bulin, C.; Ma, Z.; Guo, T.; Li, B.; Zhang, Y.; Zhang, B.; Xing, R.; Ge, X. Magnetic graphene oxide nanocomposite: One-pot preparation, adsorption performance and mechanism for aqueous Mn(II) and Zn(II). *Journal of Physics and Chemistry of Solids* 2021, *156*, 110130.

(53) Xu, X.; Zou, J.; Zhao, X.-R.; Jiang, X.-Y.; Jiao, F.-P.; Yu, J.-G.; Liu, Q.; Teng, J. Facile assembly of three-dimensional cylindrical egg white embedded graphene oxide composite with good reusability for aqueous adsorption of rare earth elements. *Colloids and Surfaces A: Physicochemical and Engineering Aspects* 2019, *570*, 127–140.

(54) Hu, L.; Yang, Z.; Cui, L.; Li, Y.; Ngo, H. H.; Wang, Y.; Wei, Q.; Ma, H.; Yan, L.; Du, B. Fabrication of hyperbranched polyamine functionalized graphene for high-efficiency removal of Pb(II) and methylene blue. *Chemical Engineering Journal* 2016, *287*, 545–556.

(55) Okewale, A.; Babayemi, K.; Olalekan, A. Adsorption isotherms and kinetics models of starchy adsorbents on uptake of water from ethanol–water systems. *International Journal of Applied Science and Technology* 2013, *3* (1), 35–42.

(56) Noorani Khomeyrani, S. F.; Ahmadi Azqhandi, M. H.; Ghalami-Choobar, B. Rapid and efficient ultrasonic assisted adsorption of PNP onto LDH-GO-CNTs: ANFIS, GRNN and RSM modeling, optimization, isotherm, kinetic, and thermodynamic study. *Journal of Molecular Liquids* 2021, *333*, 115917.

(57) Söğüt, E. G.; Karataş, Y.; Gülcan, M.; Kılıç, N. Ç. Enhancement of adsorption capacity of reduced graphene oxide by sulfonic acid functionalization: Malachite green and Zn (II) uptake. *Materials Chemistry and Physics* 2020, *256*, 123662.

(58) Ranjan Rout, D.; Mohan Jena, H. Synthesis of novel reduced graphene oxide decorated β-cyclodextrin epichlorohydrin composite and its application for Cr(VI) removal: Batch and fixed-bed studies. *Separation and Purification Technology* 2021, *278*, 119630.

(59) Tara, N.; Siddiqui, S. I.; Bach, Q.-V.; Chaudhry, S. A. Reduce graphene oxide-manganese oxide-black cumin based hybrid composite (rGO-MnO$_2$/BC): A novel material for water remediation. *Materials Today Communications* 2020, *25*, 101560.

(60) Ho, Y.-S. Review of second-order models for adsorption systems. *Journal of Hazardous Materials* 2006, *136* (3), 681–689.

(61) Chien, S.; Clayton, W. Application of Elovich equation to the kinetics of phosphate release and sorption in soils. *Soil Science Society of America Journal* 1980, *44* (2), 265–268.

(62) Liu, S.; Ge, H.; Wang, C.; Zou, Y.; Liu, J. Agricultural waste/graphene oxide 3D bio-adsorbent for highly efficient removal of methylene blue from water pollution. *Science of the Total Environment* 2018, *628–629*, 959–968.

(63) Weber Jr, W. J.; Morris, J. C. Kinetics of adsorption on carbon from solution. *Journal of the Sanitary Engineering Division* 1963, *89* (2), 31–59.

(64) Ai, L.; Jiang, J. Removal of methylene blue from aqueous solution with self-assembled cylindrical graphene–carbon nanotube hybrid. *Chemical Engineering Journal* 2012, *192*, 156–163.

(65) Ai, L.; Zhang, C.; Chen, Z. Removal of methylene blue from aqueous solution by a solvothermal-synthesized graphene/magnetite composite. *Journal of Hazardous Materials* 2011, *192* (3), 1515–1524.

(66) Orasugh, J. T.; Ray, S. S. Nanocellulose-Graphene Oxide-Based Nanocomposite for Adsorptive Water Treatment. In *Functional Polymer Nanocomposites for Wastewater Treatment*, Hato, M. J., Sinha Ray, S. Eds.; Springer International Publishing, 2022; pp. 1–53.

(67) Sivapragasam, N. *Adsorption Kinetics and Dynamics of Small Molecules on Graphene and Graphene Oxide*; North Dakota State University, 2018.

(68) Staudt, P.; Kechinski, C.; Tessaro, I.; Marczak, L.; Soares, R. d. P.; Cardozo, N. A new method for predicting sorption isotherms at different temperatures using the BET model. *Journal of Food Engineering* 2013, *114* (1), 139–145.

(69) Wang, J.; Guo, X. Adsorption isotherm models: Classification, physical meaning, application and solving method. *Chemosphere* 2020, *258*, 127279.

(70) Ncibi, M. C. Applicability of some statistical tools to predict optimum adsorption isotherm after linear and non-linear regression analysis. *Journal of Hazardous Materials* 2008, *153* (1–2), 207–212.

(71) Boer, J. H. *Dynamical Character of Adsorption*. 2nd Edition; Oxford University Press, 1968.

(72) Ruthven, D. M. *Principles of Adsorption and Adsorption Processes*; John Wiley & Sons, 1984.

(73) Langmuir, I. The constitution and fundamental properties of solids and liquids. Part I. Solids. *Journal of the American Chemical Society* 1916, *38* (11), 2221–2295.

(74) Freundlich, H. Über die adsorption in lösungen. *Zeitschrift für Physikalische Chemie* 1907, *57* (1), 385–470.

(75) Sips, R. On the structure of a catalyst surface. *The Journal of Chemical Physics* 1948, *16* (5), 490–495.

(76) Temkin, M. I. Adsorption equilibrium and the kinetics of processes on nonhomogeneous surfaces and in the interaction between adsorbed molecules. *Zhurnal Fiziche-Skoi Khimii* 1941, *15*, 296–332.

(77) Temkin, M. Kinetics of ammonia synthesis on promoted iron catalysts. *Acta Physiochimica URSS* 1940, *12*, 327–356.

(78) Foo, K. Y.; Hameed, B. H. Insights into the modeling of adsorption isotherm systems. *Chemical Engineering Journal* 2010, *156* (1), 2–10.

(79) Ayawei, N.; Ebelegi, A. N.; Wankasi, D. Modelling and interpretation of adsorption isotherms. *Journal of Chemistry* 2017, *2017*.

(80) Limousin, G.; Gaudet, J.-P.; Charlet, L.; Szenknect, S.; Barthes, V.; Krimissa, M. Sorption isotherms: A review on physical bases, modeling and measurement. *Applied Geochemistry* 2007, *22* (2), 249–275.

(81) Petkovska, M. Discrimination between adsorption isotherm models based on nonlinear frequency response results. *Adsorption* 2014, *20* (2), 385–395.

(82) Ebadi, A.; Soltan Mohammadzadeh, J. S.; Khudiev, A. What is the correct form of BET isotherm for modeling liquid phase adsorption? *Adsorption* 2009, *15* (1), 65–73.

(83) Allen, S.; Mckay, G.; Porter, J. F. Adsorption isotherm models for basic dye adsorption by peat in single and binary component systems. *Journal of Colloid Interface Science* 2004, *280* (2), 322–333.

(84) Weber, T. W.; Chakravorti, R. K. Pore and solid diffusion models for fixed-bed adsorbers. *AlChE Journal* 1974, *20* (2), 228–238.

(85) Freundlich, H. Over the adsorption in solution. *Journal of Physical Chemistry* 1906, *57* (385471), 1100–1107.

(86) Adamson, A. W.; Gast, A. P. *Physical Chemistry of Surfaces*; Interscience Publishers, 1967.

(87) Zeldowitsch, J. Adsorption site energy distribution. *Acta Physico Chimica URSS* 1934, *1* (1), 961–973.

(88) Haghseresht, F.; Lu, G. Adsorption characteristics of phenolic compounds onto coal-reject-derived adsorbents. *Energy & Fuels* 1998, *12* (6), 1100–1107.

(89) Ho, Y.; Porter, J.; McKay, G. Equilibrium isotherm studies for the sorption of divalent metal ions onto peat: Copper, nickel and lead single component systems. *Water, Air, Soil Pollution* 2002, *141* (1), 1–33.

(90) Günay, A.; Arslankaya, E.; Tosun, I. Lead removal from aqueous solution by natural and pretreated clinoptilolite: Adsorption equilibrium and kinetics. *Journal of Hazardous Materials* 2007, *146* (1–2), 362–371.

(91) Dąbrowski, A. Adsorption: From theory to practice. *Advances in Colloid Interface Science* 2001, *93* (1–3), 135–224.

(92) Altın, O.; Özbelge, H. Ö.; Doğu, T. Use of general purpose adsorption isotherms for heavy metal–clay mineral interactions. *Journal of Colloid Interface Science* 1998, *198* (1), 130–140.

(93) Hobson, J. P. Physical adsorption isotherms extending from ultrahigh vacuum to vapor pressure. *The Journal of Physical Chemistry* 1969, *73* (8), 2720–2727.

(94) Kim, Y.; Kim, C.; Choi, I.; Rengaraj, S.; Yi, J. Arsenic removal using mesoporous alumina prepared via a templating method. *Environmental Science & Technology* 2004, *38* (3), 924–931.

(95) Vijayaraghavan, K.; Padmesh, T.; Palanivelu, K.; Velan, M. Biosorption of nickel (II) ions onto Sargassum wightii: Application of two-parameter and three-parameter isotherm models. *Journal of Hazardarous Materials* 2006, *133* (1–3), 304–308.

(96) Hill, A. V. The possible effects of the aggregation of the molecules of haemoglobin on its dissociation curves. *Journal of Physiology* 1910, *40*, 4–7.

(97) Koopal, L.; Van Riemsdijk, W.; De Wit, J.; Benedetti, M. Analytical isotherm equations for multicomponent adsorption to heterogeneous surfaces. *Journal of Colloid Interface Science* 1994, *166* (1), 51–60.

(98) Ringot, D.; Lerzy, B.; Chaplain, K.; Bonhoure, J.-P.; Auclair, E.; Larondelle, Y. In vitro biosorption of ochratoxin A on the yeast industry by-products: Comparison of isotherm models. *Bioresource Technology* 2007, *98* (9), 1812–1821.

(99) Redlich, O.; Peterson, D. L. A useful adsorption isotherm. *Journal of Physical Chemistry* 1959, *63* (6), 1024–1024.

(100) Prasad, R. K.; Srivastava, S. Sorption of distillery spent wash onto fly ash: Kinetics and mass transfer studies. *Chemical Engineering Journal* 2009, *146* (1), 90–97.

(101) Ng, J.; Cheung, W.; McKay, G. Equilibrium studies of the sorption of Cu (II) ions onto chitosan. *Journal of Colloid Interface Science* 2002, *255* (1), 64–74.

(102) Gimbert, F.; Morin-Crini, N.; Renault, F.; Badot, P.-M.; Crini, G. Adsorption isotherm models for dye removal by cationized starch-based material in a single component system: Error analysis. *Journal of Hazardous Materials* 2008, *157* (1), 34–46.

(103) Wong, Y.; Szeto, Y.; Cheung, W.; McKay, G. Adsorption of acid dyes on chitosan—Equilibrium isotherm analyses. *Process Biochemistry* 2004, *39* (6), 695–704.

(104) Jossens, L.; Prausnitz, J.; Fritz, W.; Schlünder, E.; Myers, A. Thermodynamics of multi-solute adsorption from dilute aqueous solutions. *Chemical Engineering and Science* 1978, *33* (8), 1097–1106.

(105) Sips, R. Combined form of Langmuir and Freundlich equations. *Journal of Chemical Physics* 1948, *16* (429), 490–495.

(106) Pérez-Marín, A.; Zapata, V. M.; Ortuno, J.; Aguilar, M.; Sáez, J.; Lloréns, M. Removal of cadmium from aqueous solutions by adsorption onto orange waste. *Journal of Hazardous Materials* 2007, *139* (1), 122–131.

(107) Toth, J. State equation of the solid-gas interface layers. *Acta Chimica Hungarica* 1971, *69*, 311–328.

(108) Koble, R. A.; Corrigan, T. E. Adsorption isotherms for pure hydrocarbons. *Industrial & Engineering Chemistry* 1952, *44* (2), 383–387.

(109) Khan, A.; Ataullah, R.; Al-Haddad, A. Equilibrium adsorption studies of some aromatic pollutants from dilute aqueous solutions on activated carbon at different temperatures. *Journal of Colloid Interface Science* 1997, *194* (1), 154–165.

(110) Khan, A. R.; Al-Waheab, I.; Al-Haddad, A. A generalized equation for adsorption isotherms for multi-component organic pollutants in dilute aqueous solution. *Environmental Technology* 1996, *17* (1), 13–23.

(111) Emmett, P. Gases in multimolecular layers. *Journal of American Chemical Society* 1936, *60*, 309–319.

(112) Hill, T. L. Theory of physical adsorption. In Frankenburg, W. G.; Komarewsky, V. I.; Rideal, E. K., *Advances in Catalysis*, Vol. 4; Elsevier, 1952; pp. 211–258.

(113) McMillan, W.; Teller, E. The assumptions of the BET theory. *The Journal of Physical Chemistry* 1951, *55* (1), 17–20.

(114) Pellenz, L.; da Silva, L. J. S.; Mazur, L. P.; Figueiredo, G. M. D.; Borba, F. H.; Ulson de Souza, A. A.; Guelli Ulson de Souza, S. M. A.; da Silva, A. Functionalization of graphene with nitrogen-based groups for water purification via adsorption: A review. *Journal of Water Process Engineering* 2022, *48*, 102873.

5 Factors Influencing the Removal of Wastewater Contaminants on Graphene-Based Adsorbent Materials (GBAMs) Surface(s)

5.1 INTRODUCTION

Adsorption is a phenomenon whose final Q is determined by several regulating parameters. Various factors influence contaminants adsorption capability of GBAMs, notably GBAM characteristics, initial solution pH (the degree of ionization and the surface properties on the adsorbent change because of the solution's pH changing), time of contact, contaminants' initial concentration as well as adsorbent dosage, temperature, coexisting ions, and organic acid ligands, including competitive adsorption. This chapter is focused on discussing the factors affecting the adsorption of GBAMs, as presented below.

5.2 SURFACE FUNCTIONAL GROUPS

Several molecular dynamics investigations show that terminal functionalities on GBAMs play an essential role in the increased exclusion of metal ions from aqueous streams.[1,2] For the exclusion of organic as well as inorganic contaminants from wastewater, GO could be grafted or functionalized with various polymeric assemblages. For instance, due to the huge number of terminal functional groups on the GO surface, polyamidoamine dendrimers (PAMAMs) have been reportedly researched with regards to their potential to graft well onto GO as well as potential applications.[3,4] These PAMAM dendrimers are significantly larger but also more versatile in terms of terminal functionalities, like carboxylic, amino, as well as hydroxyl groups.

Yuan et al. discovered that grafted GO with dendrimers (GO-PAMAM) has a high heavy metal ion adsorption capability of 1.0 mmol/gm.[5] Zhang et al. have also reportedly investigated the use of GO-PAMAM adsorbent to eliminate heavy metal ions from aqueous media.[6] These experiments demonstrated that the metal-ion Q on the GO surface improves following polymer modification.[6]

DOI: 10.1201/9781032621302-5

Another study explored the molecular mechanisms underlying the adsorption of metal ions, notably Pb^{2+}, on G, GO, but also polyamidoamine (PAMAM) dendrimers. To better understand the molecular interaction between metal ions as well as PAMAM dendrimers, varied functionalities were grafted on the G and GO surfaces. G sheet (GS), GS/PAMAM, GS-PAMAM-COO$^-$, GS/PAMAM-OH, GO), GO/PAMAM, GO/PAMAM-COO$^-$, as well as GO-PAMAM-OH were created using the base materials.[7] Using the base materials G, GO, dendrimer with NH_2 terminal groups (PAMAM), dendrimer (PAMAM-COO$^-$) with COO$^-$ terminal groups, and dendrimer with OH terminal groups (PAMAM-OH), eight distinct surfaces were produced; that is to say, G (GS), GS/PAMAM, GS/PAMAM-COO$^-$, GS/PAMAM-OH, GO (GO), GO/PAMAM, GO/PAMAM-COO$^-$, as well as GO/PAMAM-OH. Adsorption of lead (II) ions upon the GO/PAMAM-COO$^-$ sheets proved to be meaningfully greater compared to the other fabricated surfaces with respect to the entire dosages studied in their investigated materials, as depicted in Figure 5.1. At roughly 6 mmol/L, the highest Q of GO/PAMAM for Pb^{2+} ion is reported to be 568.18 mg/g. The investigational results of adsorption Pb^{2+} order of metal ions on the distinct surfaces GO/PAMAM> GO > GS were qualitatively similar to their findings.[7] The Q_{max} of Pb^{2+} ion on various surfaces estimated using the LM equation are in the following order: GO/PAMAM-COO$^-$ > GO/PAMAM-OH > GO/PAMAM > GO > GS/PAMAM-COO$^-$ > GS/PAMAM-OH > GS/PAMAM > GS/PAMAM > GS/PAMAM > GS/PAMAM > GS/PAMAM > GS/P. It was also observed by these researchers that, with consideration to five concentrations, the Q of Pb^{2+} ions on the GO grafted dendrimers' surface were significantly greater than the virgin GO, pristine GS, as well as GS-g-dendrimer surfaces. The mechanism of metal ion adsorption on eight different surfaces was described using microscopic interactions between metal ions along with solid surfaces. Furthermore, they investigated the self-diffusion coefficient as well as the residence time of Pb^{2+} ions close to the surfaces. It is interesting to know that these authors discovered that the interaction between the Pb^{2+} ion, as well as dendrimers,

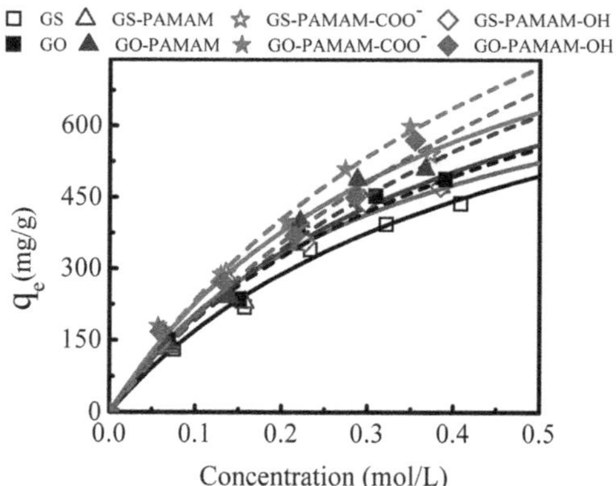

FIGURE 5.1   Langmuir adsorption isotherm showing the variation of the total amount of Pb^{2+} ion adsorbed (q_e) onto different surfaces. Open symbols for different terminal groups (–NH_2, –COO$^-$, and –OH) of dendrimer grafted to GS surface and filled symbols for different terminal groups (–NH_2, –COO$^-$, and –OH) of dendrimer grafted to GO surface. (Reproduced with permission from Kommu et al.[7] Copyright 2017, American Chemical Society.)

is important in strengthening their attachment to the grafted dendrimers' surfaces. The results reveal that employing -C=O groups of dendrimers grafted on the GO surface improves the adsorption ability of the Pb^{2+} ion substantially.[7]

The insufficiency of time may not permit us to exhaustively discuss all G-based literature available with regards to their findings on the effect of surface functionalities on their adsorption performance, though we can conclude from the available reviewed literature that surface functionalities are the major players that decide and control all other factors affecting G-based adsorbents adsorption characteristics. This implies that, researchers aiming to design as well as fabricate G-based adsorbent materials must consider the careful and predetermined engineering of surface functionalities capable of interacting well with the adsorbate(s) molecules targeted.

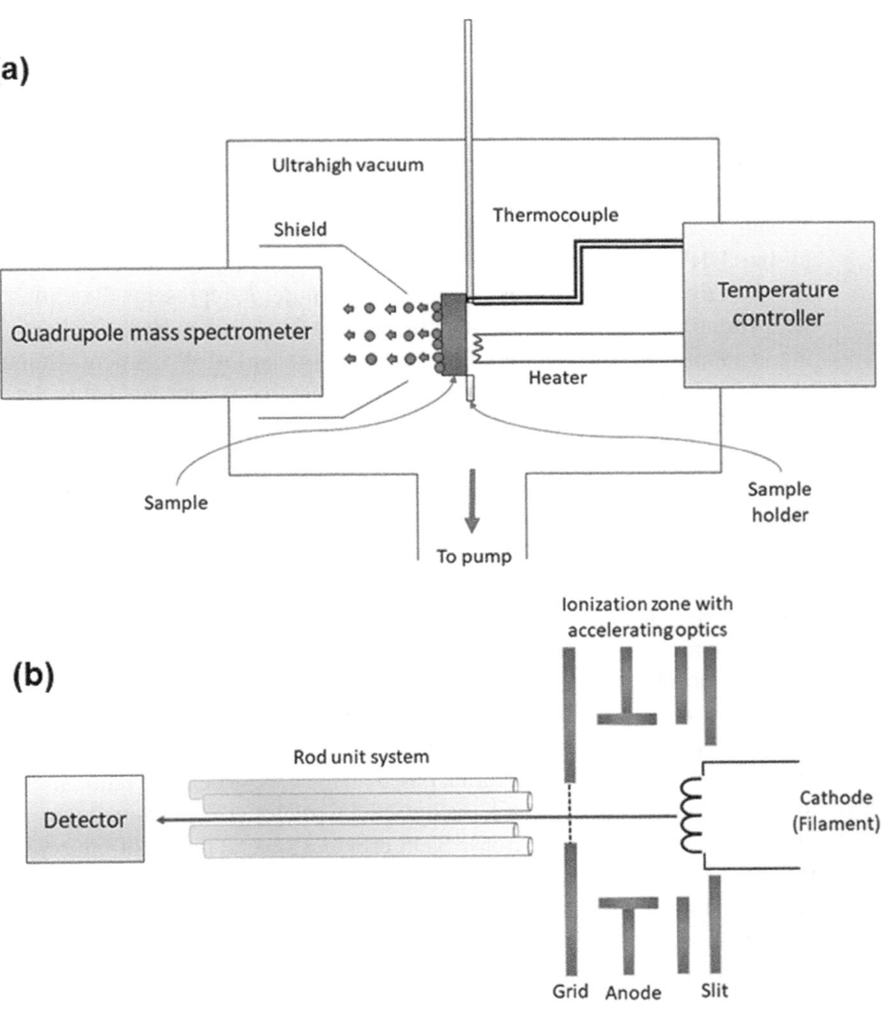

FIGURE 5.2   (a) TDS Schematic representation and (b) QMS components.

5.3   SURFACE KINETICS

The surface area of the adsorbent greatly influences its surface properties, which is also relative to the surface kinetics. The surface kinetic process of adsorbent materials is normally investigated by tracking the gas phase molecules desorbed that were earlier attached to the concerned surface by the adsorption process.

Figure 5.2a depicts the Thermal Desorption Spectroscopy (TDS) experimental set-up, which is a set-up for analyzing the surface kinetics of adsorbent materials. The experimental criteria for studying adsorption kinetics using TDS are as follows:

i. *Leak valve*: permits the fluid (gas) of interest to be dosed into the ultrahigh vacuum technique (UHV) chamber. The dosage of the gas is measured in Langmuir (L) units, with 1 L equaling 1×10^{-6} mbar/sec. If the adsorption probability 24 is 1, dosing 1 L of the gas of interest should result in a 1 monolayer (1 ML) coverage; in most circumstances, a monolayer coverage is observed at several Langmuir doses. The final pressure of the gas that adsorbs on the surface is the sum of the background pressure and the dosed pressure $\left(P_{final} = P_{background} + P_{dose}\right)$; because the $P_{background}$ (1×10^{-10} mbar) is less than the P_{dose} (mbar), the P_{final} can be regarded as the exact dose of the gas pressure of interest.

ii. *Sample heating*: The sample is placed in the manipulator, which is attached to the UHV chamber. Typically, the sample is ramped linearly, with $T_t = T_0 + \beta t$, where T_t denotes the final temperature, T_0 represents the initial temperature, β denotes the heating rate, and t is the duration. Resistive heating is adopted for the heating of the sample by means of W-filament while monitoring the temperature with a thermocouple that is spot-welded to the sample. The thermal energy produced when the sample is heated will help dissolve the bonds tying the adsorbates to the surface, leading to the adsorbates' eventual desorption from the surface.

iii. *Detector*: A temperature ramp allows a quadrupole mass spectrometer to track the desorbed molecules from the surface (QMS). The QMS assists in detecting a mass-selective pressure rise. Figure 5.2b depicts the QMS's three components (ion source, quadrupole analyzer, and electron multiplier). The QMS's W-filament is heated electrically, resulting in free electrons. The free electrons accelerate between the filament and the grid, followed by ionization between the grid and the aperture. The quadrupole analyzer is made up of four electrically connected, symmetrically aligned cylindrical rods. The analyzer permits particle separation based on the mass-to-charge ratio (m/e). To obtain the multiplied signal, the detected particles are transmitted via an electron multiplier (channeltron).[8] TDS provides several pieces of information, including a number of binding sites, surface coverage, reaction order, desorption temperature, but also desorption energy (E_{des}). Incorporating the TDS peak yields surface coverage. The binding site number could be calculated by taking into account the number of peaks in TDS. Burghaus,[9] in a particular work, reported benzene adsorption on CNTs as an example from the literature.[10] TDS in the referred work reveals that CNTs have three distinct binding sites: internal, groove, but also the exterior. The high temperature desorption peak was attributed to the inside, whereas the low temperature desorption peak was assigned to the exterior.[9] The assignments of peaks are determined by the benzene molecule's binding strength at various locations. Furthermore, a condensation peak was reportedly observed at about 145 K as a result of benzene condensation (multilayer) onto CNTs.

Surface science, via the adoption of TDS, can be used to perform both qualitative and quantitative examinations. The contours of the TDS curve can be used to identify the reaction order. Depending on the TDS test, curves for the first, second, or zeroth orders may be shown. The leading edge is aligned, and the desorption temperature shift increases with increasing coverage, according to the zeroth-order analysis. This phenomenon, which results in a shared leading edge, is observed in molecules where the intermolecular contacts are strong (for instance, hydrophobic interactions of water on G; strong intermolecular connection between water molecules). Likewise, expanding coverage and eventual multilayer development are responsible for the rise in desorption temperature. Due to coverage independence, 1st-order kinetics displays an asymmetric peak with a constant desorption temperature in contrast to zeroth-order kinetics. For non-dissociative molecule adsorption as well as desorption, 1st-order desorption is normally the case found. However, in 1st-order kinetics, the desorption temperature can decrease with increased coverage; repulsive lateral interaction of the adsorbate molecules can lead to adsorbate destabilization, which can result in a decrease in desorption temperature with increasing coverage. Furthermore, for molecules that dissolve on the surface and desorb, a 2nd-order kinetics is seen, and therefore, the desorption temperature lowers with growing coverages. Generally, the peak form in TDS permits one to determine the order of the reactions.

The TDS could be utilized to compute the binding energies, adopting one of the approaches listed below:[8]

i. Complete analysis
ii. Leading edge analysis
iii. Redhead analysis
iv. Varying heating rate analysis

Thorough analysis: this assessment is performed with respect to the expressions below, desorption rate,

$$-d\theta/dt = k_d \theta^m \tag{5.1}$$

heating rate,

$$\beta = \frac{dT}{dt} \tag{5.2}$$

inserting (5.2) in (5.1)

$$-d\theta/dT = k_d\theta^m/\beta \tag{5.3}$$

From Arrhenius model

$$k_d = ve^{-Ea/RT} \tag{5.4}$$

inserting (5.4) in (5.3)

$$-d\theta/dT = \theta^m ve^{-Ea/RT}/\beta \tag{5.5}$$

if, $-d\theta/dT = r$, then

$$r = \theta^m ve^{-Ea/RT}/\beta \tag{5.6}$$

Equation (5.6) natural logarithm results in,

$$\ln r = \ln\left(v\theta^m/\beta\right) - E_a/RT \tag{5.7}$$

A plot of $\ln r$ vs $\dfrac{1}{T}$ can be plotted for each coverage (θ), but also the gradient results in E_a (desorption energy) for each coverage. Because the coverage (θ) and order of the reaction (m) can be known from the TDS curves, the pre-exponential component (v) may be computed from the intercept. Although this procedure is accurate, it is rarely utilized due to its intricacy.

Leading edge analysis:

This approach is applied to the TDS curve at the leading edge, as reported in.[9] Equation (5.7) is used to compute the desorption energy (E_a). However, near the leading edges, the signal-to-noise ratio $\left(\dfrac{S}{N}\right)$ is extremely low. As a result, this approach is rarely utilized to calculate desorption energies.

Utilizing expression (5.5), $-d\theta/dT = \theta^m ve^{-Ea/RT}/\beta$, the Redhead evaluation is as follows. When $T = T_p$, where T_p is the desorption temperature at maximum, the rate of desorption likewise achieves a maximum, resulting in $\dfrac{d^2\theta}{dT^2} = 0$. As a result, equation (5.5) can be written as follows:

$$E_a/RT^2 = v/\beta\, m\theta^{m-1}e^{-Ea/RT} \tag{5.8}$$

Redhead predicts 1st-order desorption, therefore $m = 1$, resulting in

$$E_a/RT^2 = v/\beta \, e^{-E_a/RT} \tag{5.9}$$

With the use of natural logarithm,

$$\ln\left(E_a/RT\right) = \ln\left(vT/\beta\right) - E_a/RT \tag{5.10}$$

and,

$$E_a = RT\left(\ln\left(vT/\beta\right)\right) - \ln\left(E_a/RT\right) \tag{5.11}$$

Nevertheless, it is presumed that $\ln\left(E_a/RT\right) = 3.46$ and thus.

$$E_a = RT\left(\ln\left(vT/\beta\right) - 3.46\right) \tag{5.12}$$

For tiny molecules, the desorption energy could be computed by presuming the preexponential factor $(v) = 1013\,\mathrm{s}^{-1}$. This approach is commonly used, although it has limitations owing to the hypotheses applied.

With the variation of heat rate analysis:

A linear heating ramp is typically used in TDS. The heating rate, on the other hand, can be changed to decide the desorption energy. Changing the expression (5.9), $E_a/RT^2 = v/\beta \, e^{-E_a/RT}$ becomes,

$$\beta/T^2 = \left(vR/E_a\right) e^{-E_a/RT} \tag{5.13}$$

With the use of natural logarithm, expression (5.13) becomes.

$$\ln\left(\beta/T^2\right) = \ln\left(vR/E_a\right) - E_a/RT \tag{5.14}$$

The desorption energy is given by a negative slope caused by a change in heating rate (β). Even though this approach is reliable, it is not generally utilized because changing the rate of heating is inconvenient.

TDS is an unavoidable approach in most surface science research, and it is known as the "beauty and the beast" surface science method. Its beauty is that the test is simpler to accomplish, whereas the beast is its intricacy in data analysis. As a result, TDS can be used extensively to obtain information regarding surface-adsorbate interactions.

5.4 EFFECT OF pH

Adjusting the pH of the medium can change the electrostatic interaction between the adsorbents with the adsorbates; hence, pH variation is frequently considered while improving the adsorption of contaminants in wastewater like antibiotic,[11] heavy metals,[12] or dyes adsorption.[13] However, as the pH rises, the absorption behavior changes due to a variety of variables. The protonation-deprotonation evolution of the surface functionalities (mostly O-containing) present by the majority of GBAMs employed for contaminants removal is primarily influenced by pH.[14] The functional groups are protonated at pH_{pzc} (point of zero charge (pzc)), the surface becomes positively charged, and anions bind due to electrostatic interactions. Furthermore, a considerable quantity of H^+ and H_3O^+ compete with cations for accessible adsorption sites. These mechanisms may explain why adsorbates such as antibiotics and some dyes have a lower adsorption capability at low pH. GBAMs surfaces gradually deprotonate as pH rises, becoming negatively charged at $pH > pH_{pzc}$. For instance, due to the cationic antibiotic ions are drawn by the GBAMs in antibiotic adsorption, the absorption capacity increases. As presented in Table 5.1, the best Q pH for diverse adsorbates is different. Though for G-based adsorbents their best pH for diverse adsorbates elimination from wastewater is between 2 and 7. For example, an experiment into the adsorption of MB from simulated wastewater is best within a pH range of 2–12 revealed that the Q increases with pH, with the maximum adsorption found at a pH of 12.[15] MNZ has a pK_a of 2.58, indicating that at pH = 2, it is nearly protonated, while at pH 4 it is in a zwitterion state.[16] Because absorbents at pH pHpzc are positive at $pH > pH_{pzc}$, the existence of electrostatic repulsive interactions between the surface of the absorbent and the MNZ cationic might explain the relatively poor MNZ Q (for CXs, 94.2 mg/g) at pH = 2. Gao et al. revealed in their work with GO adsorbent that tetracycline's (TCN) GO adsorption capabilities reduced when pH or Na^+ concentration increased. Oxytetracycline and doxycycline's adsorption isotherms on GO were discussed and contrasted even as observed by other researchers, as earlier discussed.[17]

Two GOs made using modified hummers procedures (designated as GO1 and GO2) have been reportedly examined for the influence of pH on the efficiency of phenol adsorption, and the findings are presented in Figure 5.3.[18] The amount of phenol that could be removed at its highest adsorption rate without the help of UV was also demonstrated by the researchers. The two utilized adsorbents (GO1 along with GO2) achieved their maximum percentage of adsorbate removal at pH 2, and as pH increased, this percentage decreased. Figure 5.3 illustrates how the adsorptive removal percentages for GO1, but also GO2, that were not exposed to UV light fell from 93.5 percent to 77.2 percent and 80.95 percent to 70.63 percent, respectively.[18] Working at pH 2, which was the ideal pH in their work, was said to be neither practical nor economical from an economic standpoint; hence, the authors conducted their studies at pH 6. However, after being exposed to UV light, both of the employed adsorbents (GO1 but also GO2) had higher removal percentages, with GO1's removal percentage reaching its highest value at pH 2 and its lowest value at pH 10. The GO_2 displayed a distinct adsorption pattern that fluctuated, with its removal percentage, reaching a high of 91.4 percent at pH 6 and a low of 75.4 percent at pH 4. The

TABLE 5.1
Effect of pH on G-Based Adsorbents with Regard to Literature

Type of Adsorbate	Adsorbent Formulation	Maximum Adsorption pH	Max Dose	Max Q (Q_{max})	Contact Time for Max Adsorption (h)	Adsorption Kinetics & Isotherm (w.r.t R^2)		Ref.
						Kinetics	Isotherm	
Dyes								
MB	GO@CNCs	10	0.03 g/L	751.88 mg/g	3	Pseudo-second-order	Both Langmuir and Freundlich isotherms.	25
Malachite green (MG)	GO@cellulose	7.0	10 mg/L	30.091 mg/g	5	Pseudo-second-order	Langmuir	28
MB	Magnetic cyclodextrin–GO (MCGO)	10	100 mg/L	261.78 mg/g	0.83	Pseudo-second-order	Langmuir	29
MB	Activated carbon@GO	—	—	1000 mg/g	1	Pseudo-second-order	Freundlich	29
Safranin O (SO)	MgO decked multi-layered G (MDMLD)	2–12	4.0×10^{-4} M	3.92×10^{-4} mol/g	2	Pseudo-second-order	Langmuir	30
Pharmaceutical Residues								
TCN	GO	3.6	0.181 mg/L	313 mg/g	1.5	Pseudo-second-order	Langmuir	17
Oxytetracycline	GO	3.6	0.181 mg/L	212 mg/g		Pseudo-second-order	Langmuir	17
TCN	Fe_3O_4–rGO	5.0	40 mg/L	110.80 mg/g	0.47	Pseudo-second-order	Langmuir	31
Amoxicillin (AMX)	Fe_3O_4–rGO	6.0	50 mg/L	119.12 mg/g	0.47	Pseudo-second-order	Langmuir	31
Cefalexin (CLZ)	Fe_3O_4–rGO	7.0	40 mg/L	101.42 mg/g	0.47	Pseudo-second-order	Langmuir	31
Penicillin G (PEN G)	Fe_3O_4–rGO	5.0	30 mg/L	91.88 mg/g	0.33	Pseudo-second-order	Langmuir	31
Phenol	GO (GO1)	2.0	0.8 g/L	0.91 mg/g	24	—	Langmuir	18
	GO (GO2)	2.0	0.8 g/L	7.66 mg/g	24	—	Langmuir	18
Clonazepam	Magnetic-GO (mGO)	6.5	1.2 g/L	14.81 mg/g	3.0	Pseudo-first-order		32

(Continued)

TABLE 5.1 (CONTINUED)
Effect of pH on G-Based Adsorbents with Regard to Literature

Type of Adsorbate	Adsorbent Formulation	Maximum Adsorption pH	Max Dose	Max Q (Q_{max})	Contact Time for Max Adsorption (h)	Adsorption Kinetics & Isotherm (w.r.t R^2)		Ref.
						Kinetics	Isotherm	
Nalidixic	NiZrAl-layered double hydroxide-GO-chitosan (NiZrAl/LDH/GO/CS NC)	8.0	15 mg/L	277.79 mg/g	—	PSOM	—	24
Tetracycline (TC)	Zeolitic-imidazolate framework-8@rGO (ZIF-8@rGO)	7.68	233 mg/L	1776.26 mg/g	—	PSOM	LM	23
Toxic Metals Ions								
Cd(II) ions	GO	6.0 ± 0.1	20 mg/L	106.3 mg/g	—	—	Langmuir	33
Co(II) ions	GO	6.0 ± 0.1	30 mg/L	68.2 mg/g	—	—	Langmuir	33
Au(III) ions	Cross-linked chitosan (GCCS)	3.0–5.0	80 mg/L	1076.649 mg/g	15	Pseudo-second-order	Langmuir	34
Pd(II) ions	GCCS	3.0–4.0	40 mg/L	216.920 mg/g	15	Pseudo-second-order	Langmuir	34
Cu(II)	Sulfonated magnetic GO composite (SMGO)	4.68	50	62.73 mg/g	2	Pseudo-second-order	Langmuir	35
Radionuclides								
Am(III)	GO	3.5	0.038 g/L	35 ± 6 µmol/g	0.33	—	Freundlich	36
Eu(III)	GO	5.0	0.38	760 ± 75 µmol/g	0.33	—	Langmuir	36
U(VI)	GO	3.5	0.038	97 ± 19 µmol/g	0.33	—	Freundlich	36
U(VI)	GO	5.0	0.038	116 ± 5 µmol/g	0.33	—	Freundlich	36
Am(III)	GO	3.5	0.038	35 ± 6 µmol/g	0.33	—	Freundlich	36
Sr(II)	GO	6.5	0.038	272 ± 35 µmol/g	0.33	—	Freundlich	

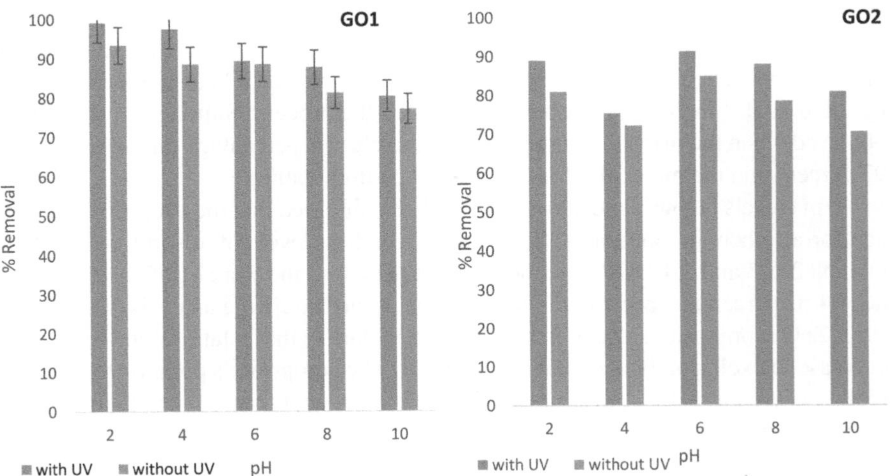

FIGURE 5.3 Effect of different pH values on the removal of phenol from aqueous media by GO1 and GO2 with and without UV. Experimental conditions: initial concentration 100 mg/L; adsorbent mass 0.02 g; volume 25 mL; temperature 25 °C; contact time 24 hr. (Reproduced with permission from Al-Ghouti et al.[18] Copyright 2022, Elsevier Science Ltd.)

structural and chemical alterations that were placed to GO, which strengthened its interaction with phenol, may be to blame for the increase in adsorption under UV. Indicating the influence of electrostatic interaction on phenol removal, GO1's better removal efficiency of phenol in comparison to GO2 was proposed to be due to weaker electrostatic repulsion. Similar to this, Hu et al.[19] developed a GO-polypyrrole (GO-PPy) composite to remove phenol from an aqueous medium and studied the variations in the adsorption process that have been brought on by altering the pH value. They demonstrated in their findings that the removal percentage decreased from 35% to 11% when the pH rose beyond 8, which may have been caused by the electrostatic attraction between the negatively charged adsorbent and the phenol, as well as the hydrophobic interaction. Additionally, in an acidic solution, the surface of the GO molecule was postulated to be positively charged when the pH was less than pH_{pzc}, and in an alkaline solution, the surface became negatively charged when the pH was greater than pH_{pzc}. Because of the enhanced electrostatic attraction between phenol and the negatively charged GO surface, the effectiveness of phenol removal at alkaline solutions is reduced. According to the authors, the electrostatic interaction might not be a deciding element in the adsorption process because there is only a negligible shift in pH from 2 to 6 when pH is less than pH_{pzc} or from 8 to 10 when pH is higher than pH_{pzc}. The hydrogen bonds between the phenol's -OH groups and the functional groups of the GO were proposed to be the primary adsorption mechanism, as the removal effectiveness dropped when the pH was raised. Additionally, the π-π bonds that form between phenol and GO may be what drives the adsorption process.[18]

To remove phenolic chemicals from the aquatic medium, Zhang et al.[20] created a sustainable composite using GO and manganese oxide. It was revealed from their study that raising the pH level from 2 to 10 results in a better adsorption capacity.

This phenomenon was reported to be likely due to the fact that oxidation potential is stronger under acidic settings than it is under alkaline conditions. The removal of phenol from an aqueous media by nitrogen-doped reduced GO has also been studied by Zhao et al.[21] in 2021 to see how pH affected the process. Similar outcomes were observed when the pH was changed, with the exclusion percentage falling from 98 to 93 percent and the molecular form of phenol predominating.[21]

At pH levels below 6 and above 7, it has been observed that the exclusion percentage for all phenolic pollutants' declines. The highest levels of adsorption for BPA, phenol, 2-CP, and 2,4-DCP were measured at pH = 6–7 and were 81.65, 86.52, 91.74, and 94.52 percent, respectively.[22] Furthermore, a surface charge analysis of the RGO-PEG-ZnO composite's Zeta potential could clarify the relationship between a decrease in exclusion % and an increase in pH. The composite's point of zero charge (pH_{pzc}) is seen at pH 7.2. From the zeta potential study, it can be seen that the composite's surface is positively charged for pH values below 7.2 and negatively charged for values more than 7.2. The hydrogen link between the adsorbate and the adsorbent weakens, resulting in a drop in the adsorption efficacy as the phenolic pollutant begins to deprotonate and generate a phenolate anion for pH values higher than 7. Additionally, because the RGO-PEG-ZnO composite surface is negatively charged (pH_{pzc} = 7.12), the phenolate anion and this surface resist one another, reducing the effectiveness of adsorption. Phenolic pollution began to protonate at pH levels lower than 7, which caused an increase in H^+ ions in the solution. The protonated phenolic compounds and the H^+ ions are more electrostatically attracted to one another as a result, depleting the adsorption process. Consequently, these authors reportedly adjusted the pH to 7 for additional studies in order to achieve high adsorption efficiency.[22]

Zeolitic-imidazolate framework-8@reduced GO (ZIF-8@rGO) has been used to study the effect of pH on TC adsorption in the pH range of 1.2 to 10.2.[23] The reported adsorbents in this report also demonstrated a time dependent adsorption as in Figure 5.4a and b, behavior. Though with emphasis on the influence of pH as seen in Figure 5.4c, the adsorption capacity increased noticeably as pH rose from 1.2 to 6.2; it peaked at pH = 6.2 before gradually declining. Because of the surface charge density of the ZIF-8@rGO aerogel and the species of TC, it was hypothesized that the TC adsorption behavior was closely associated with pH.[23] Owing to the fact that the TC molecule has two acidic functional groups and one basic functional group that can create four distinct compounds at different pH values – H_3TC^+ (pH < 3.3), H_2TC^0 (3.3 < pH < 7.68), HTC^- (7.68 < pH < 9.68), and TC^{2-} (pH > 9.68) protonation-deprotonation could take place in aqueous solution. The isoelectric point of the aerogel, which is shown in Figure 5.4c's "purple line", was reported to be 5.6. This means that when the pH was between 5.6 and higher than 5.6, the aerogel's surface was positively charged, and when it was lower than 5.6, the aerogel surface transformed into a negatively charged surface. When the pH level was <3.2, ZIF-8@rGO-2 aerogel and TC were repulsive to one another. ZIF-8@rGO-2 and H2TC0's positive or negative charge surfaces produced electrostatic attraction when the pH was between 3.2 and 7.68, increasing their ability to adsorb substances. Additionally, despite $H3TC^+$ and HTC^- having electrostatic repulsion with the adsorbent surface in both strongly acidic and strongly alkaline pH conditions, ZIF-8@rGO-2 aerogel still has

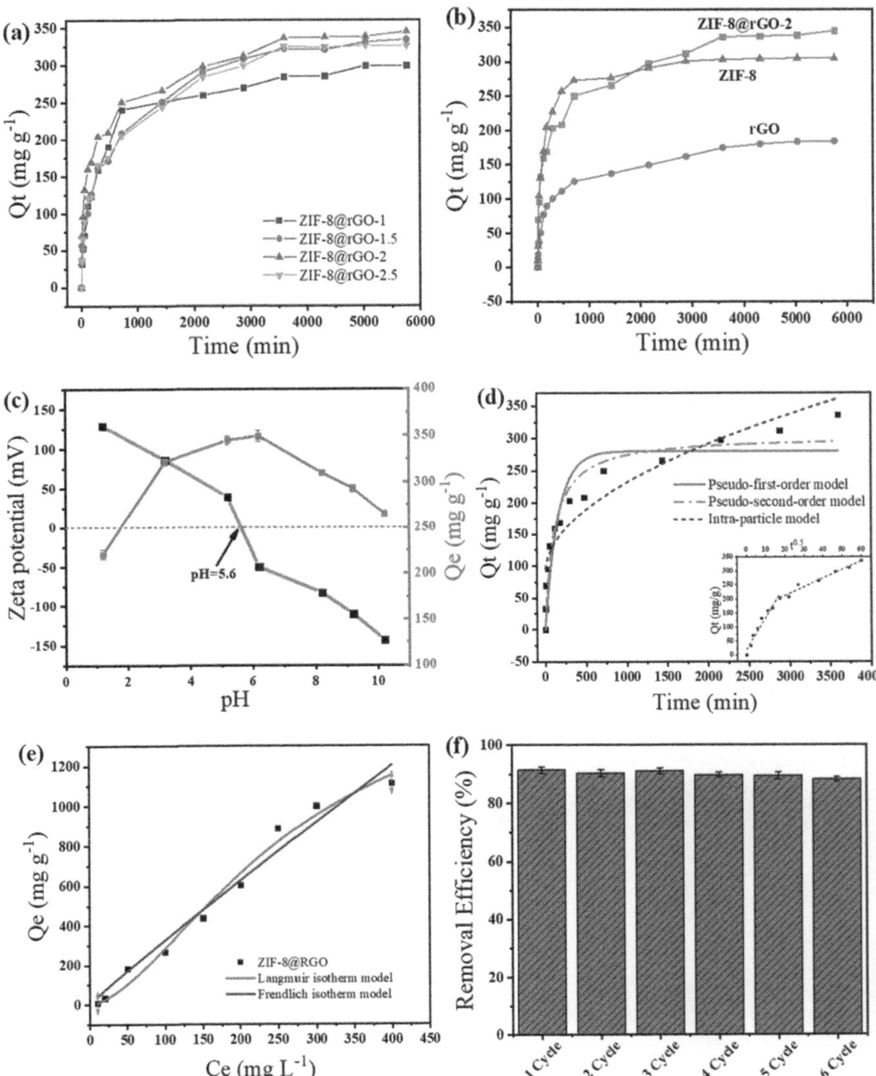

FIGURE 5.4 (a) TC adsorption properties of ZIF-8@rGO aerogel samples with different proportions of GO. (b) Comparison of Q of TC with different aerogels. Experimental conditions: amount (m) = 0.007 g, initial concentration = 100 mg L^{-1}, Volume (V) = 30 mL and temperature (T) = 30 °C. (c) Effect of initial solution pH value on TC adsorption by ZIF-8@rGO-2 (red line) and zeta potential (purple line) of ZIF-8@rGO-2 aerogel. (d) Kinetic models of TC adsorption on ZIF-8@rGO-2 aerogel. (e) Isotherm models of TC adsorption on ZIF-8@rGO-2 aerogel. (f) The reusability of ZIF-8@rGO-2 aerogel toward TC. (For interpretation of the references to color in this figure legend, the reader is referred to the web version of this article.) (Reproduced with permission from Liu et al.,[23] copyright 2022, Elsevier Science Ltd.)

a good capacity for adsorbing TC because of the interaction of π-π and cation-π bonds between benzene and imidazole rings and four aromatic rings of TC.[23] The kinetics of adsorption for the adsorbents towards TC was analyzed using PFOM, PSOM, and the intraparticle diffusion model, where they found out that adsorption kinetics followed the PSOM, indicating that the adsorption process was basically chemical adsorption, which took place on the surface of the aerogel as in Figure 5.4d.[23] In like manner, the adsorption isotherm results studied by these authors as depicted in Figure 5.4e, revealed that the oxygen-rich functional groups on ZIF crystals and rGO gave distinct adsorption sites for ZIF-8@rGO-2 aerogel, which is why the adsorption active sites on the surface of ZIF-8@rGO-2 aerogel were equally distributed and the adsorption of TC follows the multilayer adsorption process.[23] It is interesting to note that the recyclability of the adsorbent "ZIF-8@rGO-2" remained at a high level, even after 6-cycles as depicted in Figure 5.4f.

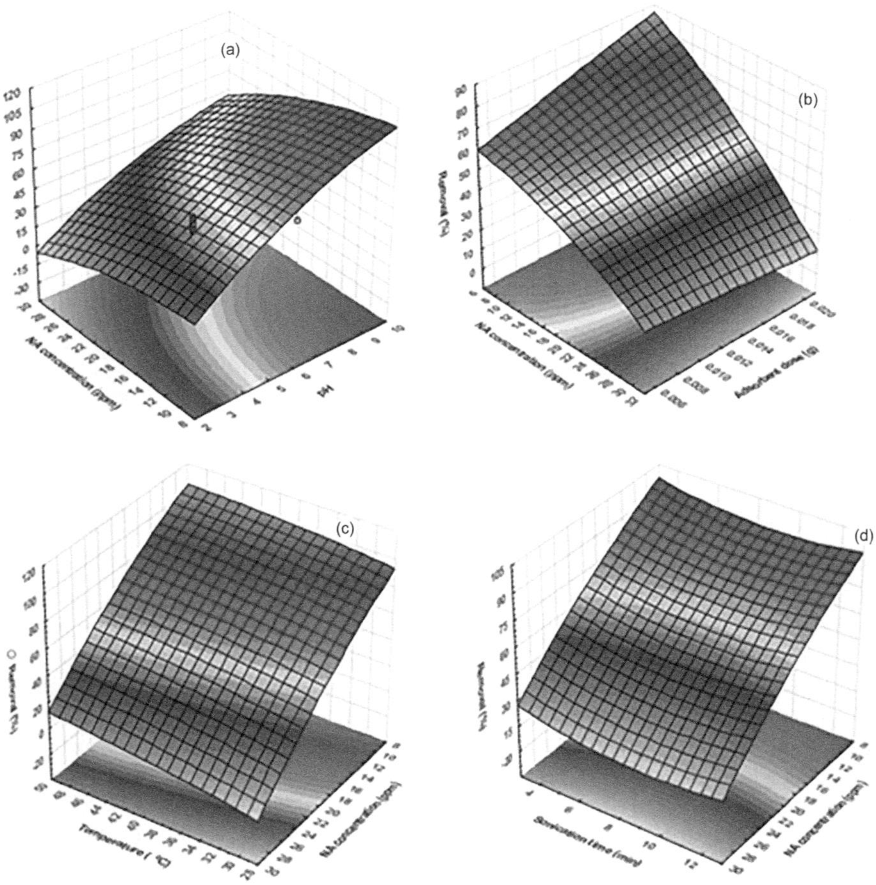

FIGURE 5.5 Visualizing the role of important variables. (Reproduced with permission from Radmehr et al.,[24] copyright 2021, Elsevier Science Ltd.)

In another instance, as presented in Figure 5.5a, it was evidently shown that both NA and pH have a significant impact on NA exclusion. It was clearly discovered that varying pH levels can affect how much the nano-adsorbent interacts with the adsorbates.[24] According to these findings, also shown in Figure 5.5a, a higher removal rate was reached when the pH level was higher than 6. Increased levels of uncharged species in the solution at pH values below the NA's pK_a (pH 6.0) increased the likelihood of a π–π interaction between the NA and the nano-adsorbent.

Radmehr et al.,[24] in their work on LDH-GO-CS NC adsorbent, as shown in Figure 5.5b, revealed that with the consideration of adsorbent dosage, the percentage removal of the adsorbate by the adsorbent at NA concentration being at the middle and/or minimum level, increased from 0.6 mg to 2 mg. They also investigated the effect of temperature on NA initial concentration while keeping other variables constant, as shown in Figure 5.5c and revealed that the removal of NA evidently increased with increasing temperature and decreased with increased concentration of NA. The

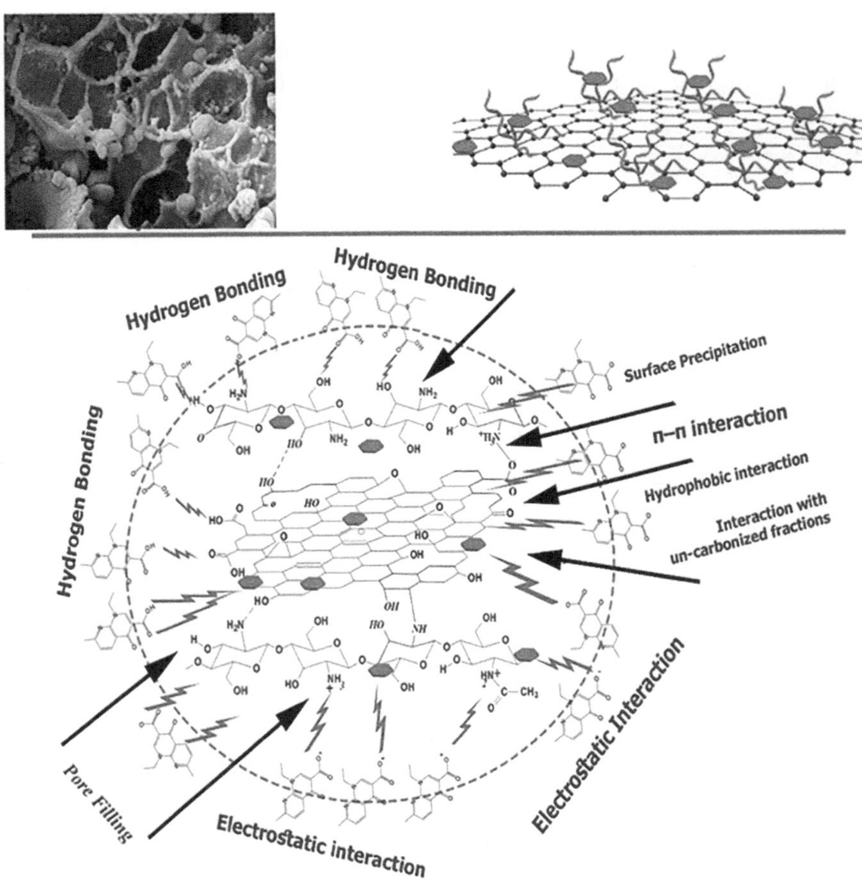

FIGURE 5.6 Adsorption mechanism of NA on the LDH-GO-CS NC. (Reproduced with permission from Radmehr et al.,[24] copyright 2021, Elsevier Science Ltd.)

influence of sonication time with respect to NAs' initial concentration was also studied by these researchers: their findings as depicted in Figure 5.5d revealed that the equilibrium sonication time was reached at ~5 min though, at higher NA concentration, removal of the adsorbate by the adsorbent was found to be <30%, which was said to signify that the adsorption process was not complete but more sonication time was needed to complete the adsorption process.[24]

Additionally, for pH values lower than 6.0, the adsorption process may be aided by electrostatic attraction and hydrogen bonding together with π–π interaction, as represented in Figure 5.6. The carboxylic acid group of NA, on the other hand, became more dissociated at pH levels above pK_a, which increased the electrostatic contact between the amino groups of CS and the carboxylic acid of NA molecules. Moreover, due to the negatively charged nature of NA molecules, electrostatic interactions may develop between LDH nanoplates and NA. LDH-GO-CS NC and NA may interact in a variety of ways, and this could explain why their exclusion performance is adequate. Based on the results discussed above, a pH of 8 was selected for the studies that followed, which is in good agreement with the pH determined by GA and DF.[24]

Another instance where the influence of pH on the adsorption behavior of GBAMs for the removal of another adsorbate is in the report by Zaman et al.[25]: these

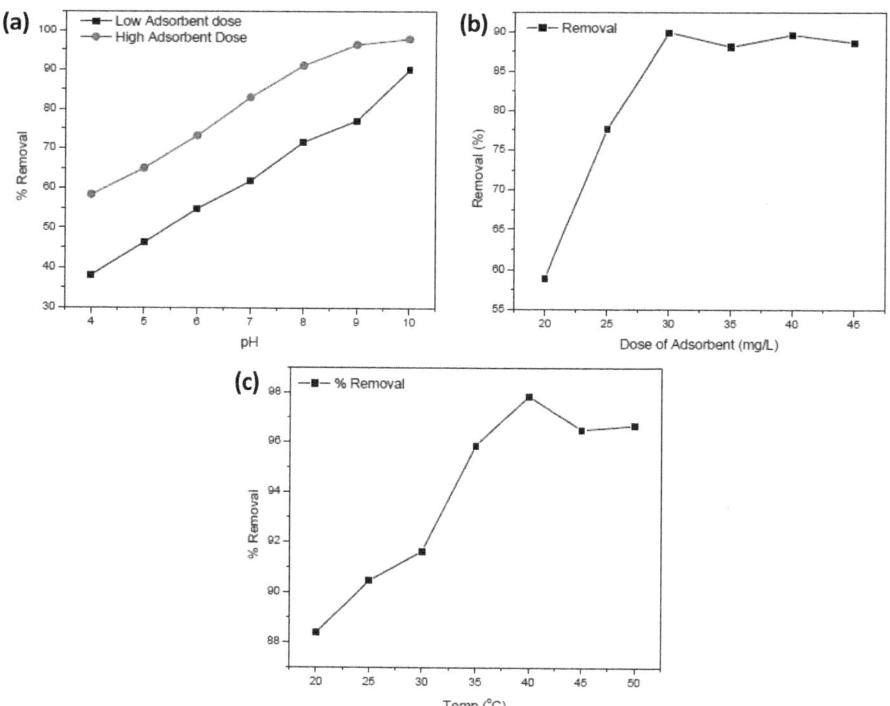

FIGURE 5.7 Adsorption of MB on GO-CNC in terms of percent exclusion with respect to (a) pH of the solution, (b) dose of adsorbent, and (c) temperature. (Reproduced with copyright permission from Zaman et al.,[25] Elsevier 2020.)

researchers selected pH range for their study was 4–10 as depicted in Figure 5.7a. They revealed that the adsorption % of the adsorbent towards MB increased with increasing pH from 2 up to 10, where 97.74% MB removal was observed revealing that the adsorption performance of GBAM favored the adsorbate removal more/better at alkaline pH. They postulated the better performance of the adsorbent in alkaline pH

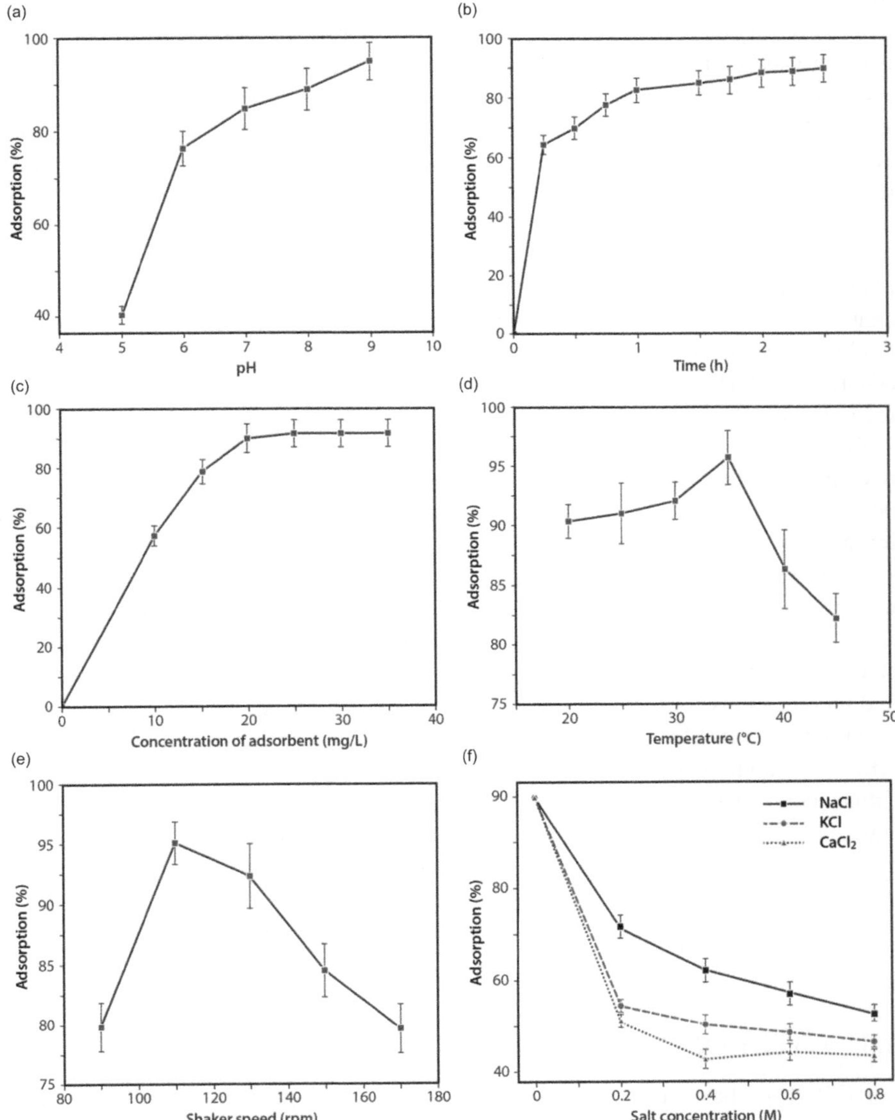

FIGURE 5.8 MB adsorption on (a) Effect of pH, (a) Effect of time, (c) Effect of concentration of adsorbent, (d) Effect of temperature, (e) Effect of shaker speed, and (f) Effect of salt concentration. (Reproduced with copyright permission from Bhattacharyya et al.,[26] Elsevier, 2021.)

to its negative surface functionalities, which where supported by the zeta potential results of the adsorbent(s). Their findings also established that the adsorption of MB by GO@CNs nanocomposite that, increasing the adsorbent dosage from 20–30 mg/L, resulted in a concurrent increase in the adsorptive exclusion of MB, while with additional use of the adsorbent beyond 30 mg/L, the adsorptive exclusion of MB decreased steadily as depicted in Figure 5.7b.[25] The phenomenon of increase in Q with increasing pH of adsorption liquor has also been supported by other researchers.[13]

Bhattacharyya et al.,[26] in their work also demonstrated that the adsorption percentage was found to be only about 40% in environments with extremely low pH levels, or acidic environments, as presented in Figure 5.8a. This is so that the carboxylic acid groups, which are primarily from GO and are present on the nanocomposite adsorbent, are not changed into their corresponding negative carboxylate ions when the pH is very low. As a result, surface adsorption mostly took place through the tiny pores on the AC surface. At low pH levels, this continued to be the only adsorption process, which has the effect of keeping the adsorption percentage very low. As the pH rises, meaning a more basic medium, where the carboxylic acid groups began to change into negative carboxylate ions, these led to the generation of a bed of negatively charged ions on the surface of the adsorbent. As a result, as the pH rises, the adsorption begins to occur in two different ways: first, through surface adsorption made possible by the pores on the surface of the AC, and second, through electrostatic attraction between the negatively charged carboxylate ion groups and the positive charge of the MB dye. At higher pH levels, both of these adsorption modalities become significant, and the adsorption percentage rises continuously, with pH leveling off at over 90% in a moderately basic media of pH 9.[26]

With respect to metallic ions adsorption, it is said that the majority of metal cations have higher adsorption capabilities on GBAMs when the pH of the original solution rises. Low Q is caused by electrostatic repulsion between the positively charged adsorbent surface and the free metal ions and competition between protons and metal cations for the few functional groups (such as the $-COO^-$ and $-O^-$ functionalities) of GBAMs in an acidic environment.[27] The surface functional groups of GBAMs become deprotonated as the pH rises, forming a negative surface charge. Consequently, surface precipitation, inner-sphere surface complexation, and electrostatic attraction all improve adsorption performance.[27] Following is a description of how divalent metal ion (M^{2+}) hydroxide develops in the solution:

$$M^{2+} \leftrightarrow -M\left(OH\right)^+ \leftrightarrow -M\left(OH\right)_2^0 \leftrightarrow -M\left(OH\right)_3^- \leftrightarrow \ldots$$

5.5 EFFECT OF CONTACT TIME VARIATION

For GBAMs, the contact time for the adsorption to reach the equilibrium state ranges from around 10 min to 24 hr.[26,37–46] With more time in contact, heavy metal adsorption grows and eventually reaches equilibrium. Due to the huge number of available empty sites and functional groups, as well as the unique layered structure of materials based on G, the adsorption occurs quickly at first.[45] It has been observed that the sluggish adsorptive process then takes place as a result of the decreased availability of the driving force and the remaining active sites.[25,45,47] As a result of the reduced availability of the still-active sites and the decline in the driving power, the sluggish adsorptive process then takes place.[40,41,48,49] According to Zhou et al. research, Cr(VI) adsorption on rGO@Fe$_3$O$_4$ started out quickly and took around 20 minutes to achieve equilibrium.[50] Pb(II) was quickly adsorbed onto low-temperature exfoliated G nanosheets within the first five minutes (approximately 85% exclusion), following which, it moved more slowly until reaching a plateau after six hours.[51] Numerous studies demonstrate that the adsorption rates of heavy metals by GBAMs are so high that they reach equilibrium faster than other carbon-based adsorbents like activated carbon and carbon nanotubes, making GBAMs desirable for heavy metal exclusion from wastewater.[52] According to Madadrang et al., the equilibrium periods for Pb(II) adsorption on EDTA@GO and GO were 20–30 min and 30–45 min, respectively, and the equilibrium state adsorption rates were substantially faster than those for activated carbon, carbon nanotubes, and other materials.[52]

It is worth nothing that different GBAMs, metal ions, adsorbent doses, and beginning adsorbate concentrations all affect how long the adsorption process takes to reach equilibrium. According to a prior study, porous CGGO monoliths and porous chitosan-gel beads took around 2 and 12 hours, respectively, for the Cu(II) adsorption to reach its maximum, whereas the inclusion of EDTA reduced the period to only 1 hr.[53] According to the different metal ions that need to be adsorbed, different adsorption times are seen. Sitko et al. found that for Zn(II), Cu(II), Cd(II), and Pb(II), respectively, the percentage adsorption on GO reached 90% after 4, 10, 14, and 15 min.[54] With increasing adsorbent supply, the contact time needed for ethylenediamine-RGO to completely remove Cr(VI) is reduced.[55] When the initial concentration of adsorbate is smaller, the adsorption equilibrium period is shorter. The Q of Pb(II) onto EDTA-GO reached 90% of its equilibrium state at initial Pb(II) concentrations of 10, 50, and 100 mg/L in 11, 15, and 18 min, respectively, according to Madadrang et al.[52] This is likely because the adsorption site adsorbed the available Pb(II) more quickly at a lower initial Pb(II) concentration. Li et al. found that reaching adsorption equilibrium only required 20 min for an initial fluoride concentration of 5 mg/g compared to 60 min for an initial fluoride concentration of 25 mg/g.[56]

It can be seen that the contact time is dependent on the pH, adsorbent dose, initial adsorbent concentration, temperature, and other factors affecting the entire adsorption process.

As depicted in Figure 5.8b, Bhattacharyya et al.,[26] observed in their work on the adsorptive removal of MB using GO@AC based adsorbent that 65% of the dye was adsorbed within 15 min. They ascribed the fast adsorption behavior of the adsorbent to both surface adsorption due to the pores and electrostatic attraction through the adsorption percentage, which went up to 80% after 60 min. However, after 90 min, most of the surface functionalities responsible for electrostatic attraction got exhausted, forcing secondary adsorption via the pores and the surface, and upon the exhaustion of these avenues, too, the adsorption becomes much slower and gets saturated with 90% removal of the dye.

5.6   ADSORBENT DOSAGE

Alongside pH as well as temperature, adsorbent dosage is a major element in causing changes in exclusion efficiency, and using an optimal G-based adsorbent dose is critical in the context of process development.[26,45,46,57]

According to Zhou et al.,[50] spanning the range of 0.25–2 g/L, the % of the adsorbed Cr^{6+} increased as the doses of GO as well as RGO-Fe_3O_4 increased. According to another group of scientists, the exclusion ratio of Cr^{6+} rose from 55.9 to 73.5 percent as the dosage of G (GN) was increased from 0.1 to 0.6 g, and from 78.3 percent to 98.2 percent when the dosage of G was modified with CTAB-GN and was boosted from 0.1 to 0.4 g.[58] Greater than 0.6 g GN as well as 0.4 g CTAB-GN, Cr^{6+} adsorptive % exclusion was indifferent.

However, other researchers found that as adsorbent dose increased, the Q, or the amount of metal ions adsorbed per unit mass of adsorbent, decreased.[58,59] When the adsorbent dosage is increased from 2 to 10 mg, the Q of Th(IV) on RGO fell from 57 to 17 mol/g, and above 10 mg, the Q was not significantly changed.[59] Wang et al.[27] found that at GO concentrations of 2, 6, and 10 mg, respectively, the amounts of Zn(II) adsorbed per unit mass of GO were 192, 81, and 77 mg/g. The strong interaction between Zn(II) and carboxyl groups is quickly responsible for the aggregation/folding of GO at higher adsorbent masses; however, GO-GO interactions may physically prevent partial active sites from adsorbing Zn(II), as well as result in electrostatic interferences that reduces adsorption. Additionally, the adsorption of metal ions is influenced by the GO content of adsorbent composites. According to research findings, the CGGO monolith's capacity to adsorb Pb(II) ions gradually increased as GO content increased because carboxyl groups from the margins of the GO sheets were incorporated into the porous substance.[53] Similar trends were evident in the exclusion of Pb(II) by polymer-based GO nanocomposite.[60] According to Li et al.,[61] the polyaniline/RGO nanocomposite's Hg(II) exclusion increased steadily up to a loading of 15 weight percent before beginning to decline. According to other researchers findings, the sorption of U(VI) by magnetic G/iron oxides was boosted by an increase in GO concentration; therefore, the composite with 80% GO was chosen, taking into account both the sorption capacity and the magnetic property of Fe_3O_4/GO.[62]

The effect of adsorbent concentration/dosage has been also studied elsewhere,[26] the authors found that the percentage of adsorption increases significantly with concentration, as shown in Figure 5.8c, since there are more functional sites available to drive both the electrostatic attraction and pore filling processes. There is no further increase in the adsorption % beyond 25 mg/L because there are not enough dye molecules to fill or bind to the adsorbing sites. As a result, the system enters equilibrium. Once it happens, even as the concentration of adsorbent is raised, the proportion of adsorption stays constant. As a result, it may be concluded that 25 mg/L is the ideal concentration for achieving maximal adsorption.

5.7   EFFECT OF TEMPERATURE

Temperature is a significant and crucial element in physical and chemical adsorption processes, and it is thought to be an effective component in influencing adsorbent capacity, as depicted in Figure 5.7c. Because temperature affects adsorption rate and reaction energy, it is frequently studied in GBAMs-adsorbate systems'.[63–65] As the temperature rises, the molecules travel faster, and the viscosity falls, increasing the effective collision between the molecules. Furthermore, increasing the temperature increases the pore volume and porosity of the adsorbent surface, enhancing the efficacy of contaminant adsorption. According to published studies, increasing the temperature in endothermic processes increases the uptake of antibiotics by GBAMs as well as the Q.[66–68] A positive ΔH^0 value, for example, represented the endothermic character of the MNZ adsorption process in a study concentrating on MNZ adsorption on CG at 20–40 °C.[69] However, extremely high temperatures do not promote MNZ adsorption on GBAMs. For instance, in research of temperature effects upon MNZ adsorption onto GO (298, 308, 318, and 328 K), an increment from 318 to 328 K was not beneficial, indicating that the adsorption was an exothermic process.[70]

With regards to toxic metals adsorption, it has been shown that the adsorption of Au(III),[34] Cd(II),[71] Co(II),[71] Cr(VI),[72] Cs(I),[73] Cu(II),[35] Eu(III),[74] Hg(II),[75] Ni(II),[76] Pb(II),[77] Pd(II),[34] Th(IV),[59] U(VI),[78] fluoride,[79] and phosphate[80] on GBAMs are enhanced with increasing temperature, indicating that these processes are endothermic. A group of researchers have revealed from their investigation that with the temperature starting from 298 to 318 K, the Q of Ni(II) on G/δ-MnO$_2$ composite increased from 46.55 to 60.01 mg/g.[76] In a different study, it was also demonstrated that as the temperature climbed from 273 to 323 K, the Q of G for fluoride increased from 12.29 to 19.59 mg/g at the fluoride equilibrium concentration of 15 mg/L. Kim and Chandra[75] The adsorption of Hg(II) by the polypyrrole-rGO composite increased as temperature increased from 5 to 70°C and remained constant at higher temperatures, as observed by Chandra and Kim. This observation can be explained by an increase in the rate at which ions diffuse or by an increase in the number of active surface sites that are available for adsorption. Another study found that iron-nanoparticle-decorated G was more effective in adsorbing Cr(VI) when it was warm, which might be attributed to the activation of the adsorbent surface and/or growth of pore size.[72] Seeing the cations are easily dissolvable at higher temperatures, the adsorption of Hg(II) and Pb(II) onto sulfur/rGO nanohybrid and few-layered GO is improved.[77] When the temperature was raised from 303 to 343 K, the ability of G to adsorb phosphate increased from 89.37 to 92.36 mg/g at a phosphate equilibrium concentration of 100 mg/L because, at higher temperatures, the interaction forces between the solvent as well as the solute are weaker than those between the adsorbent and the solute.[80]

Batch adsorption tests using rGO-PEG-ZnO composite have been reportedly conducted at 15–40 °C temperature range under ideal parameter circumstances, including solution pH 7, contact period of 45 min, initial phenolic pollutant concentration of 100 mg/L, and adsorbent dosage of 0.3 g/L[22]: The authors demonstrated the influence of temperature on the phenolic pollutant elimination is seen in Figure 5.9. The graph suggests that the adsorption process is endothermic since it indicates that as the temperature rises, the exclusion percentage also rises. It was noticed that the exclusion rate increased significantly up to 30 °C, before gradually declining after that. The ideal temperature for subsequent experiments is therefore determined to be 30°C. The acceleration of phenolic pollutants in the solution, which increases the

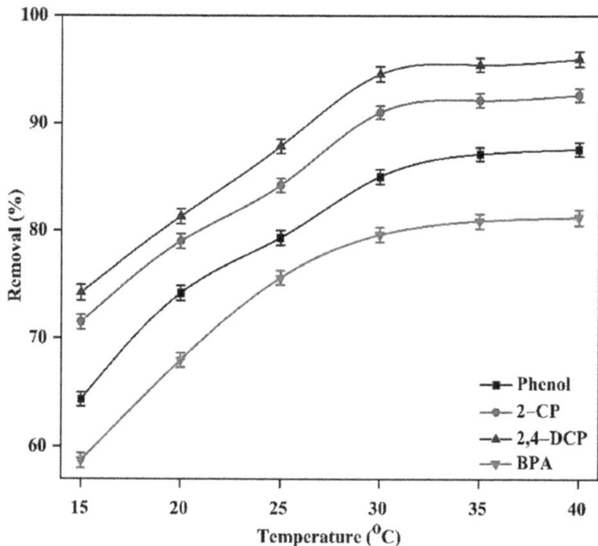

FIGURE 5.9   Temperature on phenolic pollutants removal efficiency. (Reproduced with permission from Rout et al.,[22] copyright 2022, Elsevier Science Ltd.)

pollutants' kinetic energy, is the cause of the increase in exclusion % with temperature. As a result, the rGO-PEG-ZnO composite's surface experienced an acceleration in the transfer of phenolate ions, improving adsorption efficiency. However, as the temperature rises beyond 30 °C, the attractive forces between the adsorbent surface and phenolic contaminants deteriorated, resulting in a lowering of the adsorption efficiency.[22]

Numerous studies indicate that the adsorption is more advantageous at lower temperatures, suggesting an exothermal mechanism. Wu et al. looked at how rising temperatures affected the Cr(VI) Q onto GN and CTAB-GN.[58] The Zn(II) Q of GO reduced from 246 to 225 mg/g when the temperature increased from 20 to 45°C, mostly because the attractive forces between the Zn(II) and GO surface weakened.[27] Additionally, when the temperature climbed from 10–15 to 30°C, the exclusion of As(III) and As(V) by magnetite and nanoscale zero-valent iron-modified RGO composites increased, while the exclusion of arsenic reduced as the temperature increased further.[75]

With regards to dyes, some researchers have revealed from their investigations that raising the temperature of the aqueous bath leads to a reduction in Q. For instance, the cumulative result of the adsorption MB was a reduction in the adsorption percentage with rising temperature. When the temperature is raised to 45 °C, the adsorption % dramatically decreases, dropping to ~80%. Thus, it may be inferred that 35 °C is the ideal temperature for this adsorbent, as presented at the later part of this article in Figure 5.8d.[26] In another report, the authors showed that with an increase in the temperature of the bath from 20–40 °C, the adsorption od the adsorbate(s) also increased, but upon the increase of the temperature beyond 40 °C, the Q.[25,81] Figure 5.7c depicts the observation by Zaman et al. on the effect of temperature, adsorbent dosage, as well as pH on the adsorptive exclusion of MB using GO@CNs novel adsorbent. It is believed that an ideal temperature rise can result in an increase in the porosity and, consequently, the total pore volume of the absorbent. As a result, the dye molecules would have been more likely to diffuse into the GO-CNC pores and stick to the outside surface. The fact that the percent MB elimination increased in a temperature-dependent manner also indicated that the process was endothermic. The physical links and interactions between the active adsorption sites on the molecules of the adsorbent and adsorbate might, however, become weaker at extremely high temperatures (in this example, 50 °C and above).[57]

5.8 ADSORPTION/SORPTION KINETICS

The adsorption/sorption kinetic models include pseudo-first-order, Elovich equation, pseudo-second-order, as well as intraparticle diffusion models, which could be used to explain the mechanism of adsorption for contaminants along with their rate-controlling stages, such as chemical reaction, mass transfer, as well as diffusion-controlled processes.[25] For instance, a paper demonstrated that GO had extraordinarily high sorption capacities to both CIP and SMX, as shown in Figure 5.10.[82] The sorption isotherms were very well described by the Freundlich and Langmuir models, as shown in Figure 5.10. CIP on GO had a substantially greater Langmuir maximum sorption capacity than SMX (240 mg/g), at around 379 mg/g. The CIP isotherm was reported by these authors to fit the Freundlich model somewhat better ($R^2 = 0.996$) than the Langmuir model ($R^2 = 0.968$), indicating that the heterogeneous chemisorption may be in control of the CIP's sorption on GO. This revealed in their work that CIP might have interacted with both the functionalized GO edges as well as the aromatic structure inside GO basal planes because of its two charged functional groups as well as electro deficient Pi(π) structure.[82] They also postulated

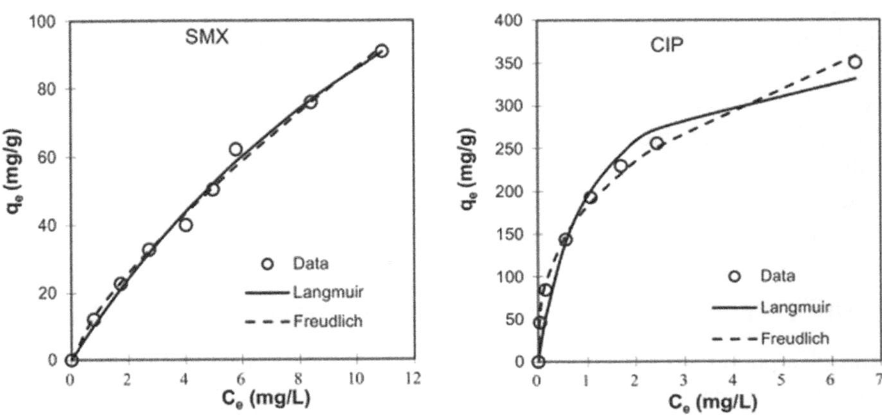

FIGURE 5.10 Sorption isotherms of sulfamethoxazole and ciprofloxacin on GOs. (Reproduced with copyright permission from Chen et al.,[82] Copyright 2015, Elsevier Science Ltd.)

that under the experimental conditions, electrostatic attraction between carboxyl or phenol groups on GO and cationic CIP species, particularly on the edges, is possible. A little pH decline that was also accompanied by CIP sorption further suggested that the GO functional groups were deprotonated during the CIP sorption process. Surface functional groups have been said to be one of the major elements regulating CIP sorption.[82] Since there was no electrostatic attraction between the adsorbent and the adsorbate, Pi-Pi (π-π) interactions were primarily responsible for SMX's sorption onto GO.[82] Electron-withdrawing sulfuric groups can attract electrons, which makes the aromatic ring of SMX a potent Pi (π)-electron acceptor. Additionally, due to its high electron negativity, the oxygen atom inside may also reduce the Pi (π)-electron density of its aromatic heterocyclic group. Because of this, extra Pi-Pi (π-π) EDA interaction with the G surfaces of carbonaceous adsorbents, such as GO, was said to be induced by the substituted benzene ring as well as an aromatic heterocyclic group of SMX.[82]

5.9   EFFECT OF AGITATION

Even though it has only been investigated by a small number of scientists, the impact of agitation is crucial for the effectiveness and rate of adsorption of GBA materials because it speeds up the transfer of adsorbate from the adsorbent to the former and vice versa.

Bhattacharyya et al. examined the effect of shaker (agitation rate) speed on the adsorption process at a temperature of 45 °C in addition to other ideal adsorption conditions as shown in Figure 5.8e.[26,45] They discovered that at a relatively low speed of 90 rpm, the percentage of adsorption remained low because the dye (MB) molecules could not get close enough to the adsorbent molecules. Increased shaker speed up to 110 rpm led to an improvement in the percentage of adsorption because the adsorbate molecules were able to come close to the GBAM's key structural components. They provided an explanation for the phenomenon, stating that, in part, as agitation speed was increased, the rate of dye diffusion from the liquid bulk towards the boundary liquid layer near the biocomposite GBA increased due to the enhancement of turbulence, leading to improved adsorption.[47] In this instance, both surface adsorption and electrostatic attraction help the adsorption process. The amount of adsorption is unaffected above 110 rpm, where the contaminants and adsorbent are sufficiently close to one another. This is mostly because an increase in rotational speed has no effect on the strength of the electrostatic bond.[45] These authors' report has been supported by their other report(s).[47]

5.10   EFFECT OF SALT ADDITION OR IN SITU COMPETING IONS

The adsorption performance efficiency is somewhat influenced by various ions that are representative of the actual effluent sample. These coexisting ions may have an impact on the hydrated particulate's electrically diffused double-layer structure and interface potential, which may then affect the binding of adsorbing species.[27] In general, ion exchange or outer-sphere surface complexation dominates the adsorption that is impacted by ionic strength, whereas inner-sphere surface complexation controls the adsorption that is affected by pH values.[62] According to the triple-layer model (TLM), o-plane sorption may take place unless the background electrolyte significantly alters the sorption process, in which case – plane sorption might be hypothesized to occur. Due to competition for the active sites on the surface of the adsorbents, various ions are said to be present in the solution, which leads to decreased Q.[50]

It has been reported in a study by Bhattacharyya et al.,[26] that Na^+ and K^+ ions, which are easily able to block the negatively charged sites on the adsorbent and then impede the interaction of the dye (MB) molecules with these sites, were created by the addition of NaCl and KCl as per Figure 5.8f. As a result, the only remaining route for dye adsorption was said to be through surface pores, and adsorption dramatically declined with continued salt addition (NaCl and KCl) to the solution. When $CaCl_2$ is added, the Ca^{2+} ions that are generated in the solution quickly combine with the OH ions to create $Ca(OH)_2$. This reduces the amount of OH ions in the solution, which in turn reduces the possibility of negative charge building up on the surface of the adsorbent. The negative ions that develop on the surface of the adsorbent are prevented from being drawn to the positive cations of the dye by the remaining Ca^{2+} that is generated in the solution. As a result, compared to the effects of NaCl or KCl, the addition of $CaCl_2$ (which results in the presence of Ca^{2+} in the solution) greatly reduces the adsorption percentage.[26]

Also, Wang et al. found that the adsorption of Zn(II) on GO decreased as the concentrations of KCl, NaCl, as well as $NaNO_3$ increased from 0.001 to 0.01 mol/L.[83] The adsorption of Zn(II) in the presence of NaCl was worse than that of KCl because Na+ had a higher affinity for the GO surface and a higher tendency for counter-ion exchange with the functional groups (II). At pH 9, the foreign cations, which were $Mg^{2+} > Na > K^+$, had a significant impact on the sorption of Co(II) on the magnetite/GO composite. By competing with HCrO4 for the active sites on the IL-EGO adsorbent, chloride, sulfate, and phosphate caused a significant reduction in the Cr(VI) adsorption, and sulfite and nitrite ions also lowered Cr(VI), lowering Cr(VI) adsorption.[84] Cd^{2+}, Co^{2+}, Zn^{2+}, Ni^{2+}, as well as Pb^{2+} did not affect the adsorption effectiveness of Cr(VI) in the case of cationic species; however, Fe^{2+} and Mn^{2+} interfered by decreasing the hexavalent chromium to trivalent state.[84] Coexisting anions, including sulfate, phosphate, carbonate, chloride, and nitrate, could compete with fluoride ions for adsorption sites on the adsorbent surface when it comes to the exclusion of fluoride.[79] A group of researchers revealed in their work that coexisting Cl, NO_3, and SO_4^2 ions had no effect on the fluoride Q of basic aluminum sulfate@G hydrogel composites.[79] However, coexisting CO_3^2 and PO_4^3 ions had a detrimental impact. Yet in another work, it has been shown that the partial reduction of Cr(VI) to Cr(III) and the presence of epoxide groups in iron nanoparticles decorated on G (G-nZVI) caused the Cr(VI) adsorption to increase with increasing NaCl concentration up to 75 mM, while higher NaCl concentrations (>75 mM) caused the Cr(VI) adsorption to decrease due to the predominance of the Cl^- ions.[72] It has also been revealed in other studies that the influence of ionic strength on the sorption of Cd(II) and Co(II) by GO nanosheets was more pronounced at low pH values than it was at high pH values. This resulted in the conclusion that the sorption of Cd(II) and Co(II) was primarily attributed to outer-sphere surface complexation or ion exchange at low pH and inner-sphere surface complexation interfering anions, including phosphate, sulfate, nitrate, fluoride, and chloride, have been studied for their effects.[71]

Numerous studies revealed that the effect of coexisting ions on the Q is quite little, indicating that such a system might be used in the field.[26,45,46,62,71] The U(VI) adsorption on GO nanosheets and Fe_3O_4/GO was discovered to be pH-dependent and ionic strength-independent, which was attributed to the inner-sphere surface complexation rather than the outer-sphere surface complexation or ion exchange.[62] The findings of control experiments carried out by Sreeprasad et al.[85] comparing the Hg(II) exclusion capacity of RGO composites in real water and distilled water revealed that the coexisting ions present in the actual water had no impact on the exclusion. The formulation the so-called SmtA-GO composites displayed extremely high selectivity for the adsorption of cadmium and excellent tolerance to coexisting metal and anionic species. For example, the tolerant concentrations of K^+, Na^+, Ca^{2+}, Mg^{2+}, Co^{2+}, Pb^{2+}, Al^{3+}, Cu^{2+}, Ni^{2+}, Zn^{2+}, Fe^{3+}, NO_3^-, Cl^-, SO_4^{2-}, HCO_3^- and $H_2PO_4^-$ were 1–800,000 fold.[86]

Because various metal ions have varying affinities for the same adsorbent, competitive adsorption is being studied by researchers globally. The competitive adsorption of binary aqueous metal cation species in mixtures has been investigated by Sitko et al.,[54] and the single and competitive adsorption experiments revealed the affinity in the following order: Pd(II) > Cu(II) >> Cd(II) > Zn(II), which was postulated to depend on the metal electronegativity and the initial stability constant of the associated metal hydroxide. Pb(II), Ni(II), and Sr(II) had the highest maximum adsorption capacities on GOs in all single, binary, and ternary metal ion adsorption systems, according to experimental and theoretical results obtained,[87] and the reasons for this were discussed using both competitive adsorption experiments and density functional theory (DFT) studies.

As far as we are aware, there is a dearth of understanding on the competitive adsorption of metal ions and several other wastewater contaminants by GBAMs. For many adsorption systems, the impact of coexisting ions on the exclusion of metal ions is complex, and the underlying mechanism is not entirely understood. Therefore, another investigation is required to conduct a competitive study for the adsorption systems containing two or more metal ions and to analyze the adsorption mechanism using data on the ionic radius, complexation, precipitation, reduction-oxidation reactions, and other related phenomena.

5.11   ORGANIC ACID LIGANDS

We are aware that a ligand is an ion or molecule (functional group) in coordination chemistry that binds with a primary metal atom to produce a coordination complex.

As organic acid ligands (bridging acids) present in wastewater have the ability to change the surface electrical characteristics of the adsorbent as well as the speciation of metal ions in solution, they can have an impact on how well an adsorbent performs. The influences of six organic acid ligands (formate, acetate, benzoate, oxalate, tartrate, and edetate) were examined by Liu et al.[88] in single- and multiligand systems for the adsorption of Cu^{2+} using Fe_3O_4/GO-supported 1,2-diaminocyclohexanetetraa cetic acid system "Fe_3O_4/GO/DCTA". The outcomes demonstrated that the oxalate, tartrate, and edetate ligands greatly modified the adsorption processes in the single-ligand systems, but the inclusion of formate, acetate, and benzoate only marginally altered the Cu^{2+} adsorption property. With regards to multiligand structures, the order of the core impact was presented as edetate greater than oxalate greater than benzoate greater than formate greater than tartrate greater than acetate, as well as the order of the two-factor interaction influence was instituted to be "formate@tartrate" greater than "acetate@edetate" greater than "formate@edetate" greater than "acetate@oxalate" greater than "formate@benzoate" greater than "formate@oxalate" greater than "formate@ acetate".

Organic ligands have been proven to enhance the adsorption of wastewater contaminants that are mainly anionic in nature, though the report was on carbon-coated ligands for the adsorption of both MO as well as Cr^{6+} ions.[89] This approach, if utilized for G-based adsorbent fabrication, will work just fine.

The effect of organic ligands on the adsorption performance of G-based adsorbents has been insufficiently studied, and we encourage researchers to consider this as a top priority when dealing with fabricated G-based adsorbents designed to adsorb biological-related adsorbates or contaminants.

5.12 CONCLUSION

This chapter has presented an inclusive, though not exhaustive, discussion of the factors that influence adsorption as a process. Knowing that research investigations can never be exhaustive, researchers can do their best to develop effective G-based adsorbents aimed at the exclusion of wastewater contaminants, especially in real life scenarios where there exists a mixture of diverse contaminants. There is a need for researchers to develop new but also effective adsorbents that will function excellently when applied for the treatment of wastewater. Investigational studies involving deep knowledge of the interaction between the adsorbent(s) and adsorbate(s) using state of the art characterization is required, since most mechanisms presented by researchers to date is mainly based on hypothesis. We suggest that researchers should seriously consider looking into factors like the morphology and structural architecture of the graphene-derived adsorbents as well as the adsorbates in question, in addition to the conventionally studied factors affecting the adsorption of contaminants by these adsorbents as mentioned above.

REFERENCES

1. Kommu, A.; Namsani, S.; Singh, J. K., Removal of heavy metal ions using functionalized graphene membranes: A molecular dynamics study. *RSC Advances* 2016, *6* (68), 63190–63199.
2. Anitha, K.; Namsani, S.; Singh, J. K., Removal of heavy metal ions using a functionalized single-walled carbon nanotube: A molecular dynamics study. *The Journal of Physical Chemistry A* 2015, *119* (30), 8349–8358.
3. Tomalia, D. A.; Baker, H.; Dewald, J.; Hall, M.; Kallos, G.; Martin, S.; Roeck, J.; Ryder, J.; Smith, P., A new class of polymers: Starburst-dendritic macromolecules. *Polymer Journal* 1985, *17* (1), 117–132.

4. Tomalia, D. A.; Baker, H.; Dewald, J.; Hall, M.; Kallos, G.; Martin, S.; Roeck, J.; Ryder, J.; Smith, P., Dendritic macromolecules: Synthesis of starburst dendrimers. *Macromolecules* 1986, *19* (9), 2466–2468.

5. Yuan, Y.; Zhang, G.; Li, Y.; Zhang, G.; Zhang, F.; Fan, X., Poly (amidoamine) modified graphene oxide as an efficient adsorbent for heavy metal ions. *Polymer Chemistry* 2013, *4* (6), 2164–2167.

6. Zhang, F.; Wang, B.; He, S.; Man, R., Preparation of graphene-oxide/polyamidoamine dendrimers and their adsorption properties toward some heavy metal ions. *Journal of Chemical & Engineering Data* 2014, *59* (5), 1719–1726.

7. Kommu, A.; Velachi, V.; Cordeiro, M. N. I. D.; Singh, J. K., Removal of Pb (II) ion using PAMAM dendrimer grafted graphene and graphene oxide surfaces: A molecular dynamics study. *The Journal of Physical Chemistry A* 2017, *121* (48), 9320–9329.

8. Sivapragasam, N., *Adsorption Kinetics and Dynamics of Small Molecules on Graphene and Graphene Oxide*; North Dakota State University, 2018.

9. Burghaus, U., Effect of carbon nanotubes' crystal structure on adsorption kinetics of small molecules: An experimental study utilizing ultra-high vacuum thermal analysis techniques. *Journal of Thermal Analysis and Calorimetry* 2011, *106* (1), 123–128.

10. Komarneni, M.; Sand, A.; Goering, J.; Burghaus, U.; Lu, M.; Veca, L. M.; Sun, Y.-P., Possible effect of carbon nanotube diameter on gas–surface interactions–The case of benzene, water, and n-pentane adsorption on SWCNTs at ultra-high vacuum conditions. *Chemical Physics Letters* 2009, *476* (4–6), 227–231.

11. Li, M.-F.; Liu, Y.-G.; Liu, S.-B.; Zeng, G.-M.; Hu, X.-J.; Tan, X.-F.; Jiang, L.-H.; Liu, N.; Wen, J.; Liu, X.-H., Performance of magnetic graphene oxide/diethylenetriamine-pentaacetic acid nanocomposite for the tetracycline and ciprofloxacin adsorption in single and binary systems. *Journal of Colloid and Interface Science* 2018, *521*, 150–159.

12. Lin, H.; Duan, Y.; Zhao, B.; Feng, Q.; Li, M.; Wei, J.; Zhu, Y.; Li, M., Efficient Hg(II) removal to ppb level from water in wider pH based on poly-cyanoguanidine/graphene oxide: Preparation, behaviors, and mechanisms. *Colloids and Surfaces A: Physicochemical and Engineering Aspects* 2022, *641*, 128467.

13. Temane, L. T.; Orasugh, J. T.; Ray, S. S., Adsorptive Removal of Pollutants Using Graphene-based Materials for Water Purification. In *Two-Dimensional Materials for Environmental Applications*, Kumar, N.; Gusain, R.; Sinha Ray, S., Eds.; Springer International Publishing, 2023; pp. 179–244.

14. Zhang, X.; Shen, J.; Zhuo, N.; Tian, Z.; Xu, P.; Yang, Z.; Yang, W., Interactions between Antibiotics and Graphene-Based Materials in Water: A Comparative Experimental and Theoretical Investigation. *ACS Applied Materials and Interfaces* 2016, *8* (36), 24273–24280.

15. Segovia-Sandoval, S. J.; Pastrana-Martínez, L. M.; Ocampo-Pérez, R.; Morales-Torres, S.; Berber-Mendoza, M. S.; Carrasco-Marín, F., Synthesis and characterization of carbon xerogel/graphene hybrids as adsorbents for metronidazole pharmaceutical removal: Effect of operating parameters. *Separation and Purification Technology* 2020, *237*, 116341.

16. Carrales-Alvarado, D.; Ocampo-Pérez, R.; Leyva-Ramos, R.; Rivera-Utrilla, J., Removal of the antibiotic metronidazole by adsorption on various carbon materials from aqueous phase. *Journal of Colloid and Interface Science* 2014, *436*, 276–285.

17. Gao, Y.; Li, Y.; Zhang, L.; Huang, H.; Hu, J.; Shah, S. M.; Su, X., Adsorption and removal of tetracycline antibiotics from aqueous solution by graphene oxide. *Journal of Colloid and Interface Science* 2012, *368* (1), 540–546.

18. Al-Ghouti, M. A.; Sayma, J.; Munira, N.; Mohamed, D.; Da'na, D. A.; Qiblawey, H.; Alkhouzaam, A., Effective removal of phenol from wastewater using a hybrid process of graphene oxide adsorption and UV-irradiation. *Environmental Technology & Innovation* 2022, *27*, 102525.

19. Hu, R.; Dai, S.; Shao, D.; Alsaedi, A.; Ahmad, B.; Wang, X., Efficient removal of phenol and aniline from aqueous solutions using graphene oxide/polypyrrole composites. *Journal of Molecular Liquids* 2015, *203*, 80–89.

20. Zhang, B.; Zhao, R.; Sun, D.; Li, Y.; Wu, T., Sustainable fabrication of graphene oxide/manganese oxide composites for removing phenolic compounds by adsorption-oxidation process. *Journal of Cleaner Production* 2019, *215*, 165–174.

21. Zhao, R.; Li, Y.; Ji, J.; Wang, Q.; Li, G.; Wu, T.; Zhang, B., Efficient removal of phenol and p-nitrophenol using nitrogen-doped reduced graphene oxide. *Colloids and Surfaces A: Physicochemical and Engineering Aspects* 2021, *611*, 125866.

22. Rout, D. R.; Jena, H. M., Polyethylene glycol functionalized reduced graphene oxide coupled with zinc oxide composite adsorbent for removal of phenolic wastewater. *Environmental Research* 2022, *214*, 114044.

23. Liu, Y.; Fu, J.; He, J.; Wang, B.; He, Y.; Luo, L.; Wang, L.; Chen, C.; Shen, F.; Zhang, Y., Synthesis of a superhydrophilic coral-like reduced graphene oxide aerogel and its application to pollutant capture in wastewater treatment. *Chemical Engineering Science* 2022, *260*, 117860.

24. Radmehr, S.; Hosseini Sabzevari, M.; Ghaedi, M.; Ahmadi Azqhandi, M. H.; Marahel, F., Adsorption of nalidixic acid antibiotic using a renewable adsorbent based on Graphene oxide from simulated wastewater. *Journal of Environmental Chemical Engineering* 2021, *9* (5), 105975.

25. Zaman, A.; Orasugh, J. T.; Banerjee, P.; Dutta, S.; Ali, M. S.; Das, D.; Bhattacharya, A.; Chattopadhyay, D., Facile one-pot in-situ synthesis of novel graphene oxide-cellulose nanocomposite for enhanced azo dye adsorption at optimized conditions. *Carbohydrate Polymers* 2020, *246*, 116661.

26. Bhattacharyya, A.; Ghorai, S.; Rana, D.; Roy, I.; Sarkar, G.; Saha, N. R.; Orasugh, J. T.; De, S.; Sadhukhan, S.; Chattopadhyay, D., Design of an efficient and selective adsorbent of cationic dye through activated carbon – Graphene oxide nanocomposite: Study on mechanism and synergy. *Materials Chemistry and Physics* 2021, *260*, 124090.

27. Wang, H.; Yuan, X.; Wu, Y.; Huang, H.; Zeng, G.; Liu, Y.; Wang, X.; Lin, N.; Qi, Y., Adsorption characteristics and behaviors of graphene oxide for Zn (II) removal from aqueous solution. *Applied Surface Science* 2013, *279*, 432–440.

28. Zhang, X.; Yu, H.; Yang, H.; Wan, Y.; Hu, H.; Zhai, Z.; Qin, J., Graphene oxide caged in cellulose microbeads for removal of malachite green dye from aqueous solution. *Journal of Colloid and Interface Science* 2015, *437*, 277–282.

29. Li, L.; Fan, L.; Duan, H.; Wang, X.; Luo, C., Magnetically separable functionalized graphene oxide decorated with magnetic cyclodextrin as an excellent adsorbent for dye removal. *RSC Advances* 2014, *4* (70), 37114–37121.

30. Rotte, N. K.; Yerramala, S.; Boniface, J.; Srikanth, V. V. S. S., Equilibrium and kinetics of Safranin O dye adsorption on MgO decked multi-layered graphene. *Chemical Engineering Journal* 2014, *258*, 412–419.

31. Arabkhani, P.; Asfaram, A., The potential application of bio-based ceramic/organic xerogel derived from the plant sources: A new green adsorbent for removal of antibiotics from pharmaceutical wastewater. *Journal of Hazardous Materials* 2022, *429*, 128289.

32. Nascimento, B. F.; Silva, L. F. O.; Araújo, C. M. B.; Silva Santos, R. K.; Gomes, B. F. M. L.; Silva Santos, P. R.; Cavalcanti, J. V. F. L.; Dotto, G. L.; Schnorr, C. E.; Motta Sobrinho, M. A., Synthesis and application of ferromagnetic graphene oxide nanocomposite as an effective adsorbent for Clonazepam: Batch experiments, modeling, regeneration, and phytotoxicity. *Journal of Environmental Chemical Engineering* 2022, *10* (5), 108331.

33. Zhao, G.; Wen, T.; Chen, C.; Wang, X., Synthesis of graphene-based nanomaterials and their application in energy-related and environmental-related areas. *RSC Advances* 2012, *2* (25), 9286–9303.

34. Liu, L.; Li, C.; Bao, C.; Jia, Q.; Xiao, P.; Liu, X.; Zhang, Q., Preparation and characterization of chitosan/graphene oxide composites for the adsorption of Au (III) and Pd (II). *Talanta* 2012, *93*, 350–357.

35. Hu, X.-J.; Liu, Y.-G.; Wang, H.; Chen, A.-W.; Zeng, G.-M.; Liu, S.-M.; Guo, Y.-M.; Hu, X.; Li, T.-T.; Wang, Y.-Q.; Zhou, L.; Liu, S.-H., Removal of Cu(II) ions from aqueous solution using sulfonated magnetic graphene oxide composite. *Separation and Purification Technology* 2013, *108*, 189–195.

36. Romanchuk, A. Y.; Slesarev, A. S.; Kalmykov, S. N.; Kosynkin, D. V.; Tour, J. M., Graphene oxide for effective radionuclide removal. *Physical Chemistry Chemical Physics* 2013, *15* (7), 2321–2327.

37. Chen, X.; Zhou, S.; Zhang, L.; You, T.; Xu, F., Adsorption of Heavy Metals by Graphene Oxide/Cellulose Hydrogel Prepared from NaOH/Urea Aqueous Solution. *Materials (Basel)* 2016, *9* (7).

38. Chiori, A., Urban water planning in Lagos, Nigeria: An analysis of current infrastructure developments and future water management solutions. 2018.

39. Dimiev, A. M.; Alemany, L. B.; Tour, J. M., Graphene oxide. Origin of acidity, its instability in water, and a new dynamic structural model. *ACS Nano* 2012, *7*, 576–588.

40. Du, Q.; Sun, J.; Li, Y.; Yang, X.; Wang, X.; Wang, Z.; Xia, L., Highly enhanced adsorption of congo red onto graphene oxide/chitosan fibers by wet-chemical etching off silica nanoparticles. *Chemical Engineering Journal* 2014, *245*, 99–106.

41. Fan, L.; Luo, C.; Sun, M.; Li, X.; Qiu, H., Highly selective adsorption of lead ions by water-dispersible magnetic chitosan/graphene oxide composites. *Colloids and Surfaces B: Biointerfaces* 2013, *103*, 523–529.

42. Gu, H.; Gao, C.; Zhou, X.; Du, A.; Naik, N.; Guo, Z., Nanocellulose nanocomposite aerogel towards efficient oil and organic solvent adsorption. *Advanced Composites and Hybrid Materials* 2021, 1–10.

43. Gupta, S. S.; Sreeprasad, T. S.; Maliyekkal, S. M.; Das, S. K.; Pradeep, T., Graphene from Sugar and its Application in Water Purification. *ACS Applied Materials and Interfaces* 2012, *4* (8), 4156–4163.

44. Gupta, V. K.; Eren, T.; Atar, N.; Yola, M. L.; Parlak, C.; Karimi-Maleh, H., CoFe$_2$O$_4$@ TiO$_2$ decorated reduced graphene oxide nanocomposite for photocatalytic degradation of chlorpyrifos. *Journal of Molecular Liquids* 2015, *208*, 122–129.

45. Bhattacharyya, A.; Banerjee, B.; Ghorai, S.; Rana, D.; Roy, I.; Sarkar, G.; Saha, N. R.; De, S.; Ghosh, T. K.; Sadhukhan, S.; Chattopadhyay, D., Development of an auto-phase separable and reusable graphene oxide-potato starch based cross-linked bio-composite adsorbent for removal of methylene blue dye. *International Journal of Biological Macromolecules* 2018, *116*, 1037–1048.

46. Bhattacharyya, A.; Mondal, D.; Roy, I.; Sarkar, G.; Saha, N. R.; Rana, D.; Ghosh, T. K.; Mandal, D.; Chakraborty, M.; Chattopadhyay, D., Studies of the kinetics and mechanism of the removal process of proflavine dye through adsorption by graphene oxide. *Journal of Molecular Liquids* 2017, *230*, 696–704.

47. Bhattacharyya, A.; Ghorai, S.; Rana, D.; Roy, I.; Sarkar, G.; Saha, N. R.; Orasugh, J. T.; De, S.; Sadhukhan, S.; Chattopadhyay, D., Design of an efficient and selective adsorbent of cationic dye through activated carbon – Graphene oxide nanocomposite: Study on mechanism and synergy. *Materials Chemistry and Physics* 2021, *260*.

48. Dutta, K.; Das, B.; Orasugh, J. T.; Mondal, D.; Adhikari, A.; Rana, D.; Banerjee, R.; Mishra, R.; Kar, S.; Chattopadhyay, D., Bio-derived cellulose nanofibril reinforced poly(N-isopropylacrylamide)-g-guar gum nanocomposite: An avant-garde biomaterial as a transdermal membrane. *Polymer* 2018, *135*, 85–102.

49. Farghali, A. A.; Bahgat, M.; El Rouby, W. M. A.; Khedr, M. H., Preparation, decoration and characterization of graphene sheets for methyl green adsorption. *Journal of Alloys and Compounds* 2013, *555*, 193–200.

50. Zhou, L.; Deng, H.; Wan, J.; Shi, J.; Su, T., A solvothermal method to produce RGO-Fe3O4 hybrid composite for fast chromium removal from aqueous solution. *Applied Surface Science* 2013, *283*, 1024–1031.

51. Huang, Z.-H.; Zheng, X.; Lv, W.; Wang, M.; Yang, Q.-H.; Kang, F., Adsorption of lead (II) ions from aqueous solution on low-temperature exfoliated graphene nanosheets. *Langmuir* 2011, *27* (12), 7558–7562.

52. Madadrang, C. J.; Kim, H. Y.; Gao, G.; Wang, N.; Zhu, J.; Feng, H.; Gorring, M.; Kasner, M. L.; Hou, S., Adsorption behavior of EDTA-graphene oxide for Pb (II) removal. *ACS Applied Materials & Interfaces* 2012, *4* (3), 1186–1193.

53. Zhang, N.; Qiu, H.; Si, Y.; Wang, W.; Gao, J., Fabrication of highly porous biodegradable monoliths strengthened by graphene oxide and their adsorption of metal ions. *Carbon* 2011, *49* (3), 827–837.

54. Sitko, R.; Turek, E.; Zawisza, B.; Malicka, E.; Talik, E.; Heimann, J.; Gagor, A.; Feist, B.; Wrzalik, R., Adsorption of divalent metal ions from aqueous solutions using graphene oxide. *Dalton Transactions* 2013, *42* (16), 5682–5689.

55. Ma, H.-L.; Zhang, Y.; Hu, Q.-H.; Yan, D.; Yu, Z.-Z.; Zhai, M., Chemical reduction and removal of Cr (VI) from acidic aqueous solution by ethylenediamine-reduced graphene oxide. *Journal of Materials Chemistry* 2012, *22* (13), 5914–5916.

56. Li, Y.; Zhang, P.; Du, Q.; Peng, X.; Liu, T.; Wang, Z.; Xia, Y.; Zhang, W.; Wang, K.; Zhu, H., Adsorption of fluoride from aqueous solution by graphene. *Journal of Colloid and Interface Science* 2011, *363* (1), 348–354.

57. Banerjee, P.; Sau, S.; Das, P.; Mukhopadhayay, A., Optimization and modelling of synthetic azo dye wastewater treatment using graphene oxide nanoplatelets: Characterization toxicity evaluation and optimization using artificial neural network. *Ecotoxicology and Environmental Safety* 2015, *119*, 47–57.

58. Wu, Y.; Luo, H.; Wang, H.; Wang, C.; Zhang, J.; Zhang, Z., Adsorption of hexavalent chromium from aqueous solutions by graphene modified with cetyltrimethylammonium bromide. *Journal of Colloid and Interface Science* 2013, *394*, 183–191.

59. Pan, N.; Deng, J.; Guan, D.; Jin, Y.; Xia, C., Adsorption characteristics of Th (IV) ions on reduced graphene oxide from aqueous solutions. *Applied Surface Science* 2013, *287*, 478–483.

60. Musico, Y. L. F.; Santos, C. M.; Dalida, M. L. P.; Rodrigues, D. F., Improved removal of lead (II) from water using a polymer-based graphene oxide nanocomposite. *Journal of Materials Chemistry A* 2013, *1* (11), 3789–3796.

61. Li, R.; Liu, L.; Yang, F., Preparation of polyaniline/reduced graphene oxide nanocomposite and its application in adsorption of aqueous Hg (II). *Chemical Engineering Journal* 2013, *229*, 460–468.

62. Zong, P.; Wang, S.; Zhao, Y.; Wang, H.; Pan, H.; He, C., Synthesis and application of magnetic graphene/iron oxides composite for the removal of U (VI) from aqueous solutions. *Chemical Engineering Journal* 2013, *220*, 45–52.

63. Ramesha, G.; Kumara, A. V.; Muralidhara, H.; Sampath, S., Graphene and graphene oxide as effective adsorbents toward anionic and cationic dyes. *Journal of Colloid and Interface Science* 2011, *361* (1), 270–277.

64. Sharma, P.; Das, M. R., Removal of a cationic dye from aqueous solution using graphene oxide nanosheets: Investigation of adsorption parameters. *Journal of Chemical & Engineering Data* 2013, *58* (1), 151–158.

65. Vijayaraghavan, K.; Padmesh, T.; Palanivelu, K.; Velan, M., Biosorption of nickel (II) ions onto Sargassum wightii: Application of two-parameter and three-parameter isotherm models. *Journal of Hazardous Materials* 2006, *133* (1–3), 304–308.

66. Song, W.; Yang, T.; Wang, X.; Sun, Y.; Ai, Y.; Sheng, G.; Hayat, T.; Wang, X., Experimental and theoretical evidence for competitive interactions of tetracycline and sulfamethazine with reduced graphene oxides. *Environmental Science: Nano* 2016, *3* (6), 1318–1326.

67. Yuan, X.; Wu, Z.; Zhong, H.; Wang, H.; Chen, X.; Leng, L.; Jiang, L.; Xiao, Z.; Zeng, G., Fast removal of tetracycline from wastewater by reduced graphene oxide prepared via microwave-assisted ethylenediamine–N, N'–disuccinic acid induction method. *Environmental Science and Pollution Research* 2016, *23* (18), 18657–18671.

68. Berkani, M.; Vasseghian, Y.; Dragoi, E.-N.; Khaneghah, A. M., The Fenton-like reaction for Arsenic removal from groundwater: Health risk assessment. *Environmental Research* 2021, *202*, 111698.

69. Manjunath, S.; Kumar, S. M.; Ngo, H. H.; Guo, W., Metronidazole removal in powder-activated carbon and concrete-containing graphene adsorption systems: Estimation of kinetic, equilibrium and thermodynamic parameters and optimization of adsorption by a central composite design. *Journal of Environmental Science and Health, Part A* 2017, *52* (14), 1269–1283.

70. Balarak, D.; Mostafapour, F. K.; Azarpira, H.; Joghataei, A., Mechanisms and equilibrium studies of sorption of metronidazole using graphene oxide. *British Journal of Pharmaceutical Research* 2017, *19* (4), 9.

71. Zhao, G.; Li, J.; Ren, X.; Chen, C.; Wang, X., Few-layered graphene oxide nanosheets as superior sorbents for heavy metal ion pollution management. *Environmental Science & Technology* 2011, *45* (24), 10454–10462.

72. WooáLee, J.; BináKim, S., Enhanced Cr (VI) removal using iron nanoparticle decorated graphene. *Nanoscale* 2011, *3* (9), 3583–3585.

73. Yang, H.; Sun, L.; Zhai, J.; Li, H.; Zhao, Y.; Yu, H., In situ controllable synthesis of magnetic Prussian blue/graphene oxide nanocomposites for removal of radioactive cesium in water. *Journal of Materials Chemistry A* 2014, *2* (2), 326–332.

74. Sun, Y.; Wang, Q.; Chen, C.; Tan, X.; Wang, X., Interaction between Eu (III) and graphene oxide nanosheets investigated by batch and extended X-ray absorption fine structure spectroscopy and by modeling techniques. *Environmental Science & Technology* 2012, *46* (11), 6020–6027.

75. Chandra, V.; Kim, K. S., Highly selective adsorption of Hg^{2+} by a polypyrrole–reduced graphene oxide composite. *Chemical Communications* 2011, *47* (13), 3942–3944.

76. Ren, Y.; Yan, N.; Wen, Q.; Fan, Z.; Wei, T.; Zhang, M.; Ma, J., Graphene/δ-MnO_2 composite as adsorbent for the removal of nickel ions from wastewater. *Chemical Engineering Journal* 2011, *175*, 1–7.

77. Zhao, G.; Ren, X.; Gao, X.; Tan, X.; Li, J.; Chen, C.; Huang, Y.; Wang, X., Removal of Pb (II) ions from aqueous solutions on few-layered graphene oxide nanosheets. *Dalton Transactions* 2011, *40* (41), 10945–10952.

78. Zhao, G.; Wen, T.; Yang, X.; Yang, S.; Liao, J.; Hu, J.; Shao, D.; Wang, X., Preconcentration of U (VI) ions on few-layered graphene oxide nanosheets from aqueous solutions. *Dalton Transactions* 2012, *41* (20), 6182–6188.

79. Chen, Y.; Zhang, Q.; Chen, L.; Bai, H.; Li, L., Basic aluminum sulfate@ graphene hydrogel composites: Preparation and application for removal of fluoride. *Journal of Materials Chemistry A* 2013, *1* (42), 13101–13110.

80. Vasudevan, S.; Lakshmi, J., The adsorption of phosphate by graphene from aqueous solution. *RSC Advances* 2012, *2* (12), 5234–5242.

81. Zaman, A.; Ali, M. S.; Orasugh, J. T.; Banerjee, P.; Chattopadhyay, D., Biopolymer-based nanocomposites for removal of hazardous dyes from water bodies. In *Innovations in Environmental Biotechnology*, Arora, S.; Kumar, A.; Ogita, S.; Yau, Y.-Y., Eds.; Springer Nature Singapore, 2022; pp. 759–783.

82. Chen, H.; Gao, B.; Li, H., Removal of sulfamethoxazole and ciprofloxacin from aqueous solutions by graphene oxide. *Journal of Hazardous Materials* 2015, *282*, 201–207.

83. Wang, J.; Tang, B.; Tsuzuki, T.; Liu, Q.; Hou, X.; Sun, L., Synthesis, characterization and adsorption properties of superparamagnetic polystyrene/Fe$_3$O$_4$/graphene oxide. *Chemical Engineering Journal* 2012, *204*, 258–263.

84. Kumar, A. S. K.; Rajesh, N., Exploring the interesting interaction between graphene oxide, Aliquat-336 (a room temperature ionic liquid) and chromium (VI) for wastewater treatment. *RSC Advances* 2013, *3* (8), 2697–2709.

85. Sreeprasad, T. S.; Maliyekkal, S. M.; Lisha, K. P.; Pradeep, T., Reduced graphene oxide–metal/metal oxide composites: Facile synthesis and application in water purification. *Journal of Hazardous Materials* 2011, *186* (1), 921–931.

86. Yang, T.; Liu, L.-H.; Liu, J.-W.; Chen, M.-L.; Wang, J.-H., Cyanobacterium metallothionein decorated graphene oxide nanosheets for highly selective adsorption of ultra-trace cadmium. *Journal of Materials Chemistry* 2012, *22* (41), 21909–21916.

87. Yang, S.; Chen, C.; Chen, Y.; Li, J.; Wang, D.; Wang, X.; Hu, W., Competitive adsorption of PbII, NiII, and SrII ions on graphene oxides: A combined experimental and theoretical study. *ChemPlusChem* 2015, *80* (3), 480–484.

88. Liu, S.; Wang, H.; Chai, L.; Li, M., Effects of single-and multi-organic acid ligands on adsorption of copper by Fe3O4/graphene oxide-supported DCTA. *Journal of Colloid and Interface Science* 2016, *478*, 288–295.

89. Li, M.; Wu, G.; Liu, Z.; Xi, X.; Xia, Y.; Ning, J.; Yang, D.; Dong, A., Uniformly coating ZnAl layered double oxide nanosheets with ultra-thin carbon by ligand and phase transformation for enhanced adsorption of anionic pollutants. *Journal of Hazardous Materials* 2020, *397*, 122766.

6 Application of Graphene (G)-Based Materials for Water Purification

6.1 INTRODUCTION

The supplementation of chemical compounds between two adjoining interfaced phases, like solid/liquid, solid/solid, liquid/gas, and liquid/liquid phases, can generally be referred to as adsorption as a material surface behavior and is reliant upon molecular diffusion. The substance being adsorbed is known as the adsorbate, while the substance performing the adsorption is known as the adsorbent. Depending on how the adsorbent surface and the adsorbate interact, adsorption can be classified as either physical or chemical. Several types of adsorption include surface modification, pore adsorption, dispersion force, electrostatic force, dipole interaction, electrostatic force, quadrupole interaction, and interaction based on charge transfer. Typically, solid/liquid and solid/gas interfaces are the sites of physical adsorption. Intermolecular gravitation, or van der Waals (vdW) forces, are primarily responsible for the physical adsorption process. Physical adsorption doesn't entail breaking or creating chemical bonds and has the advantages of being relatively quick, having a weak binding force, producing little heat during adsorption, not being selective to the adsorbed substance, and being simple to desorb. Chemical adsorption, which differs from physical adsorption, is a process in which chemical bonds are formed between the adsorbent surface and the adsorbate. It is characterized by a relatively slow adsorption rate, high adsorption heat, irreversible adsorption, a certain selectivity to the adsorbed object, and challenging desorption.

G-based adsorbent materials (GBAMs) for wastewater purification have been broadly used in the niche of adsorption chemistry where, in most cases, such adsorbents may act as both catalysts as well as adsorbents. This section will cover, to a large extent, the diverse application areas of these innovative materials. Figure 6.1 depicts some of these application areas schematically.

6.2 BIOMEDICAL APPLICATION AS WELL AS PHARMACEUTICAL TRACES REMOVAL FROM WASTEWATER USING GBAMs

In the majority of situations, leftover pharmaceutical drug residue from animal or human urine or excrement is washed into water bodies, causing toxicity to the water. In order to effectively treat wastewater for reuse, it is highly advised that these dangerous pharmaceutical residues be removed in water treatment facilities.

The ancient people used materials based on carbon, like activated carbon, to remove toxins/poisons from the human cytoplasmic fluid even before Christ (BC).

DOI: 10.1201/9781032621302-6

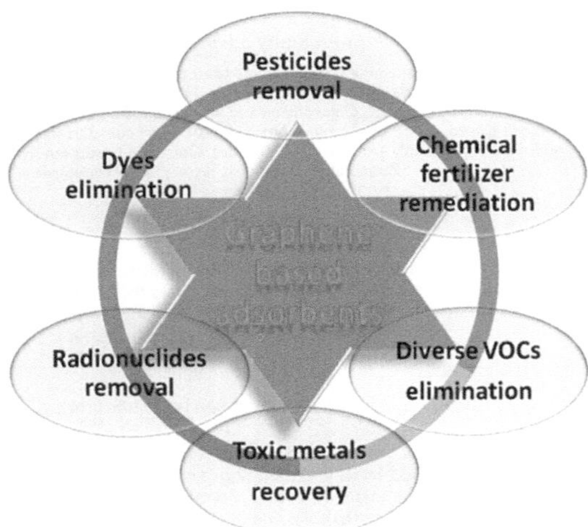

FIGURE 6.1 G-based adsorbent application niches.

For instance, German hospitals treated 178,425 cases of intoxication (poisoning) in 2016, according to the German Federal Statistical Office. In that year, there were 268,787 cases of poisoning reported to the poison control centers in the German-speaking nations, and activated charcoal use was advised in 4.37 percent of those cases. Activated charcoal application is crucial to both primary and secondary detoxification.[1] This is why for moderately severe to life-threatening poisoning, it is advised to treat with activated charcoal. Timed-release medications could be administered up to 6 h later for ingestion; however, it should be given as soon as feasible, ideally within the first hour. Diminished consciousness combined with the aspiration risk in a patient whose airway hasn't yet been guaranteed remains a significant contraindication. In situations of poisoning with bases/acids, organic solvents, alcohols, metals, inorganic salts, etc., activated charcoal is useless or partially effective. The recommended dosage is 10 to 40 times the amount of the intoxicating chemical, or 50 g for adults and 0.5–1 g/kg body weight for children. Repeated administration is advised for intoxications caused by substances that linger longer in the stomach and for intoxications brought on by timed-release medications or medications with pronounced enterohepatic or entero-enteric circulation.[1]

Due to GBMs' potential for use in the detection and treatment of complicated disorders, adsorption or enrichment has been a crucial and essential component of biomedical engineering. The existing research in this field is severely lacking in exceptional adsorptive material exploration. Due to their integrated properties, carbon-based materials have shown to be a superior choice for adsorption or enrichment carriers in the biomedical field. In this section, we talk about the use of adsorptive G-based materials' applications within the biomedical niche. Applications include breath analysis, glycopeptide, phosphopeptide enrichment, and purification of the blood/cytoplasmic fluid system. We conclude by providing a succinct assessment and a forecast for this field. Figure 6.2 presents pictorially the kind of drugs or toxins that are recognized to be adsorbed by activated carbon.

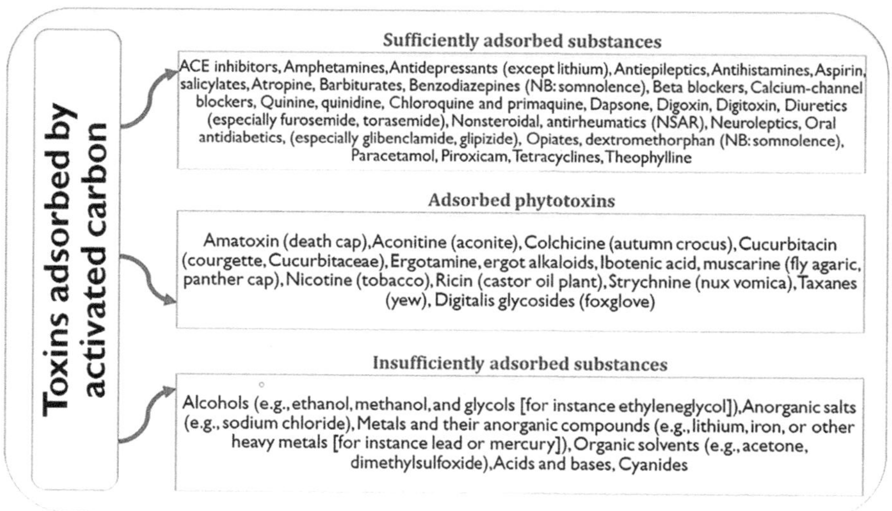

FIGURE 6.2   Pictorial representation of drugs or toxins that are recognized to be adsorbed by activated carbon.

Substances that are adsorbed by activated carbon could be simply categorized into; sufficiently adsorbed substances (antidepressants (except lithium), ace inhibitors, benzodiazepines (nb: somnolence), amphetamines, antiepileptics, aspirin, salicylates, antihistamines, barbiturates, atropine), beta blockers, calcium-channel blockers, quinidine, quinine, chloroquine and primaquine, dapsone, digoxin, digitoxin, diuretics (especially furosemide, torasemide), nonsteroidal, antirheumatics (NSAR), neuroleptics, oral antidiabetics, (like glibenclamide, glipizide), adsorbed phytotoxins (amatoxin (death cap), piroxicam, dextromethorphan (NB: somnolence), opiates, tetracyclines, paracetamol, theophylline), cucurbitacin (courgette, cucurbitaceae), (autumn crocus), colchicine aconitine (aconite), ergot alkaloids, ergotamine, muscarine (fly agaric, panther cap), ibotenic acid, (castor oil plant), nicotine (tobacco), ricin strychnine (nux vomica), taxanes (yew), digitalis glycosides (foxglove)); as well as insufficiently adsorbed substances (alcohols (for instance., methanol, ethanol, as well as glycols (e.g., ethyleneglycol)), anorganic salts (e.g., sodium chloride), metals and their inorganic compounds (e.g., lithium, iron, or other heavy metals (e.g., Pb or Hg)), organic solvents (like dimethyl ketone, (Methanesulfinyl) methane), acids as well as bases, cyano radical (–CN) containing organic compounds). Many harmful compounds, including pharmaceuticals, phytotoxins, as well as dangerous chemicals, adsorb on the surface of activated charcoal, preventing the gastrointestinal tract from absorbing them. It prevents a potential enteroenteric circulation and/or enterohepatic circulation as a secondary cleansing method.[1]

Researchers worldwide have extensively examined the use of G, its oxides, and its derivatives and composites/hybrids for the adsorptive exclusion of pharmaceutical drugs from wastewater/effluents.[2–4] It may be assumed from evidence from the literature that pharmaceuticals adsorbed onto G-based materials primarily follow PSOM and the LM, as shown in Table 6.1. The adsorption mechanism of G-based materials

TABLE 6.1

GBAs and Their Application for Pharmaceutical Trace Removal

Formulation	Pharmaceutical Traces Removed	Adsorbent Dosage	pH for Max. Adsorption	Temp. (K)	Maximum Adsorption Capacity ((q_m or Q_m) (mg/g))	Adsorption Isotherm & Kinetics		
						Isotherm	Kinetics	Ref.
G (G)								
G	Sulfamethoxazole (SMX)	60 mg/L	8.0	—	210.08	LM	PSOM	3
	Tenolol (ATL)	250 mg/L	7.5	~278	71.81	SM & TM	PSOM	5
	Carbamazepine (CBZ)	250 mg/L	7.5	~278	154.251	SM & TM	PSOM	5
	Ciprofloxacin (CIP)	250 mg/L	7.5	~278	370.11	SM & TM	PSOM	5
	Diclofenac (DCF)	250 mg/L	7.5	~278	75.801	SM & TM	PSOM	5
	Gemfibrozil (GEM)	250 mg/L	7.5	~278	109.710	SM & TM	PSOM	5
	Ibuprofen (IBP)	250 mg/L	7.5	~278	49.508	SM & TM	PSOM	5
$G\text{-}Fe_3O_4$	Pararosaniline	0.5 g/L	6.6	RT	198.23	LM	PSOM	6
G	Ciprofloxacin (CIP)	500 mg/L	7.0	298	145.9	LM	PSOM	7
Activated G(G-KOH)	Ciprofloxacin (CIP)	500 mg/L	7.0	298	194.6	LM	PSOM	7
G	Aspirin	20 mg/L	8.0	296	13.02	—	PSOM	8
	Caffeine	20 mg/L	8.0	296	19.72	—	PSOM	8
	Acetaminophen	20 mg/L	8.0	296	18.06	—	PSOM	8
G	Naphthalene	25 mg/L	7.0	298	127.7	LM	PSOM	9
	Phenanthrene	25 mg/L	7.0	298	136.4	LM	PSOM	9
	Pyrene	25 mg/L	7.0	298	170.2	LM	PSOM	9
G (GNA)	Phenanthrene	—	7.0	293.15	208.3	FM	—	10
	Biphenyl	—	7.0	293.15	102.6	FM	—	10

(Continued)

TABLE 6.1

GBAs and Their Application for Pharmaceutical Trace Removal

Formulation	Pharmaceutical Traces Removed	Adsorbent Dosage	pH for Max. Adsorption	Temp. (K)	Maximum Adsorption Capacity ((q_m or Q_m) (mg/g))	Adsorption Isotherm & Kinetics		
						Isotherm	Kinetics	Ref.
GO								
GO	IBP	0.50–1.25 g/L	6	308	9.165	LM	PSOM	11
GO	Carbamazepine (CBZ)	1 g/L	2.0	308.15	9.2 mg/g	LM	PSOM	2
	Acetaminophen (ACM)	60 mg/L	6.0	—	56.21	LM	PSOM	3
GO	Metformin	40 mg/L	6.0	288–303	50.47	FM	PSOM	12
GO	Acetaminophen	200 mg	8.0	296–3323	18.06	Type IV isotherm with an H3-type hysteresis loop	PSOM	8
	Caffeine	200 mg	8.0	77	19.72	Type IV isotherm with an H3-type hysteresis loop	PSOM	8
	Aspirin	200 mg	12.0	77	13.02	Type IV isotherm with an H3-type hysteresis loop	PSOM	8
GO	Tetracycline	0.181 mg/mL	3.6	298.15	313	LM	PSOM	13
GO	Trimethoprim (TMP)	15 mg/L			204.08	LM	PSOM	4
	Isoniazid (INH)	15 mg/L			13.89	LM	PSOM	4
GnO/MIL-101	Naproxen (NAP)	50 mg/L	7.0	—	155	LM	PSOM	14
	Ketoprofen (KTP)	50 mg/L	5.4	—	171	LM	PSOM	14
Agar-GO	Chloroquine	20 mg/L		298	31	FM	PSOM	15
GO-calcium alginate	Ciprofloxacin	2 g/L	5.9	293	39.06	FM	PSOM	16
	Chloroquine diphosphate	20 mg/L	Neutral	277.15	63	FM	—	17
	Safranin-O	20mg/L	Neutral	277.15	100	SM	—	17

Reduced GO

rGO@magnetite (rGO-M)	Ciprofloxacin (CIP)	0.2 g/L	6.2	298	18.22	LM	PSOM	18
	Norfloxacin (NOR)	0.2 g/L	6.2	298	22.20	LM	PSOM	18
rGO–Fe$_3$O$_4$	Phenazopyridine	0.8 g/L	6.0	298	14.06	LM	PSOM	19
3D rGO	Naproxen (NPX)	1000 mg/L	2	298	357	LM	PSOM	20
	Ibuprofen (IBP)	1000 mg/L	2	298	500	LM	PSOM	20
	Diclofenac (DFC)	1000 mg/L	2	298	526	LM	PSOM	20
Zeolitic-imidazolate framework-8@reduced GO (ZIF-8@rGO)	Tetracycline (TC)	233.3 mg/L	6.2	—	1776.26	LM	PSOM	21
rGO	Tetracycline (TC)	233.3 mg/L	6.2	—	1161.90	LM	POSM	21
Maltodextrin-rGO	Diclofenac	0.05 g/L	7.0	293.15	9.785	LM	—	22
	Amoxicillin	0.05 g/L	7.0	293.15	12.165	LM	—	22
Maltodextrin-rGO-CuO	Diclofenac	0.05 g/L	7.0	293.15	12.836	LM	—	22
	Amoxicillin	0.05 g/L	7.0	293.15	526.31	LM	—	22
G-based hybrids								
Fe-BiOBr/rGA	Phenol	600 mg/L	7.0	303.15	26.7067	LM	POSM	23
rGO	Phenol	600 mg/L	7.0	303.15	9.138	LM	POSM	23
Montmorillonite (Mont)	Enrofloxacin	80 mg/L	6.0	295.15	310.6	Dubunin-Radushkevich	PSOM	24
GO/rGO-polyethylene glycol methyl ether (PEG)								
MIL-68(In)–NH$_2$/GrO	Amoxicillin	—	3	298.15	—	—	PFOM	25
GO-D(4,4′-Diaminoocta-fluorodiphenyl)Fe	Sulfadiazine (SD)	160 mg/L	5	298.15	380.67 μmol/g	LM	PSOM	26
	Carbamazepine (CBZ)	160 mg/L	5.0	298.15	350.70 μmol/g	LM	PSOM	26
GO/halloysite nanotube@PANI	diclofenac	1000 mg/L	3.0	298	633.680 mg g^{-1}	LM	PFOM	27
β-CD/rGO-MWCNTs	naproxen	10 mg/L	2.0	RT	132.09 mg g^{-1}	FM	PSOM	28

towards pharmaceuticals has been proposed to be mainly physiosorptionally controlled by forces such as Vander Waal forces and π-π interactions, including electrostatic interfacial interactions.[5] The following dominant adsorption mechanisms were inferred by the authors from the enthalpy values (absolute) in thermodynamic studies, the low values of the main adsorption energy in the D-R model, but also the analysis of PG spent utilizing FTIR analysis along with Raman spectroscopy at various points based on the aforementioned discussion(s). First off, PG had a variety of oxygen-containing functionalities, as well as carboxylic (–COOH), hydroxyl (–OH), but also epoxy (–C–O–C–), as well as active sites produced by the numerous porous architected channels that covered a significant portion of its surface. Taking this into account, porous configuration, the high surface area, and availability of major transformation sites were dependent on the affinity in the process of adsorption between PG and pollutants having varied sizes. Second, the adsorption process is greatly influenced by the surface charge as well as the protonation-deprotonation transition of surface functionalities upon PG surface active sites for various pH settings. The following equations show how pharmaceuticals having both primary amines (–NH$_2$) and secondary amines (–NH) groups, like ATL and CBZ, could react with PG:

$$\text{Graphene} - \text{COO}^- + \text{Drug} - \text{NH}_3^+ \rightarrow \text{Graphen} - \text{COO}^-\text{NH}_3^+ - \text{Drug} \qquad (6.1)$$

$$\text{Graphene} - \text{OH} + \text{Drug} - \text{NH}_2 \rightarrow \text{Graphen} - \text{OH}\ldots\text{NH}_2 - \text{Drug} \qquad (6.2)$$

$$\text{Graphene} - \text{OH} + \text{Drug} - \text{NH} \rightarrow \text{Graphen} - \overset{\text{H}}{\text{O}}\ldots\overset{\text{H}}{\text{NH}} - \text{Drug} \qquad (6.3)$$

$$\text{Graphene} - \text{OH} + \text{Drug} - \text{NH} \rightarrow \text{Graphen} - \text{OH}\ldots\text{NH} - \text{Drug} \qquad (6.4)$$

$$\text{Graphene} - \text{OH} + \text{Drug} - \text{NH} \rightarrow \text{Graphen} - \overset{\text{H}}{\text{O}}\ldots\text{NH} - \text{Drug} \qquad (6.5)$$

According to the equation, electrostatic interactions can take place between the protonated amine functionalities –NH$_3^+$ along with negatively charged PG (such as –COO$^-$) functionalities (6.1). Equations (6.2), (6.3), and (6.4), correspondingly, show that surface-bridging via H-bonding may also occur between the H-atoms of the –OH groups of the PG as well as the N atoms of the amine functional groups of the drugs, or between the O atoms of the -OH groups of the PG and the hydrogen atoms of the amine groups of the drugs. Additionally, all medicines and PG may form the following hydrogen bonds (equations (6.6) and (6.7)):

$$\text{Graphene} - \text{COOH} + \text{Drug} = \text{O} \rightarrow \text{Graphen} - \text{COOH}\ldots\text{O} = \text{Drug} \qquad (6.6)$$

$$\text{Graphene} - \text{COOH} + \text{Drug} = \text{O} \rightarrow \text{Graphen} - \text{OH}\ldots\text{O} = \text{Drug} \qquad (6.7)$$

For example, it is clear how medications like CBZ and CIP can be quickly absorbed by PG (Figure 6.3). For instance, the CBZ structure enabled more than three forms of interactions due to the inclusion of trio aromatic rings, a primary

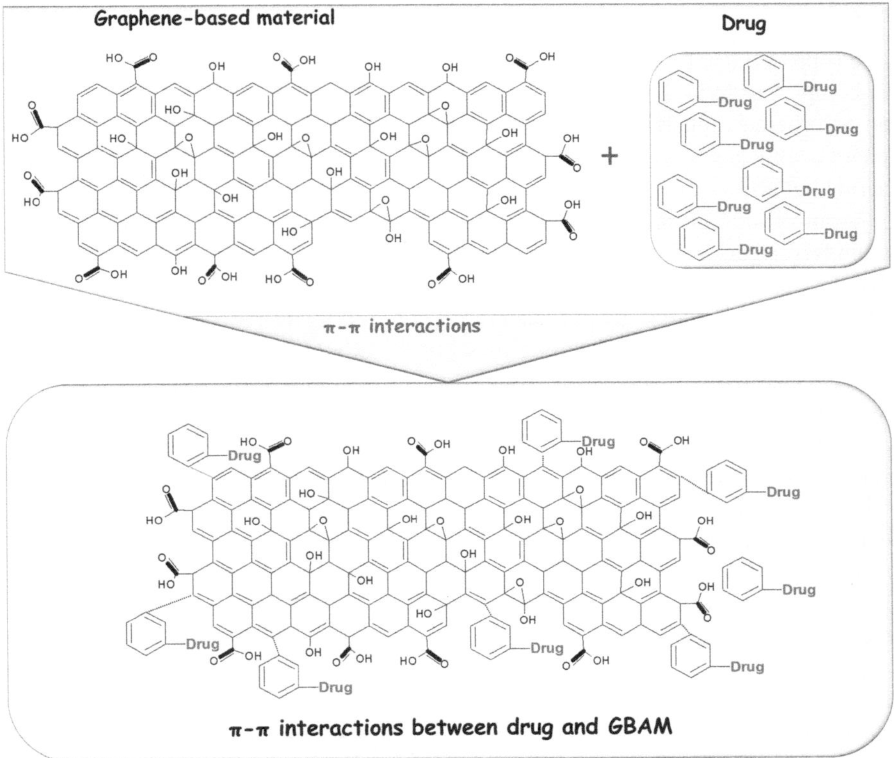

FIGURE 6.3 Postulated mechanism of aromatic drugs adsorption onto G-based adsorbent materials through π-π interactions.

amine functional group, along with a C=O group.[5] Due to the simplicity of $-NH_2$ protonation, drug molecules existed in a positively charged state, particularly under acidic environments (pH < pK_a, acidity constant pKa of CBZ = 13.9). At all measured pH levels, PG showed, as previously mentioned, a negative charge state for the particles. As a result, the adsorption mechanism includes H-bonding, vdW forces, electrostatic as well as π-π interactions, as well as electrostatic interactions. This leads to rapid absorption by PG and quick kinetics. Through interactions betwixt molecules of the drug, PG layers, but also water, the drug's hydrophobic character also plays a major role in adsorption.

Due to the nearly identical adsorption behavior/mechanism that carbonaceous compounds (PG, GO, and GI) show for the targeted pollutants, electrostatic interactions, H-bonds, π-π interactions, as well as vdW forces may all play a role in the adsorption process, albeit to varying degrees. The quantity of OCFGs present on the surface of the adsorbent determines this degree (high with regards to GO or else very low for GI). Thus, between hydrophobic organic functionalities/molecules of the contaminants but also carbon sheets of PG, electrostatic and π-π interactions may predominate.

6.2.1   G

G has proven itself as an effective adsorbent material either by itself or in the form of composites/hybrids, though herewith we focus on its pristine form application as an adsorbent for pharmaceutical contaminants: G's mechanism for the adsorption of pharmaceutical contaminants has been proposed to be mainly via electrostatic interaction but also π-π interaction.[9]

Apul et al.[10] investigated and compared the adsorption of two synthetic compounds of organic origin (phenanthrene and biphenyl) using two pristine G nanosheets "GNS" and one GO with that of activated carbon (HD4000) derived from coal, a SWCNT (single-walled carbon nanotube), a MWCNT (multi-walled carbon nanotube), as well as both in distilled as well as deionized water. Carbon nanotubes (CNTs) and granular activated carbon (GAC) were outperformed by Gs in terms of their ability to adsorb NOM. All adsorbents had their SOC (synthetic organic compound) uptake decreased by the presence of NOM. Contrary to CNTs and activated carbons, Gs were more sensitive to NOM's effects on SOC adsorption. The SOC with the flexible molecular configuration also experienced less NOM preloading damage than the SOC with the planar and stiff molecular architecture. The outcomes showed that Gs can be used as an alternate adsorbent to remove SOCs from water. They will, however, also adsorb organic pollutants if discharged into the environment, which will affect their fate and environmental effects.[10]

To explore the potential adsorptive locations and chemical processes, the adsorption of phenanthrene, naphthalene (Nap), but also pyrene onto G nanosheets as well as GO nanosheets, was studied.[9] Elemental analysis was performed by XPS, FTIR, and Raman, while SEM, and TEM were adopted to analyze the morphology and microstructure of GNS and GO. While GO adsorption was dramatically reduced after oxygen-containing functionalities were added to GNS surfaces, G exhibited an excellent affinity for polycyclic aromatic hydrocarbons (PAHs). The relationship between the adsorption coefficients (K_d) and the PAH equilibrium concentrations showed an unanticipated peak. G and GO were more difficult to adsorb due to the PAHs' hydrophobic characteristics and large molecular sizes.[9] The powerful grooved regions created by the wrinkles on GNS surfaces and the sieving effect of the π-π interactions to the flat surface account for the high affinities of the PAHs to GNS. In contrast, the adsorptive sites of GO moved to the carboxyl groups attached to the edges of GO as a result of the disappearance of the groove regions and the limitation of the π-π interactions by the polar nanosheet surfaces. The TEM and also SEM images originally showed upon PAH inclusion, the configuration as well as aggregation of G and GO nanosheets were substantially altered, which justified the annotations that the possible GNS, as well as GO sites of adsorption, were transformed during the process of adsorption.[9]

With the help of a simple one-pot solvothermal process, a superparamagnetic G-Fe$_3$O$_4$ nanocomposite (G-Fe$_3$O$_4$) was created.[6] The newly developed method's nanocomposite G/Fe$_3$O$_4$ was initially applied as an adsorbent to exclude the dye from water contamination. The newly generated G/Fe$_3$O$_4$ has a greater dye adsorption effectiveness in comparison to G-Fe$_3$O$_4$ produced via in situ chemical coprecipitation. The organic dye pararosaniline was used as the adsorbate to test the nanocomposite adsorbent's adsorption properties. The effectiveness of pararosaniline removal was examined, along with the adsorption kinetics, Q, dose, and pH of the solution. Using the FM and LM for adsorption isotherm, the ability of G-Fe_3O_4 to adsorb pararosaniline was assessed. The scientists hypothesized that the G-Fe_3O_4 nanocomposite might be easily controlled in a magnetic field for anticipated separation, resulting in a simple removal of the dye from contaminated water. In eliminating organic dyes from contaminated water, the G-Fe_3O_4 nanocomposite has a lot of potential.[6]

A one-step alkali-activated technique was used to create activated G adsorbents (G-KOH), which have a significant amount of micropores and a high specific surface area (SSA). As a result, the final product's SSA significantly rises to 512.6 m^2/g from 138.20 m^2/g. First, ciprofloxacin (CIP) was removed from aqueous solutions using the resultant G-KOH as an adsorbent. According to experimental findings, G-KOH has a strong Q of 194.6 mg/g. The inclusion of oxygen-functionalities to G-surface KOH's during the alkali-activation process helped the compound be more effective at adsorbing CIP. A PSOM better captured the adsorption kinetic, according to the results of kinetic regression. Both intra-particle diffusion and external mass transfer worked together to influence the overall adsorption process, with intra-particle diffusion taking the lead. When predicting the adsorption of CIP on G-KOH, a FM performed worse than a LM. Numerous adsorption interaction mechanisms, including H-bonds, π-π interactions based on electron donor-acceptor, as well as electrostatic interactions, are responsible for CIP's exceptional ability to adsorb onto G-KOH.[7]

Another study investigated the elimination of aspirin "ASA", acetaminophen (APAP), but also caffeine "CAF" as instances of dangerous pharmaceutical contaminants from an aqueous mixture by G nanoplatelets "GNPs". The GNPs' characterization revealed a transparent, layered assembly having a smooth surface but also many creases, along with a specific surface area of 635.2 m^2/g. The influence of adsorption time, GNPs mass, ionic strength, pH, as well as the temperature of the solution were investigated and then optimized. PFOM and PSOM were used to examine how temperature affected the adsorption kinetics, and the PSOM provided a good fit for the investigational data. Additionally, the mechanism of adsorption was investigated by adopting the intra-particle diffusion along with liquid film diffusion models. However, the results showed that neither model was the rate-determining step. The Gibbs free energy change $\left(\Delta G°\right)$, enthalpy change $\left(\Delta H°\right)$, and entropy change $\left(\Delta S°\right)$ of the adsorption were calculated after thermodynamic analysis of the process. At all temperatures, the $\Delta G°$ values were negative, showing that the adsorption of ASA, APAP, and CAF by GNPs from aqueous solution occurred spontaneously.[8]

To this end, researchers have utilized pristine G as an adsorbent for pharmaceuticals-based contaminants from wastewater, though the focus has been more on their modified forms via chemical functionalization and composite/hybrids for effective application.

6.2.2 GO

For dyes, metals, and inorganic contaminants, GO is said to be an efficient adsorbent. Different processes, including exfoliation (thermal, mechanical, and chemical), thermal breakdown, chemical vapor deposition, etc., are used to create GO from graphite. Among them, chemical oxidation is frequently used to produce huge quantities of GO using acids and oxidizing agents, and it is a flexible and easy way to make both polar and nonpolar GOs.

When employing an agar-GO "A-GO" hydrogel created through a straightforward one-step jellification procedure, the medicines *N4*-(7-Chloro-4-quinolinyl)-*N¹*,*N1*-dimethyl-1,4-pentanediamine diphosphate salt as well as safranin-O were selectively excluded from wastewater.[17] In batch experiments, the shape of the A-GO biocomposite was described, and Freundlich (Chloroquine) and Sips (Safranin-O) adsorption isotherms were shown to fit them satisfactorily ($R^2 > 0.98$). Kinetic data modeling was done using driving force models as well as Fick's diffusion equation, and a good fit was made. The findings of selective adsorption experimentations in batches revealed that when both components are mixed in a water solution, competitive adsorption takes place. The adsorptive capacities of each component decreased by about 10 mg/g, with safranin-O remaining at 41 mg/g and chloroquine remaining at 31 mg/g. Fixed-bed curves for safranin-O, as well as chloroquine using an adsorption column, revealed adsorption capacities of over 63 mg/g and 100 mg/g, correspondingly, while also displaying exceptional regenerative potentials. The biocomposite made from GO demonstrated itself to be a practical and environmentally responsible substitute for continually removing both pollutants from water.[17]

Wet spinning was used to create a novel class of biocomposite fibers called GO-doped calcium alginate (GO-CA). By examining the impact of variables like pH, dose, and contact time, the adsorption characteristics of ciprofloxacin onto the GO-CA fibers were studied. The ideal pH for ciprofloxacin to be absorbed by GO-CA fibers is 5.9. With a 2 g/L adsorbent dosage and a 6 percent GO loading, the removal rate of the GO/CA fibers is 78.9 percent. The FM was able to adequately fit the adsorption data. When the GO loading in the fibers was increased from 0% to 6% at the ciprofloxacin equilibrium concentration of 60 mg/L, the Q rose from 18.45 mg/g to 39.06 mg/g. The data from kinetic adsorption also matched.[16]

Table 6.1 presents more elaborate instances of adsorptive removal of pharmaceutical contaminants from water/wastewater as per literature.

We believe that the potential of GO towards the adsorption of pharmaceutical contaminants is inexhaustive, and more is yet to come: researchers must think beyond the ordinary.

6.2.3   rGO

Pharmaceutical wastewater treatment is crucial, as we are all aware, because of the limited water supply and expanding industry. In order to completely eradicate all pharmaceutical contaminants, an effective approach must be developed. To augment the process of eliminating pharmaceutical pollutants from the ecosystem, as well as from aquatic and industrial effluents, synthetic nanocomposite was suggested in this regard.[22] FTIR, TEM, FESEM-EDS, DLS, and XRD were utilized to construct as well as characterize binary (maltodextrin/GG composite) and ternary (maltodextrin/GG composite/CuO) composites. Drugs like diclofenac and amoxicillin were taken off the market thanks to nanocomposites. The greatest removal rate of 86 percent is achieved in 10 minutes with amoxicillin at 30 mg/L concentration, an adsorbent dosage of 0.05 g, pH of 7.4, as well as a 20 °C optimum temperature. The best removal effectiveness of 94 percent is achieved by diclofenac with nano-adsorbents when manufactured under ideal circumstances, which include an initial concentration of 20 mg/L, a 0.05 g adsorbent dose, an adsorption duration of 7 min, at 20 °C, along with a pH of 7. The produced nanocomposites provide an alternative to traditional adsorbents for removing pharmaceuticals from water, according to their findings.[22]

More complex examples of adsorptive removal of pharmaceutical pollutants from water/wastewater using GO, according to literature, are shown in Table 6.1.

6.2.4   G-Based Hybrids

The utilization of hybrid systems as adsorbents is mainly interesting to researchers and industrialists from the viewpoint that the amalgamation of three or more functional components into a single system will result in an efficient adsorbent towards single as well as mixed pollutants adsorption. Though the result is not always the same in all instances,[23] researchers must be careful in the selection, design, and fabrication of G-based hybrid adsorbents aimed at the remediation of wastewater.

In this regard, a group of researchers designed a hybrid adsorbent based on Fe-BiOBr/rGA having excellent Q 26.7067 mg/g while the Q of G aerogel 9.138 mg/g makes the hybrid system Q to be 2.92 times more than G aerogel.[23] The authors postulated that the inclusion of Fe-BiOBr in the rGA resulted in enhanced surface area and, consequently, the adsorption active functional sites of the adsorbent, thus the improvement in adsorption.[23] The hybrid adsorbent component responsible for its outstanding adsorption property performance was said to be the G aerogel via $\pi - \pi$ interaction between the aromatic phenol ring and G as the Fe-BiOBr component acted more as the catalyst portion.

A unique MIL-68(In)-NH_2/GO (GrO) composite has been created as a visible-light-driven method for amoxicillin adsorption and photocatalyst degradation (AMX). The photocatalytic activity of the MIL-68(In)-NH_2/GrO composite was significantly higher than that of the pure MIL-68(In)-NH_2. After irradiation for 120 as well as 210 min, it was possible to achieve 93 percent degradation and 80 percent TOC removal toward AMX using 0.6 g L1 MIL-68(In)-NH_2/GrO, and pH = 5. GrO was used in the modification of MIL-68(In)-NH_2/GrO, which contributed to the increased activity by acting as a sensitizer to increase the visible light absorption as well as an electron transporter for reducing photogenerated carrier recombination. Additional research demonstrated that the pH of the solution significantly affected the photodegradation of AMX. Additionally, MIL-68(In)-NH_2/GrO had good stability and reusability. H^+ and O_2^- were found to be the primary functional species in the photocatalytic system after the improved photocatalytic process over MIL-68(In)-NH_2/GrO was also anticipated.[25]

By means of intercalating chemically 4,4′-Diaminooctafluorodiphenyl (DFP) into the sheets of GO, a new magnetic spongy GO (GO-DFe) was created by budding Fe_3O_4 using an impregnation approach within porous GO (GO-DFP). To create the GO-DFe5 adsorbent having outstanding adsorption as well as segregation attributes, the concentrations of Fe^{2+}/Fe^{3+}/Fe^{3+} were tuned in this study. Adsorption kinetic models (the PSOM, Elovich, as well as intra-particle models) along with isotherm models (the LM and FM) were used to analyze the rate of adsorption but also the capacity for the magnetic GO-DFe$_5$ in the presence of sulfadiazine (SD) then carbamazepine (CBZ).[26] Further clarifications were made on how pH, ions (Cl$^-$ and NO^{3-}), and co-existing substances in municipal wastewater effluent affected the SD and CBZ's adsorption behavior on GO-DFe$_5$. Additionally, the effectiveness of GO-DFe$_5$ for cyclic reuse in SD and CBZ was assessed, and a theoretical calculation was used to suggest a probable adsorption mechanism. For SD (380.67 μmol/g) as well as CBZ (350.70 μmol/g), the GO-DFe$_5$ adsorbent demonstrated high adsorption capabilities. As per the PSOM, the amounts of GO-DFe$_5$ that two medicines adsorb were roughly 3.0 and 2.1 times greater than those of GO. The adsorption of SD and CBZ on GO-DFe$_5$ was mediated by H-bonding, electrostatic interaction, but also π-π interaction. Within six cycles, GO-DFe$_5$ in water showed good reuse performance for SD and CBZ and was simple to recover using a magnet.[26]

We also know that, in recent years, pharmaceutical pollutants have been consistently found in wastewater bodies all over the world. In response to this urgent need for the control and exclusion of pharmaceutical pollutants from water, a team of researchers recently presented a GO@halloysite nanotubes @polyaniline (GO-HNT-PANI) nanohybrid that was synthesized and used as adsorbent for the exclusion of diclofenac sodium (DS).[27] The pH fluctuation, contact time, concentration, and temperature changes, among other experimental parameters, were optimized for the elimination of DS. SEM, EDX, HR-TEM, XRD, BET, as well as FTIR analytical methods, were used to characterize the produced materials. The fastest decontamination of DS (633.680 mg/g) was accomplished at pH 3.0 in about 60 minutes. The kinetic data were best fitted to the PSOM. The results showed that the LM is the best-fitting isotherm. The adsorption equilibrium data were fitted to non-linear LM, FM, as well as TM. A spontaneous endothermic adsorption mechanism was seen in the thermodynamic investigation. In contrast to other reported adsorbents, the nanohybrid displayed greatly increased adsorptive absorption. Electrostatic interactions, π-π interactions, as well as H-bonding all contributed to the adsorption of DS. In addition, a straightforward wash with 0.1 mol/L (aq) ethanol allowed the nanohybrid to demonstrate its ability to regenerate up to five times without suffering any substantial loss. Their outcomes showed that GO-HNT-PANI nanohybrid was a highly effective material for the elimination of new pharmaceutical contaminants in water through H-bonding, π-π interaction, but also electrostatic interactions (Figure 6.4).[27] In their research, Feng et al.[28] created a brand-new β-cyclodextrin restrained 3D macrostructured rGO and MWCNTs (β-CD-rGO-MWCNTs), which was then used as an efficient adsorbent for naproxen removal from aquatic surroundings. The β-CD-rGO-MWCNTs safeguarded a large interior space that was favorable to the maintenance of adsorption sites as well as naproxen diffusion merit to their increased specific surface area (SSA) along with superior porous architecture. At room temperature, the highest naproxen adsorption capacity (q_m or Q_m) of β-CD@rGO@ MWCNTs was 132.09 mg/g, which was higher than that of the majority of regularly used naproxen adsorbents. Hydrophobic, electrostatic, π/n-π, and hydrogen bonding interactions were all implicated in the enormous amount of adsorption. The naproxen adsorption isotherm and kinetics on the adsorbents more closely resemble the

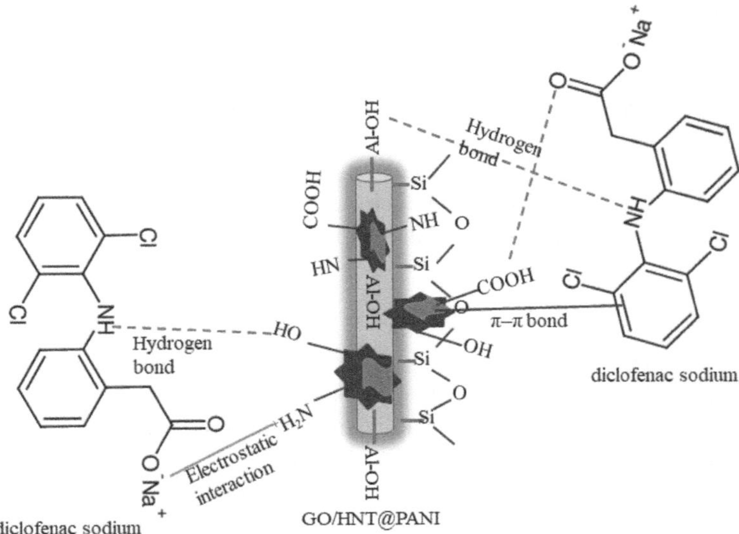

FIGURE 6.4 A proposed diclofenac sodium binding mechanism onto GO/HNT@PANI nanohybrid. (Reproduced with permission from Higgins et al.[27] Copyright 2022, Elsevier Science Ltd.)

Freundlich model and the PSOM. The thermodynamic characteristics confirmed that naproxen adsorption on β-CD@rGO@MWCNTs was a beneficial, thermodynamically viable, but also exothermic process. The adsorbent's robust 3D structural configuration gave it exceptional regeneration performance and real-world application potential.[28]

More and more G-based adsorbent hybrid systems have been designed and fabricated toward the adsorptive exclusion of pharmaceutical contaminants from water/wastewater due to their outstanding property performance when compared to pristine or composite G-based adsorbents.

6.3 ELIMINATION OF TOXIC/HEAVY METALS FROM WASTEWATER UTILIZING GO AND ITS COMPOSITES/HYBRIDS

Through the interaction of chemical or physical bonds between heavy metal ions and the adsorbent, heavy metal ions are isolated and separated during the heavy metal ions adsorption process. GO is a good example of an adsorbent for heavy metal ions, and the mechanism of adsorption mechanism involves coordination chelation (surface complexation), H-bonding electrostatic attraction, but also ion exchange. Amongst heavy metal adsorption processes, the most frequent adsorption process for heavy metals is coordination chelation. The majority of the V_A-VII_A group elements, namely O, N, as well as S, which have lone pair electrons, could be employed as chelation adsorption coordination. Coordination groups and common coordination atoms are both used. A typical carbon adsorbent is one that is built on G. In order to increase the G-based adsorbent's ability to bind heavy metal ions, coordination groups, including coordination atoms, such as $-COOH$, $-OH$, $-C = O$, $-C - O$, $-C = N$, $-C - N$, $-NH_2$, $-C - S$, $-NH$, $-C = S$, $-S = O$, $-S - S$, can be added.

G-based materials are being used in the environment more frequently as a new generation of environmental materials. The adsorption of environmental contaminants like organic pollutants and heavy metal ions, but also composite pollution, has been better demonstrated by G, GO, their related modified systems, as well as their composite/hybrid nanomaterials, than other materials.

6.3.1   G

The remediation of wastewater using G has been explored intensely by materials science researchers and industrialists within the last four to five decades. Herein, we aim to present a few instances of the utilization of this novel material for the adsorptive exclusion of toxic/heavy metals from effluents.

A two-dimensional "2D" molecular brush made of GO (and polyacrylamide was created in one study to remove Pb^{2+} and Cu^{2+} from aqueous media effectively. Due to the abundance of O and N functionalities in the composite structure, the synthesized GO 2D molecular brush possessed a Q_m of 268.4 mg/g for Pb^{2+} as well as 127.2 mg/g for Cu^{2+}, correspondingly. In addition, the interspaces between the stacked, 2D molecular brushes offer quick routes for the migration of heavy metal ions. It follows that a GO-based 2D molecular brush can approach the adsorption equilibrium in 60 minutes. These findings showed that the synthetic GO-based 2D molecular brush is an effective adsorbent for removing heavy metal ions from wastewater.[33]

In a different study, G was made using a modified Hummers' process, and cetyltrimethylammonium bromide was chosen to alter it. TEM, SEM, XPS, and FTIR were used to analyze the properties of G and modified G. The adsorption characteristics of Cr^{6+} onto G and modified G were examined in relation to parameters such as pH, contact time, temperature, and dosage. In batch studies, G and modified G were used as the adsorbent under various conditions to assess the adsorbance of Cr^{6+} from an aqueous solution.[32] The findings showed that 293 K was the best-suited temperature and that an optimal pH for the adsorption was around 2. Within the first five minutes, the adsorption processes moved quickly, reaching equilibrium in around forty minutes. The PSOM and the adsorption kinetics were well-matched. The LM predicted a Cr^{6+} Q of 21.57 mg/g at 293 K for modified G. According to the thermodynamic characteristics, Cr(VI) adsorption onto modified G was an exothermic and spontaneous process.[32]

It has been demonstrated that G nanosheets, a new nanoadsorbent, can be further tuned to maximize the adsorption capabilities for different contaminants. Sulfonated G (GS) was created by a diazotization process utilizing sulfanilic acid in order to circumvent the structural limitations of G (aggregation) as well as GO (hydrophilic surface) in aqueous media. It was shown that GS possessed sulfonic acid groups but also partial original O-containing functionalities that were highly attractive to positively charged pollutants, in addition to recovering a relatively complete hybridized sp^2 plane with a strong affinity for aromatic contaminants. The saturation adsorption capacities of GS were substantially greater than the equivalent values for rGO and GO, being 400 mg/g for phenanthrene, 906 mg/g for MB, and 58 mg/g for Cd^{2+}. The many adsorption sites in GS, including the conjugate π region sites and the functional group sites, contribute significantly to its outstanding Q toward a variety of contaminants and fast adsorption kinetic rate. Additionally, the sulfonic acid groups give GS high dispersibility and one or a few nanosheets, which ensure that adsorption activities will take place. By controlling their microstructures, surface characteristics, and water dispersion, it is quite possible to reveal the adsorption sites of G nanosheets for contaminants in water.[31]

By adding lignosulfonate (LS) to GO, a green and simple production method for creating lignosulfonate-G porous hydrogel (LGPH) was developed (GO). With LS acting as a surface functionalization agent, this approach was accomplished via a straightforward self-assembly method at low temperatures. The produced LGPH hydrogel revealed 3D interconnected pores and showed an exceptional ability to adsorb Cr^{6+} (601.2 mg/g) ions that were dissolved in the water, thanks to the numerous functional groups of LS and the substantial GO surface areas. The freestanding LGPH was easily separated from the water following the adsorption procedure, and the Cr^{6+} adsorption capacities onto LGPH were maintained at 439.1 mg/g after 5 adsorption-desorption cycles. LGPH is a promising material for eliminating heavy metals from wastewater due to its affordability and environmental friendliness.[30]

New three-dimensional (3D) G@δ-MnO_2 aerogels have purportedly been created by a group of authors through the self-assembly as well as reduction of GO, trailed by in situ solution-phase deposition of ultrathin δ-MnO_2 nanosheets. The resulting G@δ-MnO_2 architectures displayed a connected 3D network microstructure with a significant amount of ultrathin birnessite MnO_2 nanosheets uniformly deposited over the G framework. The generated 3D aerogels displayed a fast adsorption kinetic rate and a better capacity for the adsorption of heavy metal ions due to their distinctive structural properties. The LM calculated that the saturated adsorption capacities of G/δ-MnO_2 aerogels were as high as 643.62 mg/g for Pb^{2+}, 250.31 mg/g for Cd^{2+}, and 228.46 mg/g for Cu^{2+}, outperforming the comparable pristine 3D G and δ-MnO_2 nanosheets by a significant margin. It was discovered that the heavy metal ions could not only adsorb on the surface of G@ δ-MnO_2 but also intercalate into the interlayer gaps of birnessite MnO_2. This finding displayed a synergistic effect of surface complexation, electrostatic attraction, as well as ion exchange between the heavy metal ions and pre-intercalated K^+, supported by the expansion of the basic crystal structural architecture of layered MnO_2 after adsorption. It was also noteworthy that the regenerated aerogels maintained their initial shape and could be used again and again for ≥8 cycles without clearly degrading performance, achieving the sustainability of the absorbents. This was accomplished by first treating them with HCl and then treating them with KOH. What's more, hybrid aerogels are simple to separate and do not produce secondary pollutants. These hybrid aerogels are the best choice for decontaminating heavy metal ions in real-world applications due to their outstanding regeneration and reusability, high removal efficiency, quick adsorption kinetics, and simplicity of separation operation.[29]

6.3.2 GO

Among the G materials family, GO-based adsorbents have been utilized the most for toxic/heavy metals exclusion from wastewater. We here discuss some instances where researchers have utilized this novel material for the adsorptive remediation of heavy/toxic metals from water.

To ascertain the adsorption characteristics of Pb^{2+} from aqueous solutions, GO and GO-modified methyl-cyclodextrin, called GO@mCD, have been produced and employed as adsorbents.[41] The characteristic results of FTIR, XRD, Raman spectroscopy as well as SEM showed that mβCD was successfully physically bound to GO to form the GO@mβCD composite. The adsorption equilibrium and adsorption kinetics of the adsorbents were correspondingly described by the LM and the PSOM. Pb^{2+} is the maximum. It was discovered that the Q of GO@mCD (at pH = 6, at room temperature) was 312.5 mg/g, which was much higher than the Q of GO (217.39 mg/g). This suggests that the Q of GO is increased by the alteration of GO with mβCD. According to the desorption investigations, the adsorbent GO@mCD may be utilized for at least 5 cycles without significantly losing its initial Q for Pb^{2+} ions.[41]

It has been claimed that GO/cellulose membranes have been produced to conduct the efficient adsorption of heavy metal ions (Co, Ni, Cu, Zn, Cd, and Pb). Specifically, pressed and non-pressed membranes of two different sorts were created. The test demonstrated that pressed membranes are extremely robust over a range of pH levels, even in basic solutions, and that they may be used to separate or remove heavy metal ions during strong shaking in an aqueous solution. Despite being less stable than pressed membranes, non-pressed membranes can nevertheless be used effectively at high flow rates for filtration. According to these authors' batch experiment findings and measurements made using inductively coupled plasma atomic emission spectroscopy (ICP-OES), pH 4 to 8 is the optimal range for adsorption. Adsorption isotherms and kinetic investigations revealed that metal ion sorption on membranes happens in a monolayer covering. Therefore, chemical adsorption involves the intense surface complexation of metal ions with the oxygen-containing groups on the surface of GO. At a pH of 4.5, GO/cellulose membranes have maximal adsorption capacities for Co^{2+}, Ni^{2+}, Cu^{2+}, Zn^{2+}, Cd^{2+}, as well as Pb^{2+} of 15.5, 14.3, 26.6, 16.7, 26.8, and 107.9 mg/g, correspondingly. The results of the competitive adsorption experiments showed that the affinity of the generated membranes for the metal ions is Pb > Cu > Cd > Zn Ni-Co. There is consistency between the affinity order and the initial stability constant of the associated metal hydroxide and acetate. The GO/cellulose membranes' adsorption properties, reusability (more than ten cycles), and endurance in aqueous solutions make it possible to remove heavy metals from water solution. In the realm of analytical chemistry, it is also feasible to preconcentrate and/or separate trace and ultratrace metal ions using membranes.[40]

Creating a green, highly effective adsorbent for wastewater treatment is a task that needs to be completed immediately. To do this, a team of researchers created a GO-based aerogel that had been modified with silk fibroin (SF), and they first looked at its adsorption capabilities. Its well-structured pore structure, greater d-spacing, and more oxidation groups after modification are shown by SEM, XPS, XRD, Raman, and TG characterization.[38] Adsorption tests with various dyes as well as heavy metal ions showed that GO-SF aerogels possessed a satisfactory capacity for adsorbing Ag^+ as well as MB dye (cationic dyes). Compared to other GO-biopolymer materials, the MB has a greater adsorption capability of up to 1322.71 mg/g. Also employed to describe the adsorption mechanism of the GO-SF aerogels and demonstrate the hybrid aerogels' propensity for monolayer adsorption is the LM, the PSOM, and the intraparticle diffusion model. The outcomes revealed a viable option for those who could cope with the wastewater from dyeing more effectively.[38]

Composite membranes using chitosan (CS) and GO (GO) as adsorbents are used to remove inorganic pollutants such as heavy metal ions, particularly Pb^{2+}, from aqueous solutions. The CS solution was used with the modified Hummers approach to obtain GO. By creating stable chelates with the heavy metals, the addition of ethylenediaminetetraacetic acid (EDTA) compound to the CS/GO suspension improved the CS/ability of GO's to adsorb and remove the metals from the body. Utilizing inductively coupled plasma mass spectrometry, the adsorption behavior of Pb^{2+} from aqueous solutions using the produced membranes was assessed. The membranes were studied using FTIR, SEM, as well as FTIR. Examining the concentration of Pb^{2+} against the adsorption duration at an initial content of the adsorbent allowed researchers to better understand the adsorption performance of Pb^{2+} ions. The greatest Pb^{2+} metal ion adsorption efficiency was 767 mg/g for CS/EDTA/GO at 0.1 percent, 889 mg/g for CS/EDTA/GO at 0.3 percent, 970 mg/g, 853 mg/g for CS, and 1526 mg/g for GO. These results suggest that CS/EDTA/GO membranes could be used as efficient adsorbents to remove heavy metal ions from water.[39]

A wide-spectrum adsorbent for wastewater remediation has been investigated using an environmentally friendly GO-chitosan (GC) nanocomposite hydrogel column (GCCHC).[49] Methylene blue (MB) and rhodamine B (RhB) are cationic dyes, whereas methylene orange (MO) and congo red (CR) are anionic dyes. The GCCHC demonstrated a high removal capability for all of these pollutants. Additionally, the samples can go through multiple adsorption and washing cycles without losing their ability to remove contaminants.[49] The GC sponge has a 9 percent CTS content and 275.5 mg/g MBs' Q. The MB adsorption by GC sponges is caused by both electrostatic attraction but also hydrophobic interactions. The Q of MB solution at the same preliminary concentration is around 300 mg/g with GC (10:1, wt./wt.) and GC (20:1, wt./wt.) hydrogels as adsorbents, but it drops to 150 mg/g with GC (5:1, wt./wt.) as an adsorbent. The exclusion rate of MB along with RhB gradually increases as the GCCHC's GO content increases, whereas the exclusion rate of MO as well as CR gradually decreases as the GCCHC's CTS content increases (Figure 6.5).[49]

The composition of the composite hydrogel affects the dye removal capacity; for instance, increasing the GO content increases the capacity for the exclusion of MB and RhB cationic dyes, whereas increasing the CTS content increases the capacity for the exclusion of MO and CR anionic dyes. Its ability to successfully exclude both cationic as well as anionic dyes from water depends on altering the composition and characteristics of each of the reactants.[49]

There is still room at the bottom for GO within the niche of heavy metals' adsorptive exclusion form wastewater/water presently as well as in the near future.

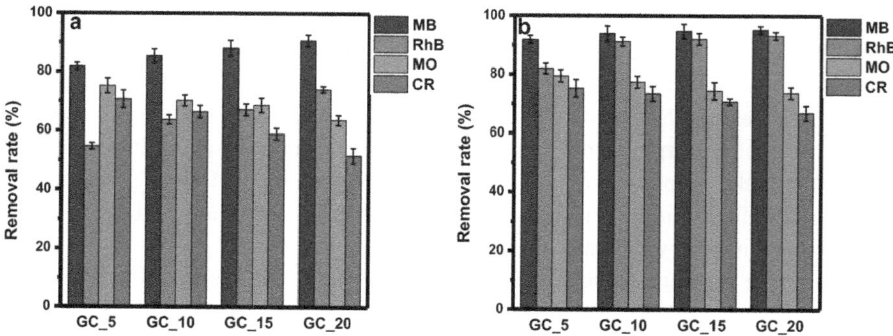

FIGURE 6.5   Dye removal rate before and after filtration through the GCCHCs via (a) direct filtration and (b) filtration after 24 h of immersion. (Reproduced with permission from Vo et al.[49] Copyright 2020, Springer Nature.)

6.3.3   rGO

Another G-based adsorbent material that has been largely utilized for the adsorptive remediation of wastewater/water is rGO, both in its pristine form of composite. This subsection presents instances of rGO utilization as an efficient adsorbent for the exclusion of toxic metals or their complexes from water as per literature.

Fu and Huang[43] have developed a method to remove the heavy metal ions Cu^{2+}, Cd^{2+}, Pb^{2+}, as well as Hg^{2+} from synthetic wastewater. developed a dithiocarbamate (DTC)-modified magnetic reduce GO ($rGO - PDTC/Fe_3O_4$). The $rGO - PDTC/Fe_3O_4$ nanocomposite was made using a unique synthesis method that also involves the bromination of GO, the nucleophilic substitution of poly(ethylenimine) (PEI), a reaction with carbon disulfide (CS_2), and the addition of Fe_3O_4 nanoparticles. XPS, FTIR, TEM, and XRD measurements of the $rGO - PDTC / Fe_3O_4$ nanocomposite reveals that DTC functionalities were chemically bound to rGO surfaces.[43] The $rGO - PDTC / Fe_3O_4$ nanocomposite demonstrated excellent adsorption capacities, rapid kinetics, and solid-liquid separation in adsorption experiments, demonstrating that it is a highly effective adsorbent for the removal of heavy metals. Using the PSOM and LM, the adsorption mechanisms were determined. The greatest adsorption capacities for the ions Cu(II), Cu^{2+}, Cd^{2+}, Pb^{2+}, as well as Hg^{2+} were 113.64, 116.28, 147.06, and 181.82 mg/g, correspondingly. After the adsorption and desorption processes, a straightforward one-step organic reaction was used to quickly recreate the used $rGO - PDTC / Fe_3O_4$ nanocomposite. Five cycles of adsorption-desorption-regeneration revealed good metal adsorption capabilities for the regenerated rGO-PDTC/Fe_3O_4 composite.[43]

MnO_2 nanotube@rGO hydrogel (MNGH) was made using a one-step hydrothermal co-assembly method in another investigation. The as-prepared MNGH had a high surface area of 156 m²/g and 3D architecture.[44] MnO_2 nanotubes with a consistent size (20 nm) were encased in porous reduced GO (rGO) sheets that were joined as a result of the morphology of MNGH being studied. The MNGH sample showed respectable adsorption performance for different heavy metal ions (Pb^{2+}, Cd^{2+}, Ag^+, Cu^{2+}, and Zn^{2+}), utilizing the 3D porous nanostructure as well as the synergetic action between rGO and MNTs. The sorption datum fits the LM and the Q of Pb^{2+} on MNGH can reach 356.37 mg/g. Furthermore, MNGH's adsorption capabilities can be preserved even after numerous regeneration tests. These findings showed that the MNGH in its as-prepared form might be used as a promising adsorbent for the exclusion of heavy metal ions.[44]

Both kinds of GO-based gels produced by González's research have been shown to be capable of adsorbing both acidic and basic dyes. The ability of chitin (Chi) as well as its hybrid gels, to adsorb contaminants is shown to be pH and Chi:nGO proportion dependent. The optimal adsorption pH for NR and RB was correspondingly 5.0 and 4.0.[50] Since Chi:nGO 3:1 has a high Q but also good mechanical properties, it was determined to be the material most suitable for the isotherm, kinetic, as well as desorption studies. The addition of nGO would be required to provide the material with the best mechanical qualities, despite the hybrid's lower capability for adsorbing an acidic dye than the Chi gel. A compromise between mechanical resistance along with adsorption character would, therefore, be implied by the amount of nGO in the hybrid. With the exception of Chi-RB, which used the Redlich-Peterson model, all systems demonstrated the SM to have a satisfactory fitting. No matter the adsorbent substance, the absorption rates of RB and NR displayed the same kinetic pattern. The Chi-RB system, however, had a better Elovich equation fit than the other systems, which had a better fit with the pseudo-first model. The optimum desorption pH values of RB and NR were found to be 9.0 and 10.0, correspondingly, for all the loaded compounds; this would be advantageous if the goal of reusing the adsorbents was pursued.[50]

This subsection has only presented readers with a few instances as a major guideline towards success in their diverse research on the utilization of rGO as a G-based adsorbent aimed at the exclusion of heavy/toxic metals form water; more instances are presented in Table 6.2.

6.3.4 G-Based Hybrids

The utilization of G-based hybrids for the adsorptive removal of toxic metals from wastewater has been one of the major attractive niches pursued by material science researchers focused on water treatment.

It might be very difficult to combine humic acid and heavy metal ions for synergistic adsorption. Their competitive adsorption onto the majority of adsorbents is substantially to blame for this.[48] In order to overcome this, a composite hierarchically structured made of polyethyleneimine-magnetic mesoporous SiO_2 as well as GO (MMSP-GO) were created. In order to chemically conjugate with the carboxyl groups on GO sheets, magnetic mesoporous silica microspheres were created and functionalized with PEI molecules, which provided a number of amine groups and increased the affinity between the pollutants and the mesoporous SiO_2.[48] Measurements made using TEM, SEM, TGA, DLS, as well as VSM were used to describe the characteristics of the composites. The simultaneous and effective removal of humic acid and heavy metal ions utilizing MMSP-GO composites as adsorbents were demonstrated by a series of adsorption experiments. By using the LM, the Q_m of MMSP-GO for Pb^{2+} and Cd^{2+} were calculated to be 333 and 167 mg/g, correspondingly. Due to their

TABLE 6.2

GBAs and Their Application for Toxic/Heavy Metals Removal

Formulation	Toxic/Heavy Removed	Adsorbent Dosage	pH for Max. Adsorption	Temp. (K)	Maximum Adsorption Capacity (mg/g)	Adsorption Isotherm & Kinetics		Ref.
						Isotherm	Kinetics	
G and composites								
G-δ-MnO$_2$	Pb^{2+}	40 mg/L	6.0	298	439.1	LM	—	29
	Cd^{2+}	40 mg/L	6.0	298	40.11	LM	—	29
	Cu^{2+}	40 mg/L	6.0	298	64.8	LM	—	29
LGPH (lignosulfonate-G porous hydrogel)	Cr^{6+}	40 mg/L	3.0	298	601.2	LM	PSOM	30
Sulfonated G (GS)	Cd^{2+}	40 mg/L	5.0	298	58	LM	PSOM	31
CTAB-GN	Cr^{6+}	8 g/L	2.0	293	21.57	LM	PSOM	32
PAMAM dendrimer-g-G (GS-PAMAM)	Pb^{2+}	1 g/L	6.0	298	268.4	LM	PFOM	33
	Cu^{2+}	1 g/L	7.0	298	127.2	LM	PFOM	33
GO and Composites								
GO	Eu^{3+}		6.0					34
Colloidal GO	Gd^{3+}		5.9					35
GO	Eu^{3+}		4.5					36
GO	La^{3+}		6.0					37
	Nd^{3+}		6.0					37
	Gd^{3+}		6.0					37
	Y^{3+}		6.0					37
GO/silk fibroin	Ag^{+}							38
CS-EDTA-GO	Pg^{2+}	133 mg/L	—	—	3.7 mmol/g	—	—	39

GO-cellulose	Co^{2+}	—	4.5	—	15.5	LM	PSOM	40
	Ni^{2+}	—	4.5	—	14.3	LM	PSOM	40
	Cu^{2+}	—	4.5	—	26.6	LM	PSOM	40
	Zn^{2+}	—	4.5	—	16.7	LM	PSOM	40
	Cd^{2+}	—	4.5	—	26.8	LM	PSOM	40
	Pb^{2+}	—	4.5	—	107.9	LM	PSOM	40
Methyl-β-cyclodextrin-GO	Pb^{2+}	—	6.0	RT	312.5	LM	PSOM	41
GO	Pb^{2+}	—	6.0 RT	RT	217.39	LM	PSOM	41
Reduced GO and Composites								
Poly(acrylamide) (PAM) polymer brushes on rGO sheets (rGO/PAM)	Pb^{2+}	150 mg/L	6	298.18	1000	LM	PSOM	42
rGO-Dithiocarbamate (DTC)/Fe_3O_4	Cu^{2+}	0.4 mg/L	7	298	113.64	LM	PSOM	43
	Cd^{2+}	0.4 mg/L	7	298	116.28	LM	PSOM	43
	Pb^{2+}	0.4 mg/L	7	298	147.06	LM	PSOM	43
	Hg^{2+}	0.4 mg/L	7	298	181.82	LM	PSOM	43
MnO_2 nanotubes@reduced GO	Pb^{2+}	20 mg/L	5.0	298	1.71 mmol/g	LM	PSOM	44
	Cd^{2+}	20 mg/L	5.0	298	1.58 mmol/g	LM	PSOM	44
	Ag^+	20 mg/L	5.0	298	1.28 mmol/g	LM	PSOM	44
	Cu^{2+}	20 mg/L	5.0	298	1.91 mmol/g	LM	PSOM	44
	Zn^{2+}	20 mg/L	5.0	298	1.28 mmol/g	LM	PSOM	44
G-based Hybrids								
Poly(AMPS-co-VI)-g-GT/PVA	Pb^{2+}	0.2–2 g/L	6	298	81.78	LM	PSOM	45
	Cu^{2+}	0.2–2 g/L	6	298	69.67	LM	PSOM	45
	CV	0.2–2 g/L	6	298	94.0	LM	PSOM	45
	CR	0.2–2 g/L	6	298	101.74	LM	PSOM	45

(Continued)

TABLE 6.2 (CONTINUED)

GBAs and Their Application for Toxic/Heavy Metals Removal

Formulation	Toxic/Heavy Removed	Adsorbent Dosage	pH for Max. Adsorption	Temp. (K)	Maximum Adsorption Capacity (mg/g)	Adsorption Isotherm & Kinetics		Ref.
						Isotherm	Kinetics	
Amino-silane 8 functionalized GO (GO-APTS) and copolymer of 2-acrylamido-2- 9 methylpropanesulfonic acid (AMPS)/maleic anhydride (MA) [poly(AMPS-co-MA)]	Pb(II)	0.4 g/L	≥6	298.15	310.10	LM	PSOM	46
	Cu^{2+}	0.4 g/L	≥6	298.15	282.25	LM	PSOM	46
	Co^{2+}	0.4 g/L	≥6	298.15	416.06	LM	PSOM	46
G-MWCNT-PDA	Cu^{2+}	50 mg/L	7.0	298.15	5.01	LM	PSOM	47
	Pb^{2+}	50 mg/L	7.0	298.15	1.69	LM	PSOM	47
Fe_3O_4@Mesoporous SiO_2-GO	Pb^{2+}	10 mg/L	7.1	298.15	333	LM	—	48
	Cd^{2+}	10 mg/L	7.1	298.15	167	LM	—	48

interactions in aqueous solutions, HA increases the adsorption of heavy metals by MMSP-GO composites. The fundamental mechanism of humic acid and heavy metal ion synergistic adsorption was addressed. The concurrent exclusion of humic acid as well as heavy metals in wastewater treatment systems has shown promise for the use of MMSP-GO nanocomposite adsorbents. Their findings with regard to the adsorption isotherm demonstrated that the LM with an R^2 value of more than 0.99 suited the data better than the FM, indicating that the adsorption of Pb^{2+} and Cd^{2+} on MMSP-GO can be regarded as a monolayer adsorption process. The nearby oxygen atoms were made available to bind metal ions by the several O-containing functionalities that were present on the GO surface. MMSP's amino groups displayed a strong affinity for metal ions concurrently, as well as the coordinate interactions might be used to explain a potential adsorption mechanism.[48]

Another example involves the very effective heavy metal ions adsorption using the ultra-lightweight but also strong 3D G/polydopamine modified MWCNT-PDA hybrid aerogels that were made using a viable green method. The hybrid aerogel was molded without the need for any extra reducing agents, significantly lowering the emissions that contribute to environmental pollution. When used appropriately, MWCNT-PDA gave hybrid aerogels a sufficient level of structural stability, successfully stopped GO sheets from stacking, and exposed more active sites for better heavy metal ion adsorption. The generated hybrid aerogels showed better heavy metal ion adsorption ability, with saturation adsorption capacities for Cu^{2+} and Pb^{2+} of 318.47 and 350.87 mg/g, correspondingly, according to the LM. Due to the substantial specific surface area, porous structure, and capable active sites, this was made achievable. Their study revealed that the primary adsorption mechanisms were the synergistic interactions of surface complexation but also chelation between active sites within the hybrid aerogels with the heavy metal ions. Hybrid aerogels are the best option for treating heavy metal ions in real-world applications owing to their high Q and simplicity of water separation after adsorption.[47]

6.4 REMOVAL OF TOXIC COLORANTS/DYES USING GO/GO-BASED COMPOSITES

Due to its special qualities, including ultra-high carrier mobility, π–π conjugated structure, rich surface functionalities for enhanced electrostatic as well as hydrogen bonding interactions, and special electrochemical properties, GO, also known as a "star molecule," is widely used in technical fields like the treatment of environmental pollutants.

The first 2D carbon crystal with a honeycomb-like structure and unbound π-π electrons is called pure graphene. Through interactions between π-π stacks and hydrophobic properties, pure graphene helps to remove contaminants. Its limited use in the removal of pollutants from water is, however, also a result of its substantial π-conjugated structure and potent hydrophobic performance. The derivatives of G, GO, and rGO, consequently, exhibit greater performance in the adsorption of various contaminants. The C:O ratio in GO is 2–4:1 and in its reduced form is 8–246:1,[51] correspondingly. G derivatives that include O-functionalities are good materials for adsorption applications. Since they are more appealing, O-containing functionalities including –COOH, O–C–O, and –OH have been widely used to remove a wide range of impurities from solid pollutants, water pollutants, and air pollution. Additionally, inorganic particles such as metal and non-metal, along with their oxides in nanoparticle form, can be adopted for GO functionalization.[51]

6.4.1 GO

GO among G-based materials has been the most utilized carbon-based as well as G-based adsorbent for the adsorptive exclusion of dyes from wastewater within recent decades. Some instances are herewith presented in this subsection.

Mesoporous C-material has drawn much attention and has been used in a variety of industries due to its exceptional features, which include a high specific surface area, a steady structure, as well as good adsorbability. In this article, we presented an easy method to make N-doped G-C (NGLC) nanosheets and examined their effectiveness in wastewater decontamination. NGLC nanosheets were produced in this study by combining restricted synthesis with low-temperature calcination under O-limited conditions. Such NGLC nanosheets had a porous structure made of a number of carbon layers, giving them a substantial specific surface area (421.85 m^2/g). During this time, the NGLC was enriched with a large number of active sites, including nitrogen species and O-contained groups, through low-temperature calcination under oxygen-limited circumstances. Due to these characteristics, the NGLC makes a superior adsorbent for the removal of organic pollutants. The highest adsorption capabilities for the cationic dyes, RhB and MB, were as high as 1272.74 mg/g and 679.55 mg/g, respectively. This indicates the promising potential for the treatment of cationic pollutants. Batch adsorption investigations revealed outstanding reusability and broad pH compatibility (2–10). The RhB removal efficiencies from bulk polydopamine produced carbon material (denoted as PDA carbon) and NGLC-450-Ar (formed at argon atmosphere) were only 17.56 percent and 62.03 percent, correspondingly, according to the results of the kinetic adsorption test. The importance of restricted synthesis with template aid and oxygen-limited oxidation was revealed by the large difference in adsorption performance. In an experiment and analysis using FTIR and XPS, where their report established that the process of adsorption involved H-bonding, pore filling as well as electrostatic interaction, but also π-π interaction.[62]

Readers are referred to Table 6.3 for more instances where GO, and its composite materials as adsorbents have been used for the exclusion of diverse dyes for water/wastewater with reference to available literature.

6.4.2 rGO

Utilizing graphite oxide along with metal ions (Fe^{3+} and M^{2+}) as starting materials, a one-pot solvothermal synthesis technique was created to synthesize rGO supported ferrite (MFe_2O_4, M stands for Co, Mn, Zn, as well as Ni) hybrids. XRD, Raman spectra, FESEM, energy dispersive X-ray spectroscopy, TEM, but also vibrating sample magnetometer were used to characterize the hybrids. The homogenous deposit of

TABLE 6.3
GBAs and Their Application for Colorants/Dyes Removal

Formulation	Colorants/Dyes Removed	Adsorbent Dosage	pH for Max. Adsorption	Temp. (K)	Maximum Adsorption Capacity (mg/g)	Adsorption Isotherm & Kinetics		Ref.
						Isotherm	Kinetics	
G								
G	MB		10	293	153.85			52
	Cationic red X-GRL			288	217.39			53
	MB			303	1.52 g.g^{-1}			54
	MB			298	184			55
	RhB			298	72.5			55
	MO			298	11.5			55
G-Fe$_3$O$_4$	MB	0.4 g/L	—	298	43.82	LM	PSOM	56
Fe$_3$O$_4$-G	MB	200 mg/L	—	298	45.27	LM	PSOM	57
	CR	200 mg/L	—	298	33.66	LM	PSOM	57
GO								
GO-SiO$_2$	MO	100 mg/L	3	RT	344.83	LM	PSOM	58
	MB	100 mg/L	10	RT	416.67	LM	PSOM	58
Chi:nGO 3:1	RB	1000 mg/L	4.0	298.15	70	Sips	Elovich	50
	NR	1000 mg/L	5.0	298.15	165	Sips	PFOM	50
GO-silk fibroin	MB	50 mg/L	9.0	303.0	1322.71	Freundlich	PSOM	38
GO-chitosan	MB	—	—	—	275.5	—	—	49
	RhB	—	—	—	92.05 ± 2%	—	—	49
	MO	—	—	—	74.45 ± 2%	—	—	49
	CR	—	—	—	70.72 ± 3%	—	—	49
GO	Acridine orange	—	4–5	RT	2158	LM	—	59
GO-chitosan	MB	500 mg/L	9.0	298	1.60 mmol/g	LM	PSOM	60
	MO	500 mg/L	3.0	298	0.80 mmol/g	Freundlich	PSOM	60
GO	MB	20 mg/L	4.3	298.15	350	LM	PSOM	61

(*Continued*)

TABLE 6.3 (CONTINUED)

GBAs and Their Application for Colorants/Dyes Removal

Formulation	Colorants/Dyes Removed	Adsorbent Dosage	pH for Max. Adsorption	Temp. (K)	Maximum Adsorption Capacity (mg/g)	Adsorption Isotherm & Kinetics		Ref.
						Isotherm	Kinetics	
N-doped G	RhB	400 mg/L	3.0	298	1272.74	LM	PSOM & PFOM	62
	MB	400 mg/L	9.0	298	679.55	LM	PSOM & PFOM	62
	MG	20 mg/L	4.3	298.15	248	LM	PSOM	61
Reduced GO								
rGO/PAM	MB	150 mg/L	6	298.18	1530	LM	PSOM	42
rGO supported ferrite (Mn/Zn/Co/Ni)Fe_2O_4	RhB	0.6 g/L	—	293.15	23	LM	PSOM	63
	MB	0.6 g/L	—	293.15	35	LM	PSOM	63
rGO-Fe_3O_4	RhB	25 mg/mL	7.55	RT	30	—	PSOM	64
	R6G	25 mg/mL	7.55	RT	>90	—	PSOM	64
	Acid blue 92 (AB92),	25 mg/mL	7.55	RT	50	—	PSOM	64
	Orange (II) (OII)	25 mg/mL	7.55	RT	88	—	PSOM	64
	Malachite green (MG)	25 mg/mL	7.55	RT	50	—	PSOM	64
	New coccine (NC)	25 mg/mL	7.55	RT	~46	—	PSOM	64
rGO	MB	6 g/L	6.4	298	100%	Freundlich	PSOM	65
	RhB	6 g/L	6.4	298	97%	Freundlich	PSOM	65
G-based hybrids								
Amino-silane 8 functionalized GO (GO-APTS) and copolymer of 2-acrylamido-2- 9 methylpropanesulfonic acid (AMPS)/maleic anhydride (MA) [poly(AMPS-co-MA)]	MB	0.4 g/L	≥6		416.06	LM	PSOM	46
	CV	0.4 g/L	8.0	298.15	440.97	LM	PSOM	46
PVA-$Ni_{0.5}Zn_{0.5}Fe_2O_4$-GO-CNTs	MB	1340 mg/L	10.0	296.15	71.03	LM	PSOM	66
	Methyl orange (MO)	1340 mg/L	—	296.15	4.26	LM	PSOM	66
	CV	1340 mg/L	—	296.15	9.02	LM	PSOM	66
GO-Chitosan-PVA	CR	6 g/L	2.0		88.17%	LM	POSM	67

monodispersed, uniform-sized MFe_2O_4 microspheres on rGO nanosheets was demonstrated.[63] On the morphology of the hybrids, the effect of metal ion concentration was looked into. The hybrids have low coercivity and remanence, along with significant saturation magnetization. It is significant that the produced hybrids work well as dye pollution adsorbents. When the hybrids are concentrated to 0.6 g/L, it has been discovered that they can eliminate more than 92% of RhB and 100% of MB with a concentration of 5 mg/L in under 2 minutes. RhB and MB degradation was additionally accelerated by the hybrids' heightened photocatalytic activity. With the use of a magnet, the hybrids may be quickly isolated from the solution due to their greater saturation magnetization. RhB and MB's rGO-$MnFe_2O_4$ hybrid's adsorption capability is less than that of pure RGO nanosheets (~37 & 54 mg/g for RhB & MB, correspondingly).[63]

With reference to the literature that is currently available, readers are referred to Table 6.3 for other examples of the usage of GO and its composite materials as adsorbents for the exclusion of various dyes for water/wastewater.

6.4.3 G-Based Hybrids

Successfully created and used in the exclusion of CR dye from its aqueous solution is the GO/Chitosan-PVA polymer composite. By using FTIR, SEM, but also TGA analysis to describe the hydrogel polymer, it was discovered that it had a porous surface, was thermally stable up to 340 °C, and then started to degrade. Based on the collected data, it is evident that the dye-containing solution can be treated using hydrogel, which has been created. In various experimental settings, the Q that depends on time was assessed. According to experimental findings, the preparation of the GO/Chitosan-PVA polymer composite, the amount of adsorbent mass, the pH in the solution, and the initial concentration of CR dye all have a substantial impact on dye adsorption.[67] It was discovered that GO/Chitosan-PVA hydrogel produced better outcomes than Chitosan-PVA hydrogel adsorbent when ultimate concentrations of 20 mg/L CR dye solution were compared at various contact times and pH6 conditions. In the instance of the GO/Chitosan-PVA hydrogel, better outcomes were obtained because the addition of GO to the biopolymer matrix (Chitosan-PVA) improves the hybrid hydrogel's surface's thermal stability and porosity, both of which increase adsorption efficiency. When the initial adsorbate concentration was raised from 10 mg/L to 25 mg/L, the final dye concentration in the solution increased from 1.56 mg/L to 5.13 mg/L. At pH 2, it was found that the removal efficiency and dye concentration at the last stage were, correspondingly, 88.17 percent and 2.9 mg/L. The efficacy of dye removal, on the other hand, was reduced by up to 78.2 percent under higher pH conditions. In order to forecast and assess the CR dye's adsorption process on GO/Chitosan-PVA hybrid hydrogel, a three layers ANN structure model employing a backpropagation algorithm was constructed and effectively designed based on the experimental data. The value of the R^2 acquired using ANN models demonstrated a strong concordance between the simulated and experimental outcomes. As a result of its high surface area and active functional groups, including O-containing groups, the polymer composite with GO produced for the experiments demonstrated very excellent adsorption characteristics.[67]

6.5 EXCLUSION OF PESTICIDES FROM WASTEWATER USING GO/GO-BASED COMPOSITES

Additionally, some research teams have looked into the use of G-based materials as adsorbents to remove pesticides from aqueous solutions.

Due to their carcinogenic effects, pesticides and organic contaminants pose a threat to the human and animal biota. The current research direction is toward using G-based adsorbents to clean up wastewater that contains pesticides. Using the literature that is currently available, we looked at the use of G-based adsorbents for the removal of pesticides in water treatment. The utilization of G, its composites, as well as hybrids for the adsorptive removal of pesticides from wastewater is presented in Table 6.4 below.

Enhancements in adsorption/desorption settings or adsorbent design may be made by better understanding the adsorption mechanism, which is a significant problem. The properties of the adsorbent have an impact on adsorption's effectiveness as well as mechanism (like surface area, porosity, but also surface functionalities), the molecular characteristics of the organic pollutants (such as molecular size, aliphatic vs aromatic composition, as well as hydrophobicity), the characteristics of the background solution (such as pH, ionic strength, natural organic matter, but also temperature), and the interactions between the functional groups of the adsorbent and adsorb Based on the intensity of the interactions between the adsorbate along with the adsorbent, the only adsorption mechanisms documented in the literature are chemisorption or physisorption. According to this theory, physisorption is caused by weak intermolecular reversible physical interactions like dipole-dipole interactions, π-π stacking, diffusion, London interactions, hydrophobic interactions, vdW forces, and H-bonds, with the formation of multiple layers of the adsorbate on the adsorbent. By exchanging or transferring electrons, chemisorption, on the other hand, entails irreversible chemical interactions in a layer between the sorbent and the sorbate, including covalent bonds, chelation, complexation, proton shift, and redox reactions. Depending on the surface functionalities and porosity of the adsorbent, the adsorption of pesticides on G-based adsorbents may be based on physisorption or chemisorption. The mechanism of sorbate removal by the sorbent with regard to G-based adsorbents and pesticides is represented in Figure 6.6 below.

6.5.1 G

G has been proven to serve as an effective adsorbent for the exclusion of pesticides based contaminants from wastewater over recent decades. In this vein, the exclusion of six triazine pesticides from wastewater, including simezine, simeton, cyprazine, ametryn, atrazine, and prometryn, was accomplished by Zhang et al. using a cellulose/G composite (CGC).[70] With the exception of cyprazine, for which the ideal pH was 11, they discovered that the Q of CGC increased with increasing pH and reached its highest at pH 9. Additionally, they claimed that CGC's adsorption capabilities outperformed those of graphitic carbon black, cellulose, primary, secondary amines, and also graphitic carbon "GCB". The outcomes demonstrate that the rGO is effective in altering the cellulose. Comparing the adsorption rate of the composite to

TABLE 6.4

GBAs and Their Application For Pesticides Removal

Formulation	Pesticides Removed	Adsorbent Dosage	pH for Max. Adsorption	Temp. (K)	Maximum Adsorption Capacity (Q_m)	Adsorption Isotherm & Kinetics		Ref.
						Isotherm	Kinetics	
G								
G glazed silica (GCS)	Phonamiphos	170 mg/L	—	—	—	—	—	69
	Phorate	170 mg/L	—	—	—	—	—	69
	Dimethoate	170 mg/L	—	—	—	—	—	69
	Parathion-methyl	170 mg/L	—	—	—	—	—	69
	Pirimiphos-methy	170 mg/L	—	—	—	—	—	69
	Malathion	170 mg/L	—	—	—	LM	—	69
	Fenthion	170 mg/L	—	—	—	—	—	69
	Isocarbophos	170 mg/L	—	—	—	—	—	69
	Chlorfenvinphos	170 mg/L	—	—	—	FM	—	69
	Profenofos	170 mg/L	—	—	—	—	—	69
	Methidathion	170 mg/L	—	—	—	—	—	69
Cellulose/G	Simeton	3 g/L	9.0	—	—	LM	—	70
	Simazine	3 g/L	9.0	—	—	—	—	70
	Atrazine	3 g/L	9.0	—	—	—	—	70
	Cyprazine	3 g/L	9.0	—	—	—	—	70
	Ametryn	3 g/L	9.0	—	—	—	—	70
	Prometryn	3 g/L	9.0	—	—	—	—	70
GO								
2-vinylpyridine-GO	Phenol	5 mg/L	7.0–8.0	298	333.33	LM	PSOM	68
	2,4-dichlorophenol	5 mg/L	7.0–8.0	298	103.092	LM	PSOM	68

(Continued)

TABLE 6.4 (CONTINUED)

GBAs and Their Application For Pesticides Removal

Formulation	Pesticides Removed	Adsorbent Dosage	pH for Max. Adsorption	Temp. (K)	Maximum Adsorption Capacity (Q_m)	Adsorption Isotherm & Kinetics		Ref.
						Isotherm	Kinetics	
C-mCS/GO	Nap	—	—	298	0.093 mmol/g	—	—	[71]
	Anthracene	—	—	298	0.089 mmol/g	—	—	[71]
GO	Phenol	1 g/L	2.0	298	48.54	LM	—	[72]
GO	Methomyl (Met)	1 mg/L	12.0	298	106.22	SM	PSOM & Elovich models (EVM)	[73]
	Acetamiprid (Ace)	1 mg/L	2.0	298	285.96	SM	PSOM & EVM	[73]
	Azoxystrobin (Azo)	1 mg/L	2.0	298	2896.84	SM	PSOM & EVM	[73]
Polyacrylic acid-modified GO (GO-PAA)	Phenol	0.2 mg/L	2.0	298.15	10	LM	—	[72]
GO	Phenol	0.2 mg/L	2.0	298	70	LM	—	[74]
Metal organic framework-GO (UiO-67/GO)	Glyphosate	0.1 mg/L	4.0	RT	482.69	LM	PSOM	[75]
N, N-bis(2-hydroxyethyl) glycine (Bicine)-GO	p-nitrophenol (PNP)	250 mg/L	3.0	298	67.16	LM	—	[76]
	1-naphthol (AN)	250 mg/L	3.0	298	16.99	LM	—	[76]
	Hydroquinone (HQ)	250 mg/L	3.0	298	40.17	LM	—	[76]
	p-aminophenol (PAP)	250 mg/L	3.0	298	96.83	LM	—	[76]
	3-nitrophenol (MNP)	250 mg/L	3.0	298	76.33	LM	—	[76]
	Tert-butyl hydroquinone (TBHQ)	250 mg/L	3.0	298	55.47	LM	—	[76]

Reduced GO								
rGO-PEG-ZnO	BPA	0.25 g/L	6.0	298	485.756	LM	PSOM	[77]
	Phenol	0.25 g/L	6.0	298	511.248	LM	PSOM	[77]
	2–CP	0.25 g/L	6.0	298	531.804	FM	PSOM	[77]
	2,4–DCP	0.25 g/L	7.0	298	570.641	FM	PSOM	[77]
N-doped-rGO	Phenol	200 mg/L	4.0	303.15	155.82	—	PSOM	[78]
	p-nitrophenol	200 mg/L	4.0	303.15	80.60	—	PSOM	[78]
Fe_3O_4-rGO	Atrazine	0.5 g/L	5.0	298	75.24%	LM	PSOM	[79]
	Ametryn,	0.5 g/L	5.0	298	93.61%	LM	PSOM	[79]
	Prometryn	0.5 g/L	5.0	298	91.34%	LM	PSOM	[79]
	Simazine	0.5 g/L	5.0	298	88.55	LM	PSOM	[79]
	Simeton	0.5 g/L	5.0	298	81.22	LM	PSOM	[79]
rGO	Chlorpyrifos (CP)	—	—	303.15	1200	—	—	[80]
	Endosulfan (ES)	—	—	303.15	1100	—	—	[80]
	Malathion (ML)	—	—	303.15	800	—	—	[80]
rGO	Methomyl (Met)	1 mg/L	12.0	298	96.86	SM	PSOM & EVM	[73]
	Acetamiprid (Ace	1 mg/L	2.0	298	357.65	SM	PSOM & EVM	[73]
	Azoxystrobin (Azo	1 mg/L	2.0	298	2818.04	SM	PSOM & EVM	[73]
G-based Hybrids								
AS/NZVI/GO	Methomyl	10 mg/L	7.0	RT	59.13	FM	PSOM	[81]
	Isoprocarb	10 mg/L	7.0	RT	21.33	FM	PSOM	[81]
	Carbaryl	10 mg/L	7.0	RT	61.91	FM	PSOM	[81]
Aminoguanidine modified magnetic-GO	Chlorpyrifos	10 mg/L	7.0	298	85.47	FM	PSOM	[82]
Polyacrylic acid-GO "GO-PAA"	Phenol	1 g/L	2.0	298	84.03	LM	—	[72]

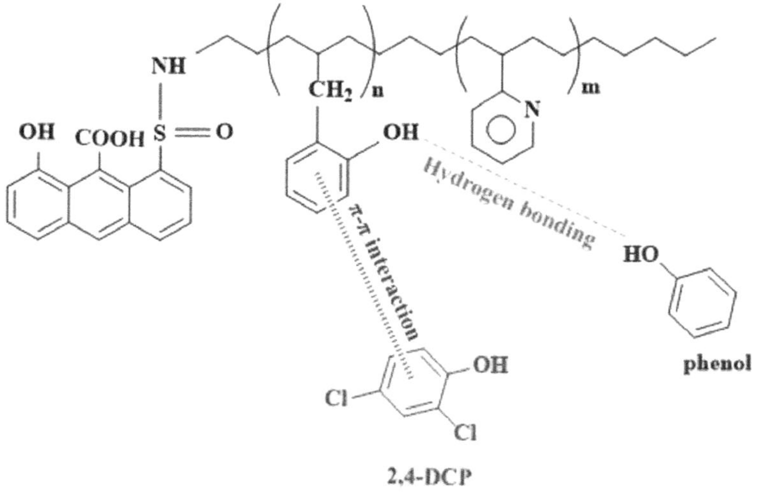

FIGURE 6.6   Adsorption mechanism of phenol and 2, 4-DCP with GO-COOH/TCL/AA/ NIPAM/2-VP. (Reproduced with copyright permission from Khedri et al.,[68] Elsevier 2022.)

that of GCB, GCB, PSA, cellulose, but also G; the rough and wrinkled surfaces with a lamellar structure offer a favorable situation and noticeably improve it. For a 10 mL solution of triazine insecticides, the adsorption process only calls for the addition of 30 mg of CGC. According to tests on adsorption isotherms, the LM more accurately captures the process of adsorption. The nature of adsorption is endothermic, spontaneous, and beneficial. Adsorption is a physical process with rising entropy, as determined by thermodynamic parameters. Furthermore, after six cycles of recycling with a straightforward organic solvent, the CGC's adsorption effectiveness remains above 85%. Therefore, the results of their study demonstrated that cellulose that has been transformed by rGO is a reliable as well as effective adsorbent for triazine herbicides in water. It is postulated that though they obtained excellent results, extra research is necessary to determine the possible Q of the cellulose.[70]

Exfoliated GO was combined with acid-treated SiO_2 to create a novel kind of G-coated SiO_2 (GCS), which was then reduced with hydrazine hydrate to coat the silica particles. This approach is easy, practical, and reliable. Utilizing optical images, Raman spectroscopy, FTIR, XRD, TGA, BET analysis, and elemental analysis, the GCS composite particles were characterized. These investigations demonstrated how successfully G was coated onto silica particles during the process. For the eleven pesticides tested, the composite particles outperform five different sorbents (graphite carbons, activated carbon, pure G, C18 silica, and silica) in terms of adsorption levels and range of applications. We explore the adsorption mechanism and think that it depends on the strong p-bonding network of the benzene rings and the electron-donating properties of the S, P, and N atoms. This study reveals that composite materials based on G could be utilized to eliminate pesticide residues in watery environments.[69] The adsorption process was accelerated and made simpler by the use of silica as a G carrier. The majority of the surface was coated in G, according to the authors, and silica had little to no impact on the adsorption of OPPs. The experimental data were most well-fitted by the LM. Profenofos, fenthion, primiphos-methyl, parathion-methyl, phorate, chlorphenvinphos, methidathion, malathion, isocarbophos, phonamiphos, as well as dimnethoate were the OPPs with the strongest adsorption, in that order. The N, S, and O atoms' strong electron-donor capacities as well as the phenyl rings' – bonding networks, supported the adsorption.[69]

6.5.2 GO

GO has been largely utilized for the adsorptive exclusion of radioactive contaminants from wastewater. We have thus presented some interesting instances within this subsection.

In one study, N, N-bis(2-hydroxyethyl)glycine (Bicine) modified GO was made using a straightforward covalent conjugation process, then was used to adsorb and separate phenols, organic dyes, and REE^{3+} (Y^{3+}, Nd^{3+}, and Eu^{3+}) from aqueous solution. Different types of characterization, such as SEM, FTIR, XPS, and BET, were used. Research was also done on the GO-Bicine composite's ability to adsorb REE^{3+}, phenols, and dyes. The adsorption capabilities for various pollutants by the GO-Bicine composite varied slightly. Compared to Nd^{3+}, and Eu^{3+}, Y^{3+}, had a Q that was 4.8 and 6.8 times higher. As opposed to 1-naphthol, hydroquinone, p-aminophenol, 3-nitrophenol, tert-butyl hydroquinone, and p-nitrophenol presented Q that was 1.4, 5.7, 2.4, 1.3, and 1.7 times higher. When it came to the adsorption of organic dyes, the GO-Bicine composite had a high Q for MB, which was four times higher than that for alizarin yellow R. While π-π stacking interactions, electrostatic interactions and also H-bonding interactions were thought to be the key mechanisms of adsorption for phenols as well as the dyes; though electrostatic interactions, complexation, and Lewis acid-base interactions were implicated in the REE^{3+} adsorption. The LM satisfactorily explained the p-nitrophenol adsorption on the mixture of GO and bicine. The GO-Bicine composite's endothermic and spontaneous adsorption of p-nitrophenol was shown by the adsorption thermodynamics. The GO-Bicine composite will be a viable adsorbent for removing or separating various contaminants from aqueous solutions due to its adsorption selectivity.[76] The adsorption mechanism of the adsorbent for the adsorbate towards REE^{3+}s ($Y3^{3+}$, Nd^{3+}, Eu^{3+}), phenols (AN, HQ, PAP, PNP, MNP, TBHQ), and dyes (MB and AYR) were explained by the authors, as illustrated in Figure 6.7 below.

Yang et al. created hybrid nanocomposites of GO and metal-organic framework (UiO-67-GO) for the elimination of organophosphorus pesticides (OPPs).[75] According to the findings, 482.69 mg/g of glyphosate was adsorbed at pH 4, being consistent with the LM and PSOM. According to XPS investigations, the glyphosate's phosphate surface functionalities interaction with the surface functionalities of UiO-67/GO, which include oxygen, is what causes the substance to adsorb favorably.[75]

A handful of thorough investigations have compared the potential of novel materials to that of conventional materials; new materials have been developed based on the ideas of green chemistry. Proanthocyanidin (PAS) and glutaraldehyde were employed

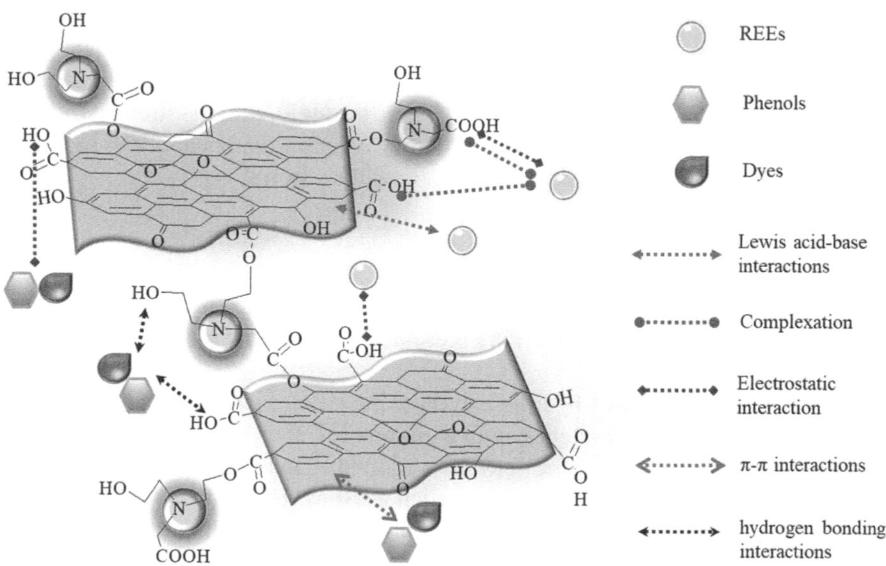

FIGURE 6.7   The schematic illustration of possible intermolecular interactions between different substances and GO-Bicine composite. (Reproduced with permission from Ren et al.,[76] copyright 2021, Elsevier Science Ltd.)

in this study to make two composites based on GO and magnetic chitosan, one using a green approach (G-mCS/GO) and the other using a traditional method (C-mCS/GO), both employing the crosslinking agents as the two different types of mCS. Following that, their efficacy was evaluated by polycyclic aromatic hydrocarbons (PAHs) exclusion from synthetic wastewater employing them as adsorbents. The magnetization plots showed increased saturation magnetization with respect to G-mCS/GO, and the XRD study revealed the crystallographic characteristics. FTIR investigations also revealed the existence of functional groups linked to GO, chitosan, as well as Fe_3O_4 particles. Better Q of G-mCS/GO was observed for Nap (0.093 mmol/g) and anthracene (0.089 mmol/g), having exclusion effectiveness of 93.55 percent and 89.25 percent, correspondingly, in adsorption assays carried out at an adsorbent dosage of 0.1 mg/mL, pH 6.5, 298 K, for 30 min, and an initial PAH concentration of 0.01 On the other hand, fluoranthene (0.076 mmol/g) had a superior removal efficiency with C-mCS/GO, at 76.07 percent. According to the results of the molecular simulation, Nap molecules are more stable than anthracene and fluoranthene molecules. This stability increases the π-π, but also hydrophobic interactions between the adsorbent surface and PAH molecules in situ, improving the Q. The findings of this investigation suggest that G-mCS/GO is a feasible and long-lasting adsorbent for the removal of PAH from wastewater.[71]

For phenol cleanup from synthetic and actual wastewater, a novel polyacrylic acid-modified GO (GO-PAA) composite material was studied. SEM, TEM, FTIR, EDX, and BET were used to physically and chemically evaluate the GO and GO-PAA (BET). It investigated how several experimental variables, such as pH, temperature, and beginning phenol concentration, affected the results. Results showed that at pH 2 and 25 °C, the best adsorption took place. The Q of GO-PAA was double that of unmodified GO, showing the value of surface adjustment, but also the addition of C=O functionalities, in enhancing the adsorption process. Furthermore, it was determined that the LM was the most appropriate isotherm to represent the phenol adsorption process. Also, the thermodynamics investigations supported the impulsiveness, but also the exothermic nature, of the adsorption phenomenon. Under ideal conditions of pH 2 and 25 °C, GO-PAA was able to remove phenol from synthetic and actual wastewater by 75% and 18%, respectively. These findings suggest that, especially in acidic environments, phenolic compound adsorption utilizing GO-PAA may be a quick and efficient way to clean up pollution.[72]

A facile approach for making 2-vinylpyridine/thermosensitive polymer functionalized-GO to be used as a nanoengineered adsorbent for the exclusion of phenol along with 2, 4-dichlorophenols' in wastewater via ultra-performance liquid chromatography was reported by Khedri et al.[68] Aimed at optimizing the exclusion process, experimental factors such as pH, contact time, temperature, the dosage of adsorbent, phenols' initial concentration, but also 2, 4-dichlorophenols' were investigated. pH=7, a contact period of 10 min, 0.005 g dosage of adsorbent, as well as a 298 K temperature were found to be the ideal conditions for phenol elimination. The 2, 4-dichlorophenol removal procedure was tested under the following conditions: pH = 8, 0.005 g of adsorbent dose, and 5 min of contact time at 298 K.[68] The experimental results supported the Langmuir isotherm model for the elimination of phenol and 2, 4-dichlorophenol from an aqueous solution and were in good agreement with the PSOM. For phenol and 2, 4-dichlorophenol, correspondingly, the maximal Langmuir sorption capacities were 333.33 and 103.092 mg/g at 298 K. The procedure was possible and spontaneous, according to the results of the thermodynamic study ΔG for phenol, and 2, 4-dichlorophenol was found to be −8.199 to −8.733 kJ mol^{-1} and −14.796 to −15.069 kJ mol^{-1} but also an exothermic nature for phenol, and 2, 4- dichlorophenol established as ΔH value of -1821.51 J mol^{-1} and −18325.7 J mol^{-1}. There was an 89 percent and a 99 percent maximum removal efficiency for 2, 4-dichlorophenol compounds, correspondingly. There was an 89 percent and a 99 percent maximum removal efficiency for 2, 4-dichlorophenol compounds, correspondingly.[68]

6.5.3   rGO

The adsorptive removal of pesticides from wastewater using G-based adsorbents have been demonstrated by several researchers globally within the recent decades. The drive toward the utilization of rGO as well as its composites/hybrids has been confirmed from literature to be based upon its large surface area and active surface functionalities, which help in active interaction with the contaminants molecules via hydrogen bonding as well as electrostatic interaction: rGO's structure also possess the potential to interact with pesticide molecules via π-π interaction mainly for pesticides containing aromatic benzene rings in their structure.

At pH 5 at 25 °C, Boruah et al. found adsorption efficiencies of 93.61, 91.34, 88.58, 81.22, as well as 75.24 percent for ametryn, prometryn, simazine, simetone, but also attrazine, correspondingly.[79] Due to the greatest electrostatic interactions, while the least amount of iron was leaching from the adsorbate at pH 5, the Q

heightened with rising pH and was maximal. The oxygen-containing functions of the Fe_3O_4/rGO nanocomposite were discovered to engage electrostatically with the $\equiv N^+$-group of the insecticide. For good adsorption, the strong π–π contacts were most important.[79]

For the treatment of phenolic compound-contaminated wastewater, conjugated π region G-based materials offer special advantages. N-doped rGO ("N-rGO") with an expanded conjugated π area along with increased hydrophobicity was created by annealing the chitosan-based composite that had been exfoliated as well as anchored within a GO matrix in order to examine the mechanisms of interaction aimed at optimizing the exclusion efficiency of phenol along with p-nitrophenol. At 200 mg/L initial concentration, pH 6, but also a temperature of 30 °C, their findings revealed that the equilibrium adsorption capacities of N-rGO for phenol as well as p-nitrophenol were 155.82 and 80.60 mg/g, correspondingly. Robust π-π and hydrophobic interactions were credited with the enhanced removal efficiency. Regeneration tests further demonstrated that N-rGO may continue to maintain a high removal effectiveness (greater than 80%) after five reuse cycles. Using an effective and reusable adsorbent, these results indicate that N-rGO is a potential method for removing phenolic chemicals from wastewater.[78]

Maliyekkal et al. investigated the adsorption of malathion, endosulfon, and chlorpyrifos onto GO, but also rGO. According to reports, the adsorption capacities of CP, ES, as well as ML were as high as 1200, 1100, and 800 mg/g, respectively. The lack of sensitivity to pH changes in the adsorption process demonstrated that the rGO surface COO^- functionalities did not directly interact with the adsorbabte (pesticide).[80] They utilized first-principles pseudopotential-based DFT to understand the atomistic mechanism of interactions, leading to their revelation that only the pesticide polar bonds altered substantially when G and water came into contact. Electrostatic interactions may be seen when the structure and charge density are shown.

The functionalization of magnetic GO by aminoguanidine (abbreviated AGu@mGO(R)) allowed a group of researchers to design and produce a G-based nanoadsorbent in an intriguing inquiry. First, using a modified version of Hammer's approach, GO was created from rice husk biomasses.[82] Then, after co-precipitation of Fe^{2+} and Fe^{3+} onto GO in primary media, magnetic GO (mGO) was produced. Aminoguanidine was then used to functionalize the mGO to create AGu@mGO(R) nanoparticles. Accordingly, the synthesized nanoadsorbent made of G was characterized. The produced nanocomposite "AGu@mGO(R)" was tested for its ability to remove chlorpyrifos, one of the most commonly used pesticides, from real samples of water and cucumber juice. In order to achieve the most effective pesticide removal, a variety of the factors that affect the adsorption process were tuned. We also analyzed the adsorption processes' isotherm, kinetic, and thermodynamic properties. Using HPLC-MS/MS, it was possible to measure the chlorpyrifos desorption efficiency from AGu@mGO(R). The outcomes demonstrated AGu@mGO(Refficient)'s absorptive capacity for removing chlorpyrifos. The remaining herbicide in agricultural wastewater would, therefore, be removed using this nonabsorbent. The measurement of pesticide residues in samples might also be done using the AGu@mGO(R).[82]

It's important to keep in mind that industrialisation and agricultural development have both had negative effects on living things recently. Community members have been worried about the environment because of the growing issue of organic pollution. These include priority harmful pollutants that are continually emitted into the environment by a variety of businesses, such as phenolic pollutants like phenol, 2-chlorophenol (2-CP), 2,4-dichlorophenol (2,4-DCP), but also bisphenol-A (BPA). In this study, a biocompatible composite of rGO and zinc oxide with polyethylene glycol functionalization (rGO-PEG-ZnO) was created and tested for the adsorptive removal of harmful phenolic contaminants from water. The solution pH optimal adsorption parameters were 7, 60 min of adsorption period, 25 °C, and 0.25 g/L of dose. The maximal uptake of BPA, phenol, 2-CP, and 2,4-DCP were 485.756, 511.248, 531.804, and 570.641 mg/g, correspondingly. The isotherms for BPA and phenol were well matched by the LM, however, for 2-CP and 2,4-DCP, Freundlich was the best-suited model. The PSOM is supported by the kinetic data for all phenolic contaminants. The process was confirmed to be spontaneous by the thermodynamic study, which reveals that the Gibb's free energy ($\Delta G°$) values for all the pollutants were negative. The aforesaid adsorption process was endothermic, according to the positive values of change in enthalpy ($\Delta H°$), which were 28.261, 37.205, 46.182, and 61.682 kJ/mol for BPA, phenol, 2-CP, and 2,4-DCP, correspondingly. With a slight decrease in the removal percentage, the composite can be utilized for up to five cycles. According to the synthetic composite's adsorption efficiency for synthetic industrial effluents, up to 86.54 percent of the waste was removed in 45 minutes. RGO-PEG-ZnO composite can be considered a successful adsorbent for treating phenolic contaminants from wastewater due to its exceptionally quick adsorption and high Q.[77]

6.5.4   G-Based Hybrids

The drive for more effective and efficient G-based adsorbent materials in recent decades has driven researchers and industrialists towards the design and assembly of advanced materials for which hybrid systems are of pivotal interest. In this subsection, we have presented G-based hybrid systems that have been utilized by researchers for the adsorptive exclusion of wastewater contaminants, as per the literature.

Pesticides have apparently been removed from sewage using a low-cost, reusable multifunctional adsorbent made from herb waste, nano-zero-valent ferrium, and GO (AS/NZVI/GO). The AS/NZVI/GO has numerous outstanding qualities that have been claimed to be helpful in achieving high adsorption efficiency, including an abundance of pore structures, many surface functions, and a high specific surface area. The highest adsorption capacities of AS/NZVI/GO for carbaryl, isoprocarb, as well as methomyl up to 59.13, 21.33, and 61.91 mg/g, correspondingly, demonstrated the high adsorption effectiveness of AS/NZVI/GO for pesticides.[81] Pore filling, electron donor-acceptor interaction (π-π), hydrogen bonds, and electrostatic interaction were the primary factors in the adsorption mechanism, as depicted in Figure 6.8. According to the results, the Freundlich isotherm model and the PSOM provided a superior fit to the experimental data, demonstrating that physical and chemical adsorption mostly drove the overall rate of the adsorption process. In addition, the

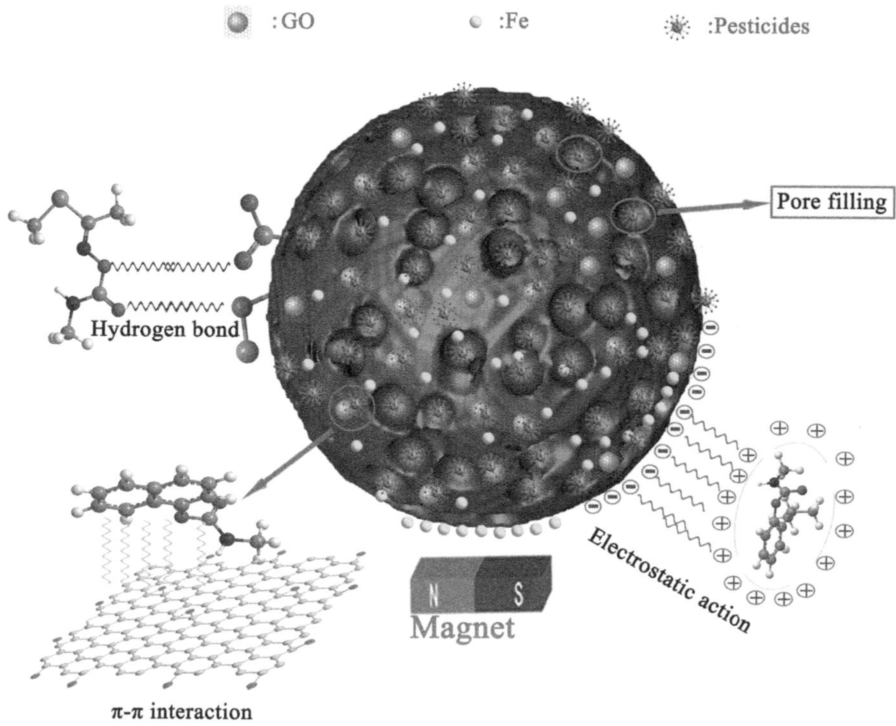

FIGURE 6.8 The schematic diagram of the possible adsorption mechanism of AS/NZVI/GO on pesticide molecules. (Reproduced with permission from Du et al.[81] Copyright 2021, Elsevier Science Ltd.)

trial on reusability showed that after five recycles, the AS/NZVI/GO had removed three pesticides with an efficiency of over 70%. In conclusion, the new AS/NZVI/GO could be employed as a powerful adsorbent to remove pesticides.[81]

In one study, magnetic GO was functionalized by aminoguanidine (abbreviated Agu/mGO®) to create G-based nanoadsorbent.[82] The GO was initially created using a modified Hammer's approach using rice husk biomasses. Then, magnetic GO (mGO) was produced by the co-precipitation of Fe^{2+} and Fe^{3+} onto GO in primary media. Aminoguanidine was then used to functionalize mGO to create Agu/mGO® nanoparticles. Vibrating sample magnetometer (VSM) analysis, SEM, TEM, AFM, DLS, XRD, and other methods were used to analyze the synthesized G-based nanoadsorbent. Chlorpyrifos, one of the most commonly used pesticides, was removed from genuine samples of water and cucumber juice using Agu/mGO® to assess its effectiveness. In order to achieve the most effective pesticide removal, a variety of adsorption-altering factors were tuned. We also analyzed the adsorption processes' isotherm, kinetic, and thermodynamic properties. Using HPLC-MS/MS, it was possible to measure the chlorpyrifos desorption efficiency from Agu/mGO®. The outcomes demonstrated Agu/mGO(Refficient)'s absorptive capacity for removing chlorpyrifos. The remaining herbicide in agricultural wastewater would, therefore, be removed using this nonabsorbent. The measurement of pesticide residues in samples might also be done using the Agu/mGO®.[82]

Modern industrial operations have led to a widespread presence of persistent organic pollutants (POPs) in the environment. In an effort to address the POPs contamination problem, a team of researchers developed an efficient POPs decontamination technique based on POPs dynamic adsorption onto the surface of rGO-coated SiO_2-Pt Janus magnetic micromotors toward their suitable final disposition. While the motors move quickly through a contaminated fluid, it has been found that POPs are effectively adsorbed in a short period of time.[83] As model POPs, polybrominated diphenyl ethers (PBDEs) and 5-chloro-2-(2,4-dichlorophenoxy) phenol (triclosan) were chosen, and the contaminants were successfully removed. Comparing rGO-coated micromotors to their contemporaneous GO-coated, static rGO-coated, as well as dynamic SiO_2 micromotor counterparts, it was found that the rGO-coated micromotors had superior adsorbent qualities. Using a variety of micromotors with magnetic properties that were employed for their collection from environmental samples, the degree of decontamination was investigated. For four reuse cycles of the micromotors, the adsorption properties were intact. The new rGO-coated SiO_2 functional material-based micromotors have shown exceptional skills for POP removal and subsequent disposition, bringing up new opportunities for the effective environmental cleanup of these dangerous compounds.[83]

Hybrid materials are nowadays used by most researchers owing to their inherently better property performance in comparison to their individual components or their composite counterparts, as discussed in several studies earlier.

6.6 REMOVAL OF RADIONUCLIDES

Radioactive waste discharge is a growing source of concern since the exposure of living organisms to radioactive wastewater can cause a variety of health issues due to chemical toxicity and radiation. In effluent from research facilities, nuclear power plants, nuclear weapons testing, and spent fuel reprocessing, large quantities of radionuclides with extended half-lives, such as ^{134}Cs, ^{135}Cs, and ^{137}Cs, are produced during fission. Cs must be removed from wastewater as a result.[84,85] It is still highly challenging to successfully remove these radioactive pollutants from wastewater due to their high affinity, good solubility in water, and ease of mobility in the environment.[85]

Ion exchange, membrane separation, chemical precipitation, adsorption, and evaporation are all relatively efficient techniques that can be used to remove radionuclides. Adsorption is thought to be a good strategy because it is inexpensive and simple to implement. Numerous different adsorbents have been developed for the treatment of radioactive effluents. The efficiency, capacity, and selectivity of the currently accessible adsorbents don't meet the demands for utilization, nevertheless, in most instances. Therefore, for the effective treatment of radioactive effluents, new GBAs must be developed.

A substantial amount of radioactive material is released into the aquatic environment as a result of mining and nuclear operations. As a result, aquatic life is at risk.[86] Through water systems, the emitted fission product gets into the food chain. Radioactive nuclides are issued into the environment by human activities such as fertilizer use, lignite burning in powder plants, and mineral processing. Therefore, an effective method of treatment is required to eliminate long-lived radionuclides. The environmental effects of radioactive waste produced during mining and nuclear energy generation are long-lasting.[85] The water system can allow the fission byproducts ^{90}Sr, ^{137}Cs, ^{235}U, and ^{129}I to enter the food chain. These radionuclides can pollute water, which can then seep into the soil, be ingested by plants, and finally reach animals and people. Ore processing, lignite burning in power plants, and fertilizer use all result in the production of radionuclides. This radioactive pollution of freshwater is a severe problem.[85]

Owing to their exceptional physicochemical properties, like their high specific surface area as well as the abundance of surface functional groups, G, GO, as well as rGO, are receiving more attention for the adsorptive exclusion of radioactive contaminants from wastewater.[84,85] The utilization of various GBA materials for the exclusion of radionuclides, as well as their performance in terms of properties, are shown in Table 6.5.

6.6.1 G

By processing GO with reducing agents, it is simple to produce rGO, which is frequently used as a substitute for G in environmental cleanup. However, rGO displays fewer OCGs along with a more aromatic structure compared to GO, while having a richer structure than G. In water filtration, rGO is highly wanted because of its unique structure, which might embrace wrinkles, faults, and functional groups. The reduction paths for rGO syntheses can be broadly categorized as biological reduction, chemical reduction, thermal reduction, as well as electrochemical reduction, according to prior investigations.

A type of 2D layer of carbon with sp^2 bonds is called a G nanosheet (GNS). Contrary to traditional carbon materials such as activated carbon or CNTs, however, their high surface area is independent of their solid state pores distribution.[87] Owing to their large surface area but also exceptional mechanical rigidity, they have a number of important advantages in the manufacture of carrier matrices. A technique using microwave assistance was used to create a specific sort of G nanosheet/δ-MnO$_2$ (GNS/MnO$_2$) composite.[87] The LM can be used to explain the adsorption isotherm. At ambient temperature, Ni (II) may saturate and adsorb itself to GNS/MnO$_2$ at a rate of 46.6 mg/g, which is 1.5 and 15 times greater than pure δ-MnO$_2$ and GNS, correspondingly. The endothermic reaction, as well as the rise in randomness at the solid-liquid interface during the Ni (II) adsorption process, are suggested by the positive values of both ΔH but also ΔS. The ΔG values are negative, indicating spontaneous adsorption and GNS/MnO$_2$ has a high recovery rate of 91 percent. The PFOM and PSOM were explored to determine the adsorption kinetics of Ni^{2+}. The GNS/MnO$_2$ composite, MnO$_2$, and GNS have correlation coefficients (R^2) of 0.8672, 0.7934, and 0.6228, correspondingly, according to the first kinetic model. High regression correlation coefficients ($R^2 > 0.99$) are discovered for the PSOM, indicating that it provides a more accurate description of the adsorption kinetics than the PFOM. For all adsorbents, the estimated (cal) values of q$_e$ from the PFOM are significantly lower

TABLE 6.5
Removal Radionuclides Using G-Based Adsorbents

Formulation	Radionuclide Removed	Adsorbent Dosage	pH	Temp. (K)	Maximum Adsorption Capacity (Q_m)	Adsorption Isotherm & Kinetics		Ref.
						Isotherm	Kinetics	
G								
G/δ-MnO$_2$	Ni^{2+}		9.0	298	46.55 mg/g	LM	PSOM	[87]
Prussian blue analog-doped reduced GO aerogels (RGOA/TPBAs)	Radioactive cesium (Cs$^+$ or Cs(I))	0.25 g/L	7.0	298.15	226.98 mg/g	LM	PSOM	[84]
MXene@GO	U(VI)	0.3–0.5 g/L	5.0	298	1003.5 mg/g	LM	PSOM	[88]
GO composite nitrogen-doped carbon (GO@NCs)	U(VI)	50 ppm	5.0	298.15	879.2 mg/g	LM		[89]
Chitosan@GO	U(VI)	160 mg/L	6.0	303	822 mg/g	LM	PSOM	[90]
GO								
Alginate/GO/layered double hydroxide (LDH@GO-5%)	Sr^{2+}	1 g/L	5.95	RT	0.91 ± 0.1	Freundlich	—	[91]
	SeO$_4$$^{2-}$	1 g/L	5.95	RT	0.37 ± 0.1	LM	—	[91]
PAM@GO	Sr^{2+}	0.2 g/L	8.5	303	2.11 mmol/g	LM	PSOM	[92]
GO/hydroxyapatite (HAP)	U (VI)	0.1 g/L	3.0	Room temp.	373.00 mg/g	LM	PSOM	[93]
PVA/GO	Sr(II)	0.5 g/L	7.0	298	20.07 mg/g	LM	PSOM	[94]
GO-PEDOT-PSS	U (VI)	0.2 g/L	4.5	298	84.51 mg/g	LM	PSOM	[95]
Phosphate-functionalized-GO (PGO)	U (VI)	0.6 g/L	4.0	303	251.7 mg/g	LM	PSOM	[96]

(*Continued*)

TABLE 6.5 (CONTINUED)

Removal Radionuclides Using G-Based Adsorbents

Formulation	Radionuclide Removed	Adsorbent Dosage	pH	Temp. (K)	Maximum Adsorption Capacity (Q_m)	Adsorption Isotherm & Kinetics		Ref.
						Isotherm	Kinetics	
rGO								
rGONF	U (VI)	0.3 mg/L	5.0	333 ± 2	200 mg/g	LM	PSOM	[97]
	Th (IV)	0.3 mg/L	5.0	333 ± 2	126.68 mg/g	LM	PSOM	[97]
	U (VI)	2 mg/100 ml	6.5	298	1532.35 mg/g	LM	PSOM	[98]
	Eu(III)	2 mg/100 ml	6.5	298	886.44 mg/g	LM	PSOM	[98]
GO-WO$_3$	Sr^{2+}	1 g/L	6–8	—	149.56 mg/g	LM	PSOM	[99]
rGO-Prussian blue	Cs$^+$				226.98 mg g	LM	PSOM	[84]
Hybrids								
M@GO@Cs	U(VI)	15 mg/L	6.0		504 mg/g	LM	PSOM	[100]
GO@nickel ferrite (GONF)	Uranium (VI) (U (VI))	0.3 mg/L	5.0	333 ± 2	135.13 mg/g	LM	PSOM	[97]
	Thorium (IV) (Th (IV))	0.3 mg/L	5.0	333 ± 2	88.49 mg/g	LM	PSOM	[97]
Ca–Mg–Al LDH-GO	Eu(III)	0.35 mg/L	6.0	303	1.12×10^{-3} mol/g	LM	PSOM	[101]

than the experimental (exp) values. The theoretical and experimental q_e values, however, match up quite well with the PSOM.[87]

The reports on the utilization of pristine G for the adsorptive removal of radioactive wastewater pollutants have been extremely understudied, based on available literature, and require careful investigation, seeing G's outstanding performance towards radionuclides removal.

6.6.2 GO

Amongst G-based materials, GO has been the most utilized material for the adsorptive removal of radionuclides from wastewater owing to its excellent surface functionalities and uniform dispersion in water.

In this regard, an interesting research has shown the co-adsorption of radioactive SeO_4^{2-} and Sr^{2+} using GO-coated alginate particles.[91] The synthetic LDH/GO alginates beads, according to the scientists, were also used to set up small-nonevent adsorption columns having loaded synthetic SeO_4^{2-} and Sr^{2+} polluted wastewater. According to the water chemistry, utilizing alginate beads dramatically increased the amount of Sr^{2+} adsorbed, which was attributable to the GO or alginic acid surface functionalities. The manufactured LDHs present in the alginate beads substantially influenced the incorporation of SeO_4^{2-}. After the creation of alginates, precisely, the Q of Sr^{2+} (0.85–0.91 mmol/g) on GO somewhat increased. So, it was concluded that this multilayer material had been partially exfoliated during manufacturing, increasing the sorption sites.[91]

According to one research investigation, Sun et al. developed iron-rGO for the adsorptive exclusion of U (VI).[102] Li et al. have investigated MXene@GO nanocomposites to eliminate U (VI)[89] recently. These two substances could both be efficiently absorbed by their produced adsorbent/nanocomposites, which could also be recycled up to five times.[103]

In an interesting report, GO-Bicine engineered nanocomposite was explored as an adsorbent for REE^{3+} selective adsorption; compared to Nd^{3+} and Eu^{3+}, the adsorption capability of Y^{3+} was 4.8 and 6.8 times greater.[76] Even after 8 regeneration cycles, the adsorption efficiency of the composite toward Y^{3+} was still 39%, which presented the adsorbent as a great material towards radioactive contaminants remediation from wastewater.[76]

In a different study, polyacrylamide-modified GO (PAM/GO) composites were created and examined using transmission and scanning electron microscopy, Fourier transform infrared spectroscopy, Raman spectroscopy, and X-ray diffraction. FITEQL 3.2 code was used to model the surface site distribution of PAM/GO. In order to remove Sr(II) from aqueous solutions, the PAM/GO composites that were produced were used. It was discovered that the adsorption conditions, including contact time, pH value, and ionic strength, depended on these variables. The highest adsorption capacity was found to be 2.11 mmol/g at pH = 8.5 ± 0.1 and T = 303 K, which was in good agreement with the LM. The adsorption of Sr(II) on PAM/GO composites was likely an endothermic, spontaneous process, according to thermodynamic characteristics determined from the temperature-dependent adsorption isotherms. PAM/GO may be a possibility due to the effective adsorption for the preconcentration and separation of Sr(II) from wastewater.[92] The authors hypothesized that the adsorption process is connected to electrostatic interaction in addition to the fact that adsorption is increased with rising pH.

By co-precipitating GO with nickel and iron salts in a single pot, inverse spinel nickel ferrite (GONF) and rGO-based inverse spinel nickel ferrite (rGONF) nanocomposite were created for the removal of U^{6+} and Th^{4+}.[97] Up to five cycles of recycling and reuse were possible with GONF and rGONF, which could be separated by an external magnetic field with no discernible decrease in adsorption ability.[97]

By directly joining GO (GO) and poly(3,4-ethylenedioxythiophene):poly(styrenesulfonate) (PEDOT:PSS) using a simple ball milling approach, a group of authors created a novel material (GO/PEDOT:PSS), which exhibits impressive performance for the immobilization of U(VI).[95] According to the batch tests, GO/PEDOT:PSS displays ionic strength-independent sorption edges and temperature-promoted sorption isotherms, showing an inner-sphere complexation with endothermic character. The PSOM, which produces a rate constant of 1.09. $\times 10^{-2}$ g/mg/min can be used to illustrate the sorption kinetics. Meanwhile, the sorption isotherms coincide with the LM, which predicts that the Q_m is measured to be 384.51 mg/g at pH 4.5 under 298 K, indicating a monolayer sorption mechanism. According to the results of the FT-IR and XPS examinations, the surface carboxyl/sulfonate group is responsible for the chelation of U(VI), indicating that the PSS moiety may be to blame for the increased sorption capacity. These results can significantly influence the design strategy for creating extremely effective adsorbents in the field of treating radioactive wastewater.[95]

One study used the Arbuzov reaction to graft triethyl phosphite onto the surface of GO (GO), producing phosphate-functionalized GO (PGO). With a Q_m of 251.7 mg/g at pH = 4.0 ± 0.1 and T = 303 K, it was examined how the PGO may be used to remove U(VI) from the aqueous solution. A study using X-ray photoelectron spectroscopy also looked into the adsorption mechanism and found that U(VI) was chemically adsorbing to the surface of PGO. Additionally, the findings of the experiments showed that U(VI) was extracted from environmental contaminants more specifically than other heavy metal ions, with greater removal efficiency toward U(VI) on PGO surface.[96] Their synthesized composite demonstrated excellent selective adsorption of U(VI) in a mixture of Sr(II), U(VI), and Co(II) as depicted in Figure 6.9 below.

6.6.3   rGO

rGO has been vastly utilized by numerous researchers globally for the adsorptive removal of wastewater contaminants in recent decades.

GONF as well as rGONF composites, which is made by co-precipitating GO with nickel along with iron salts in one pot, have been reportedly used to explore the removal of U^{6+} but also Th^{4+} from water.[97] These researchers discovered that rGONF and GONF have porous surface morphologies with mean particle sizes of 32.16 nm and 41.41 nm, correspondingly. Tests using the magnetic property measurement system were done to verify that the creation of ferromagnetic GONF and superparamagnetic rGONF was successful (MPMS). According to their research, Th(IV) and U may be adsorbate using the PSOM kinetics of adsorption (VI).[97] Their isotherm data

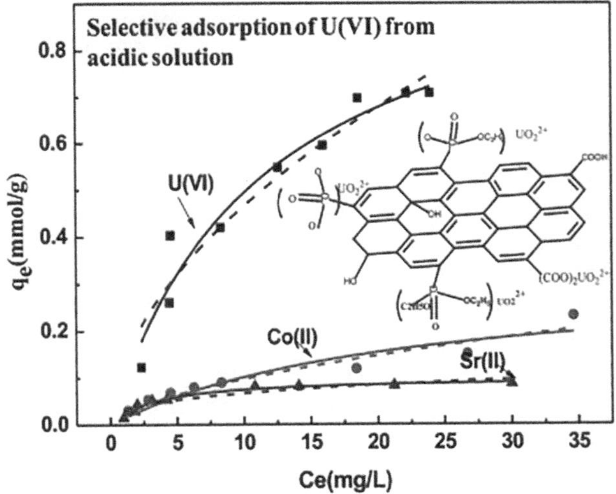

FIGURE 6.9 Adsorption isotherms of U(VI), Sr(II) and Co(II) on PGO surface, the solid line stands for the LM and the dash line stands for the Freundlich model, pH = 4.0 ± 0.1, m/V = 0.05 g/L, I = 0.01 mol L^{-1} NaClO$_4$, T = 303 K. (Reproduced with permission from Xia et al.,[96] Copyright 2015, Elsevier Science Ltd.)

for adsorption showed that Th(IV) and U were absorbed by the monolayer on the GONF and rGONF's homogeneous surface (VI). U(VI) and Th adsorptions increased when the system temperature increased from 293 to 333 ± 2 K (IV). The thermodynamic studies indicated that the adsorption of Th(IV) and U(VI) onto rGONF and GONF was endothermic. The Q of rGONF and GONF, which could be separated by an external magnetic field, was recycled and used up to five times without any material loss. rGO and WO$_3$ were combined to create 3D nanostructured composite adsorbents, according to Mu and his coworkers.[99] As a consequence of the research, it appears to have a mesoporous 3D structure with uniformly loaded WO$_3$ nanorods on the rGO surface. The advantages of GO and WO$_3$ are combined to create composites that have a higher capacity for adsorption over a wider pH range (4–11) for removing Sr^{2+} from aqueous solutions. The Q_m of 149.56 mg/g was attained, which is significantly greater than that for GO, WO$_3$, and other comparable adsorbents, and adsorption isotherms demonstrate that the data fit the LM well (R > 0.99). Within 200 minutes, equilibrium for Sr^{2+} adsorption on rGO/WO$_3$ was reached. The numerous adsorption sites made available by the scattered WO$_3$ nanoparticles on the RGO surface are primarily responsible for the rapid and high rates of adsorption of RGO/WO$_3$. The exclusion of Sr^{2+} ions by rGO/WO$_3$ was not significantly influenced by the presence of Na$^+$ ions, and the rGO/WO$_3$ adsorbent can be reused at least five times without suffering a substantial loss in Q, according to an adsorption-desorption experiment. The possibility of removing 90Sr from radioactive wastewater is therefore demonstrated by rGO-WO$_3$.[99]

The removal of radioactive cesium (Cs) from effluent has allegedly been accomplished using rGO aerogels doped with ternary Prussian blue analogues (rGOA/TPBAs), which were produced utilizing a controlled self-assembly approach. A unique 3D layered structure that makes it simpler to expose adsorption sites and, as a result, promotes the adsorption of Cs^+ is another feature of rGOA/TPBAs, in addition to their huge specific surface area and superior mechanical strength. The composite material has a good cesium adsorption performance, with a Q_m of 226.98 mg/g (dosage of 0.25 g/L, pH of 7, temperature of 298.15 K). Additionally, rGOA/TPBAs maintained a high removal rate of more than 90% while maintaining outstanding Cs+ selectivity in a solution with other ions ($m/V = 0.25$ g/L, $Mn^+ = Ca^{2+}$, Mg^{2+}, Na^+, K^+, $[Cs^+]_{initial} = 10$ mg/L). It is evident that the adsorption is homogeneous monolayer chemisorption because the proposed secondary kinetic model and the LM both suit the adsorption process well ($R^2 = 0.9982$, $R^2 = 0.9929$). The investigations with coexisting ions show that the composite has good selectivity.[84]

The preparation of high-performance adsorbent by modifying GO with magnetite and chitosan via a chemical solution mixing for the adsorption of uranium has been explored by Sharma et al.,[100] as depicted in Figure 6.10. These researchers observed that their prepared composite, which was more of a hybrid, displayed a maximum adsorption capacity of 504 mg/g: the adsorbent (M@GO@Cs) could be used 10 times while still maintaining 90% and also, and the synthesized adsorbent was proven to be effective for the adsorption of mixed radionuclides such as strontium and cesium mixture.

An effective method for getting rid of uranium has been proposed, which is photocatalysis-enabled transition of U from the soluble U^{6+} to the insoluble U^{4+}.[104] In this study, a novel aerogel composite with a 3D structure (rGO@TiO$_2$) was used to reduce U^{6+} using visible light. The addition of RGO might widen the area of light absorption and improve the effectiveness of charge separation in 3D rGO@TiO$_2$. As a result, 3D rGO@TiO$_2$-3 (where the mass ratio of GO and H-TiO$_2$ was 1:1) produced the highest U^{6+} reduction rate (0.03752 /min), which is nearly 5 times higher than that of rGO. While this was happening, U^{6+} was primarily reduced to (UO$_2$)O$_2$·2H$_2$O in air-equilibrated wastewater by a reaction involving photogenerated electrons as well as superoxide radicals ($^\cdot$O$_2^-$). This research advances our understanding of how to clean up radioactive environments and demonstrates the potential of 3D rGO@TiO$_2$ in the photocatalytic field.[104]

6.6.4 G-BASED HYBRIDS

G-based hybrid materials are among the leading systems that researchers and industrialists have focused on, owing to the excellent amalgamation of diverse materials characteristics into a single system having better overall property performance as adsorbents, photocatalysts, active reinforcing fillers, etc.[85,88] We have herewith considered a few instances where G-based hybrids have been used as effective/efficient adsorbent material for the adsorptive removal of radioactive materials from wastewater/water.

In their report, Sharma et al. described an inventive new magnetite@go@chitosan (M@GO@Cs) adsorbent for effectively removing uranium (VI) from wastewater. It was determined that uranium (VI) was bonded to the -CONH and -OH functionalities complexations in M@GO@Cs. They found that the U (VI) adsorption isotherm on the M@GO@Cs followed the Langmuir isotherm (LM), while the kinetic followed

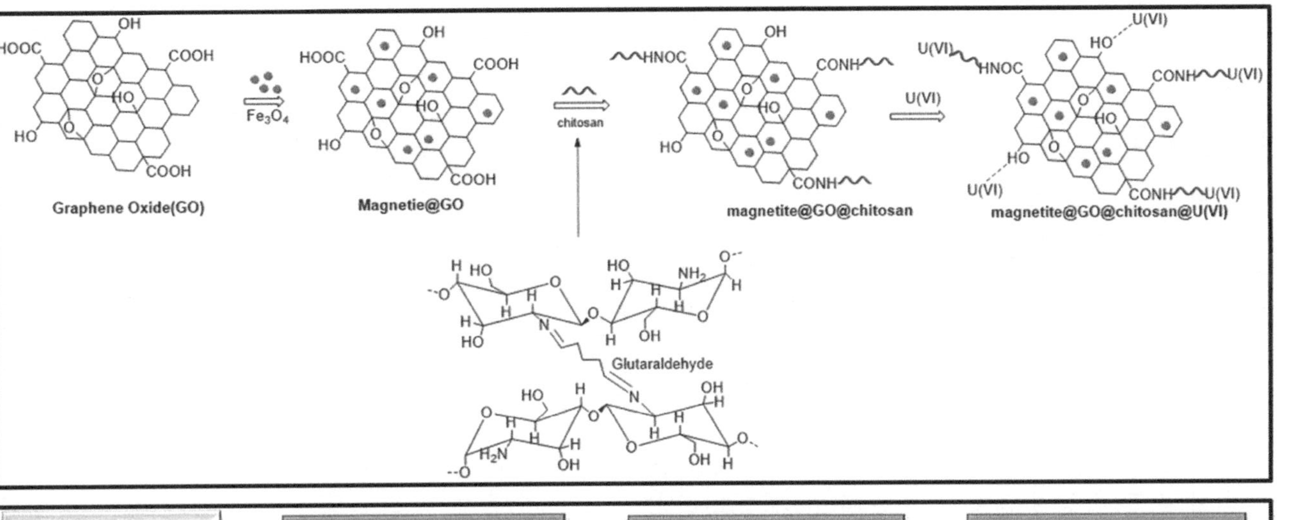
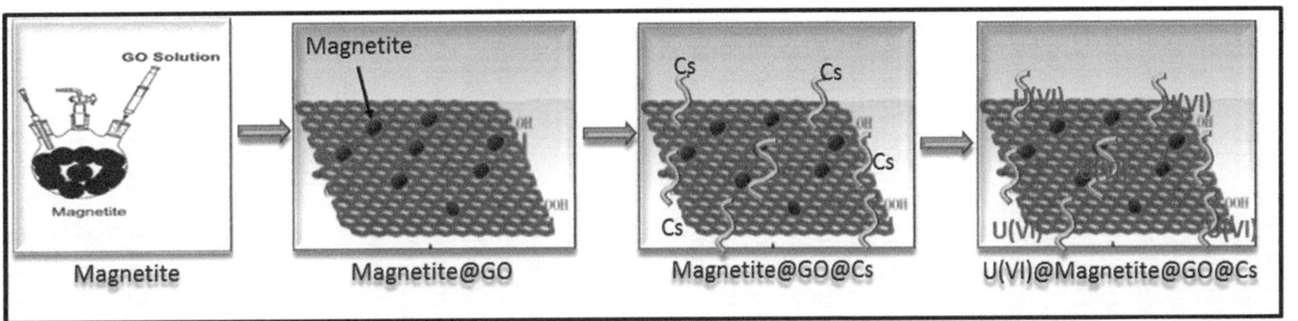

FIGURE 6.10 Adsorption of uranium on the surface of M@GO@Cs. (Reproduced with permission from Sharma et al.[100] Copyright 2022, Elsevier Science Ltd.)

the PSOM model in accordance with other findings for the removal of radionuclides from water.[85,88]

In an intriguing work, composites made of Ca-Mg-Al layered double hydroxide coated on GO (Ca-Mg-Al LDH-GO) were effectively synthesized, described, and used to decontaminate Eu(III) and fulvic acid (FA) under various experimental set-ups. Different addition sequences and concentration gradients were also achieved for Eu(III) and FA, revealing various interaction processes. According to the experimental findings, FA and Eu(III) cohabitation improved each compound's ability to adsorb in ternary systems. The obtained Ca-Mg-Al LDH-GO composites were used to remove Eu(III) and FA, and they also demonstrated outstanding chemo-physical stability and Q of 1.12×10^{-3} mol/g and 3.54×10^{-4} mol/g, correspondingly. Through kinetic procedures and temperature isotherms, the exceptional adsorption performances of Ca-Mg-Al LDH-GO were confirmed. This demonstrated that the kinetics processes were simulated using a PSOM pattern, and the adsorption isotherms were beautifully simulated using a LM pattern. By confining Eu^{3+} and FA to the surfaces of Ca/Mg/Al LDH/GO composites; these oxygen-containing groups played a substantial role, according to an examination of their XPS spectra. Based on the findings of the experiments, Ca-Mg-Al LDH-GO composites could be utilized as promising adsorbents with readily available regenerative reusability towards the decontamination of Eu^{3+} as well as FA from wastewater emanating from nuclear fuel systems.[101]

A team of researchers reportedly used GO-based inverse spinel nickel ferrite "GONF" magnetic composite with a mean particle size of 41.41 nm and its rGO counterpart "rGONF" with a mean particle size of 32.16 nm to achieve adsorptive exclusion of U^{6+} and Th^{4+}. The adsorbates' adsorption kinetics experiments on the adsorbent were in good agreement with the PSOM. However, the results of the adsorption isotherms showed that U^{6+} and Th^{4+} were adsorbed in homogeneous surface monolayers onto GONF and rGONF. The adsorption of both U(VI) and Th was enhanced in response to an increase in the system temperature (from 293 to 333 2 K) (IV). Thermodynamic analysis, on the other hand, showed that the U^{6+} and Th^{4+} adsorption onto GONF and rGONF was endothermic. The adsorbent could be recovered using an external magnetic field, then recycled and reused up to five times without suffering any appreciable loss of adsorption ability.

The niche of radionuclides adsorptive removal from water/wastewater is somewhat new, though it is growing rapidly. It would be great if researchers would develop effective/efficient G-based adsorbents whose performance is sustained even after their regeneration upon reuse, with the aim of applying the same in real-life applications.

6.7 CONCLUSIONS

The G family is made up of G, GO, but also rGO, which have the same structure in the bulk but have distinctive surface topologies. The creation of GBNs, their structural analysis, surface interactions, 2D/3D macrostructure edifices, as well as their use in water treatment have all seen considerable advancements in studies over the past few decades. New solutions to difficult environmental challenges are now possible thanks to the development of G-based materials. Thought should be given to the environmental friendliness of the entire life cycle, including synthesis and structural modification, as well as its assembly, along with the application, for practical applications in water treatment. Practical water filtration must take into account cost effectiveness, which is a crucial element. In order to protect the environment, fabrication techniques should be enhanced. Extensive production of G, GO, and rGO as well as their systems, such as composites/hybrids, calls for extremely effective synthetic processes. The ideal procedure should be quick, recyclable, and free of hazardous emissions. A bottleneck that demands further development in the near future is the manufacture of ideal large-area G nanosheets, despite the fact that significant work has been done to advance the synthetic approach. The structural characteristics of G-based adsorbent materials, rather than the synthetic process, are what make them so appealing. Their uses will also be influenced by their structures, which depend on the synthetic conditions. The molecular structure of GO or rGO has not, however, been established with absolute certainty. To put it another way, we need to pay closer attention because the original structural configuration of GO/rGO is still a mystery. The adsorption of organic and metallic pollutants, catalysis, and other applications of these materials in water treatment are very promising. On the contrary, the ecosystem and people may be at risk due to the peculiar features of these structures, such as sharp edges and their oxygen-containing functionalities. The accompanying environmental risk should, therefore be managed properly along with the performance. It seems like a good idea for practical use to structurally reconstruct isolated G-based materials into 2D/3D G-based materials. Nanoscale, but also macroscale, structures are cleverly combined in reconstructions to preserve the G-based materials original attributes, as well as to add new capabilities, including size sieving, simple recycling, as well as simple operation. Modern studies show that G-based adsorbent materials in their form as 2D membranes have a precise sieving ability for molecule/ion separation, but there are several issues that still need to be resolved. For multilayer constructed G-based membranes, it is still difficult to control the sieving holes at the angstrom scale while keeping a stable structure. Creating uniform sieving pores with high density is challenging for monolayer pristine G membranes. The two ends of balance for membrane separation, on the other hand, are flow and sieving performance. Lower flux results from having smaller sieving pores, as it did with G-based membranes. Researchers must give more effort to increasing the flow in the sieving performance maintenance premise of these materials. In addition to the superior performance of 2D membranes, 3D G-based adsorbent materials macrostructures make recycling and operation simple. We suggest that additional consideration be given to structural stability, functional modification, recycling, and reusability. Materials based on G offer fresh approaches to solving complex environmental issues. The environmental friendliness of these systems should be the primary consideration in the analysis of their entire life cycle, from fundamental manufacture to actual uses, while taking into account their applications in environmental remediation.

Herewith, we have conversed cumulatively in this chapter about the application of G-based adsorbent systems for the adsorptive remediation of wastewater, which is of great interest to emerging researchers within this niche.

REFERENCES

1. Zellner, T.; Prasa, D.; Färber, E.; Hoffmann-Walbeck, P.; Genser, D.; Eyer, F., The use of activated charcoal to treat intoxications. *Dtsch Arztebl International* 2019, *116* (18), 311–317.

2. Bhattacharya, S.; Banerjee, P.; Das, P.; Bhowal, A.; Majumder, S. K.; Ghosh, P., Removal of aqueous carbamazepine using graphene oxide nanoplatelets: Process modelling and optimization. *Sustainable Environment Research* 2020, *30* (1), 1–12.

3. Rosli, F. A.; Ahmad, H.; Jumbri, K.; Abdullah, A. H.; Kamaruzaman, S.; Fathihah Abdullah, N. A., Efficient removal of pharmaceuticals from water using graphene nanoplatelets as adsorbent. *Royal Society Open Science* 2021, *8* (1), 201076.

4. Çalışkan Salihi, E.; Wang, J.; Kabacaoğlu, G.; Kırkulak, S.; Šiller, L., Graphene oxide as a new generation adsorbent for the removal of antibiotics from waters. *Seperation Science and Technology* 2021, *56* (3), 453–461.

5. Khalil, A. M. E.; Memon, F. A.; Tabish, T. A.; Salmon, D.; Zhang, S.; Butler, D., Nanostructured porous graphene for efficient removal of emerging contaminants (pharmaceuticals) from water. *Chemical Engineering Journal* 2020, *398*, 125440.

6. Wu, Q.; Feng, C.; Wang, C.; Wang, Z., A facile one-pot solvothermal method to produce superparamagnetic graphene–Fe_3O_4 nanocomposite and its application in the removal of dye from aqueous solution. *Colloids and Surfaces B: Biointerfaces* 2013, *101*, 210–214.

7. Yu, F.; Ma, J.; Bi, D., Enhanced adsorptive removal of selected pharmaceutical antibiotics from aqueous solution by activated graphene. *Environmental Science and Pollution Research* 2015, *22* (6), 4715–4724.

8. Al-Khateeb, L. A.; Almotiry, S.; Salam, M. A., Adsorption of pharmaceutical pollutants onto graphene nanoplatelets. *Chemical Engineering Journal* 2014, *248*, 191–199.

9. Wang, J.; Chen, Z.; Chen, B., Adsorption of Polycyclic Aromatic Hydrocarbons by Graphene and Graphene Oxide Nanosheets. *Environmental Science & Technology* 2014, *48* (9), 4817–4825.

10. Apul, O. G.; Wang, Q.; Zhou, Y.; Karanfil, T., Adsorption of aromatic organic contaminants by graphene nanosheets: Comparison with carbon nanotubes and activated carbon. *Water Research* 2013, *47* (4), 1648–1654.

11. Banerjee, P.; Das, P.; Zaman, A.; Das, P., Application of graphene oxide nanoplatelets for adsorption of Ibuprofen from aqueous solutions: Evaluation of process kinetics and thermodynamics. *Process Safety and Environmental Protection* 2016, *101*, 45–53.

12. Zhu, S.; Liu, Y.-G.; Liu, S.-B.; Zeng, G.-M.; Jiang, L.-H.; Tan, X.-F.; Zhou, L.; Zeng, W.; Li, T.-T.; Yang, C.-P., Adsorption of emerging contaminant metformin using graphene oxide. *Chemosphere* 2017, *179*, 20–28.

13. Gao, Y.; Li, Y.; Zhang, L.; Huang, H.; Hu, J.; Shah, S. M.; Su, X., Adsorption and removal of tetracycline antibiotics from aqueous solution by graphene oxide. *Journal of Colloid Interface Science* 2012, *368* (1), 540–546.

14. Sarker, M.; Song, J. Y.; Jhung, S. H., Adsorptive removal of anti-inflammatory drugs from water using graphene oxide/metal-organic framework composites. *Chemical Engineering Journal* 2018, *335*, 74–81.

15. Bezerra de Araujo, C. M.; Wernke, G.; Ghislandi, M. G.; Diório, A.; Vieira, M. F.; Bergamasco, R.; da Motta, Alves Sobrinho, M.; Rodrigues, A. E., Continuous removal of pharmaceutical drug chloroquine and Safranin-O dye from water using agar-graphene oxide hydrogel: Selective adsorption in batch and fixed-bed experiments. *Environmental Research* 2022, *216*, 114425.

16. Wu, S.; Zhao, X.; Li, Y.; Zhao, C.; Du, Q.; Sun, J.; Wang, Y.; Peng, X.; Xia, Y.; Wang, Z., Adsorption of ciprofloxacin onto biocomposite fibers of graphene oxide/calcium alginate. *Chemical Engineering Journal* 2013, *230*, 389–395.

17. Bezerra de Araujo, C. M.; Wernke, G.; Ghislandi, M. G.; Diório, A.; Vieira, M. F.; Bergamasco, R.; da Motta, Alves Sobrinho, M.; Rodrigues, A. E., Continuous removal of pharmaceutical drug chloroquine and Safranin-O dye from water using agar-graphene oxide hydrogel: Selective adsorption in batch and fixed-bed experiments. *Environmental Research* 2023, *216*, 114425.

18. Tang, Y.; Guo, H.; Xiao, L.; Yu, S.; Gao, N.; Wang, Y., Synthesis of reduced graphene oxide/magnetite composites and investigation of their adsorption performance of fluoroquinolone antibiotics. *Colloids and Surfaces A: Physicochemical and Engineering Aspects* 2013, *424*, 74–80.

19. Karimi-Maleh, H.; Shafieizadeh, M.; Taher, M. A.; Opoku, F.; Kiarii, E. M.; Govender, P. P.; Ranjbari, S.; Rezapour, M.; Orooji, Y., The role of magnetite/graphene oxide nanocomposite as a high-efficiency adsorbent for removal of phenazopyridine residues from water samples, an experimental/theoretical investigation. *Journal of Molecular Liquids* 2020, *298*, 112040.

20. Umbreen, N.; Sohni, S.; Ahmad, I.; Khattak, N. U.; Gul, K., Self-assembled three-dimensional reduced graphene oxide-based hydrogel for highly efficient and facile removal of pharmaceutical compounds from aqueous solution. *Journal of Colloid and Interface Science* 2018, *527*, 356–367.

21. Liu, Y.; Fu, J.; He, J.; Wang, B.; He, Y.; Luo, L.; Wang, L.; Chen, C.; Shen, F.; Zhang, Y., Synthesis of a superhydrophilic coral-like reduced graphene oxide aerogel and its application to pollutant capture in wastewater treatment. *Chemical Engineering Science* 2022, *260*, 117860.

22. Moradi, O.; Alizadeh, H.; Sedaghat, S., Removal of pharmaceuticals (diclofenac and amoxicillin) by maltodextrin/reduced graphene and maltodextrin/reduced graphene/copper oxide nanocomposites. *Chemosphere* 2022, *299*, 134435.

23. An, W.; Yang, T.; Wang, Y.; Xu, J.; Hu, J.; Cui, W.; Liang, Y., Adsorption and in-situ photocatalytic Fenton multifield coupled degradation of organic pollutants and coking wastewater via FeBiOBr modification of three-dimensional graphene aerogel. *Applied Surface Science* 2023, *610*, 155495.

24. Akpotu, S. O.; Diagboya, P. N.; Lawal, I. A.; Sanni, S. O.; Pholosi, A.; Peleyeju, M. G.; Mtunzi, F. M.; Ofomaja, A. E., Designer composite of montmorillonite-reduced graphene oxide-PEG polymer for water treatment: Enrofloxacin sequestration and cost analysis. *Chemical Engineering Journal* 2023, *453*, 139771.

25. Yang, C.; You, X.; Cheng, J.; Zheng, H.; Chen, Y., A novel visible-light-driven In-based MOF/graphene oxide composite photocatalyst with enhanced photocatalytic activity toward the degradation of amoxicillin. *Applied Catalysis B: Environmental* 2017, *200*, 673–680.

26. Wang, W.; Zhang, J.; Xiao, G.; Liu, X.; Qu, H.; Zhou, S., Preparation of magnetic porous graphene oxide by intercalating rigid molecule and subsequent magnetization for enhancing pharmaceuticals removal from water. *Materials Today Communications* 2023, *34*, 105119.

27. Higgins, P.; Siddiqui, S. H.; Kumar, R., Design of novel graphene oxide/halloysite nanotube@polyaniline nanohybrid for the removal of diclofenac sodium from aqueous solution. *Environmental Nanotechnology, Monitoring & Management* 2022, *17*, 100628.

28. Feng, X.; Qiu, B.; Sun, D., Enhanced naproxen adsorption by a novel β-cyclodextrin immobilized the three-dimensional macrostructure of reduced graphene oxide and multiwall carbon nanotubes. *Separation and Purification Technology* 2022, *290*, 120837.

29. Liu, J.; Ge, X.; Ye, X.; Wang, G.; Zhang, H.; Zhou, H.; Zhang, Y.; Zhao, H., 3D graphene/δ-MnO$_2$ aerogels for highly efficient and reversible removal of heavy metal ions. *Journal of Materials Chemistry A* 2016, *4* (5), 1970–1979.

30. Zhang, Z.; Li, L.; Ji, X.; Chen, J.; Yang, G.; Lucia, L. A., Facile Synthesis of Lignosulfonate-graphene Porous Hydrogel for Effective Removal of Cr (VI) from Aqueous Solution. *BioResources* 2019, *14* (3), 7001–7014.

31. Shen, Y.; Chen, B., Sulfonated graphene nanosheets as a superb adsorbent for various environmental pollutants in water. *Environmental Science & Technology* 2015, *49* (12), 7364–7372.

32. Wu, Y.; Luo, H.; Wang, H.; Wang, C.; Zhang, J.; Zhang, Z., Adsorption of hexavalent chromium from aqueous solutions by graphene modified with cetyltrimethylammonium bromide. *Journal of Colloid and Interface Science* 2013, *394*, 183–191.

33. Dong, Y.; Sang, D.; He, C.; Sheng, X.; Lei, L., Graphene oxide-based two-dimensional molecular brush for efficient removal of lead and copper ions from water media. *Journal of Chinese Chemical Society* 2020, *67* (7), 1183–1188.

34. Greaves, M.; Elderfield, H.; Klinkhammer, G., Determination of the rare earth elements in natural waters by isotope-dilution mass spectrometry. *Analytica Chimica Acta* 1989, *218*, 265–280.

35. Yusan, S.; Gok, C.; Erenturk, S.; Aytas, S., Adsorptive removal of thorium (IV) using calcined and flux calcined diatomite from Turkey: Evaluation of equilibrium, kinetic and thermodynamic data. *Applied Clay Science* 2012, *67*, 106–116.

36. Li, D.; Zhang, B.; Xuan, F., The sorption of Eu (III) from aqueous solutions by magnetic graphene oxides: A combined experimental and modeling studies. *Journal of Molecular Liquids* 2015, *211*, 203–209.

37. Ashour, R. M.; Abdelhamid, H. N.; Abdel-Magied, A. F.; Abdel-Khalek, A. A.; Ali, M.; Uheida, A.; Muhammed, M.; Zou, X.; Dutta, J., Rare earth ions adsorption onto graphene oxide nanosheets. *Solvent Extraction and Ion Exchange* 2017, *35* (2), 91–103.

38. Wang, S.; Ning, H.; Hu, N.; Huang, K.; Weng, S.; Wu, X.; Wu, L.; Liu, J.; Alamusi, Preparation and characterization of graphene oxide/silk fibroin hybrid aerogel for dye and heavy metal adsorption. *Composites Part B: Engineering* 2019, *163*, 716–722.

39. Croitoru, A.-M.; Ficai, A.; Ficai, D.; Trusca, R.; Dolete, G.; Andronescu, E.; Turculet, S. C., Chitosan/graphene oxide nanocomposite membranes as adsorbents with applications in water purification. *Materials* 2020, *13* (7), 1687.

40. Sitko, R.; Musielak, M.; Zawisza, B.; Talik, E.; Gagor, A., Graphene oxide/cellulose membranes in adsorption of divalent metal ions. *RSC Advances* 2016, *6* (99), 96595–96605.

41. Nyairo, W. N.; Eker, Y. R.; Kowenje, C.; Zor, E.; Bingol, H.; Tor, A.; Ongeri, D. M., Efficient removal of lead (II) ions from aqueous solutions using methyl-β-cyclodextrin modified graphene oxide. *Water, Air, & Soil Pollution* 2017, *228* (11), 1–10.

42. Yang, Y.; Xie, Y.; Pang, L.; Li, M.; Song, X.; Wen, J.; Zhao, H., Preparation of reduced graphene oxide/poly(acrylamide) nanocomposite and its adsorption of Pb(II) and methylene blue. *Langmuir* 2013, *29* (34), 10727–10736.

43. Fu, W.; Huang, Z., Magnetic dithiocarbamate functionalized reduced graphene oxide for the removal of Cu(II), Cd(II), Pb(II), and Hg(II) ions from aqueous solution: Synthesis, adsorption, and regeneration. *Chemosphere* 2018, *209*, 449–456.

44. Zeng, T.; Yu, Y.; Li, Z.; Zuo, J.; Kuai, Z.; Jin, Y.; Wang, Y.; Wu, A.; Peng, C., 3D MnO_2 nanotubes@reduced graphene oxide hydrogel as reusable adsorbent for the removal of heavy metal ions. *Materials Chemistry and Physics* 2019, *231*, 105–108.

45. Sahraei, R.; Sekhavat Pour, Z.; Ghaemy, M., Novel magnetic bio-sorbent hydrogel beads based on modified gum tragacanth/graphene oxide: Removal of heavy metals and dyes from water. *Journal of Cleaner Production* 2017, *142*, 2973–2984.

46. Sahraei, R.; Hemmati, K.; Ghaemy, M., Adsorptive removal of toxic metals and cationic dyes by magnetic adsorbent based on functionalized graphene oxide from water. *RSC Advances* 2016, *6* (76), 72487–72499.

47. Zhan, W.; Gao, L.; Fu, X.; Siyal, S. H.; Sui, G.; Yang, X., Green synthesis of amino-functionalized carbon nanotube-graphene hybrid aerogels for high performance heavy metal ions removal. *Applied Surface Science* 2019, *467–468*, 1122–1133.

48. Wang, Y.; Liang, S.; Chen, B.; Guo, F.; Yu, S.; Tang, Y., Synergistic removal of Pb (II), Cd (II) and humic acid by Fe_3O_4@ mesoporous silica-graphene oxide composites. *PloS One* 2013, *8* (6), e65634.

49. Vo, T. S.; Vo, T. T. B. C.; Suk, J. W.; Kim, K., Recycling performance of graphene oxide-chitosan hybrid hydrogels for removal of cationic and anionic dyes. *Nano Convergence* 2020, *7* (1), 4.

50. González, J. A.; Villanueva, M. E.; Piehl, L. L.; Copello, G. J., Development of a chitin/graphene oxide hybrid composite for the removal of pollutant dyes: Adsorption and desorption study. *Chemical Engineering Journal* 2015, *280*, 41–48.

51. Zhu, Q.; Wang, J.; Zhang, L.; Yan, D.; Yang, H.; Zhang, L.; Miao, X., Research progress of graphene for the adsorption, enrichment and analysis of environmental pollutants. *Bulletin of Materials Science* 2023, *46* (1), 17.

52. Liu, T.; Li, Y.; Du, Q.; Sun, J.; Jiao, Y.; Yang, G.; Wang, Z.; Xia, Y.; Zhang, W.; Wang, K., Adsorption of methylene blue from aqueous solution by graphene. *Colloids and Surfaces B: Biointerfaces* 2012, *90*, 197–203.

53. Li, Y.; Liu, T.; Du, Q.; Sun, J.; Xia, Y.; Wang, Z.; Zhang, W.; Wang, K.; Zhu, H.; Wu, D., Adsorption of cationic red X-GRL from aqueous solutions by graphene: Equilibrium, kinetics and thermodynamics study. *Chemical and Biochemical Engineering Quarterly* 2011, *25* (4), 483–491.

54. Wu, T.; Cai, X.; Tan, S.; Li, H.; Liu, J.; Yang, W., Adsorption characteristics of acrylonitrile, p-toluenesulfonic acid, 1-naphthalenesulfonic acid and methyl blue on graphene in aqueous solutions. *Chemical Engineering Journal* 2011, *173* (1), 144–149.

55. Zhao, D.; Sheng, G.; Chen, C.; Wang, X., Enhanced photocatalytic degradation of methylene blue under visible irradiation on graphene@ TiO_2 dyade structure. *Applied Catalysis B: Environmental* 2012, *111*, 303–308.

56. Ai, L.; Zhang, C.; Chen, Z., Removal of methylene blue from aqueous solution by a solvothermal-synthesized graphene/magnetite composite. *Journal of Hazardous Materials* 2011, *192* (3), 1515–1524.

57. Yao, Y.; Miao, S.; Liu, S.; Ma, L. P.; Sun, H.; Wang, S., Synthesis, characterization, and adsorption properties of magnetic Fe_3O_4@graphene nanocomposite. *Chemical Engineering Journal* 2012, *184*, 326–332.

58. Sari Yilmaz, M., Graphene oxide/hollow mesoporous silica composite for selective adsorption of methylene blue. *Microporous Mesoporous Materials* 2022, *330*, 111570.

59. Sun, L.; Yu, H.; Fugetsu, B., Graphene oxide adsorption enhanced by in situ reduction with sodium hydrosulfite to remove acridine orange from aqueous solution. *Journal of Hazardous Materials* 2012, *203–204*, 101–110.

60. Yan, H.; Yang, H.; Li, A.; Cheng, R., pH-tunable surface charge of chitosan/graphene oxide composite adsorbent for efficient removal of multiple pollutants from water. *Chemical Engineering Journal* 2016, *284*, 1397–1405.

61. Bradder, P.; Ling, S. K.; Wang, S.; Liu, S., Dye adsorption on layered graphite oxide. *Journal of Chemical & Engineering Data* 2011, *56* (1), 138–141.

62. Huang, Q.; Jin, S.; Song, S.; Chen, Z., The nitrogen-doped graphene-like carbon nanosheets: Confined construction and oxygen-limited oxidation for higher removal efficiency toward organic contaminants. *Journal of Cleaner Production* 2022, *363*, 132604.

63. Bai, S.; Shen, X.; Zhong, X.; Liu, Y.; Zhu, G.; Xu, X.; Chen, K., One-pot solvothermal preparation of magnetic reduced graphene oxide-ferrite hybrids for organic dye removal. *Carbon* 2012, *50* (6), 2337–2346.

64. Geng, Z.; Lin, Y.; Yu, X.; Shen, Q.; Ma, L.; Li, Z.; Pan, N.; Wang, X., Highly efficient dye adsorption and removal: A functional hybrid of reduced graphene oxide–Fe_3O_4 nanoparticles as an easily regenerative adsorbent. *Journal of Materials Chemistry* 2012, *22* (8), 3527–3535.

65. Tiwari, J. N.; Mahesh, K.; Le, N. H.; Kemp, K. C.; Timilsina, R.; Tiwari, R. N.; Kim, K. S., Reduced graphene oxide-based hydrogels for the efficient capture of dye pollutants from aqueous solutions. *Carbon* 2013, *56*, 173–182.

66. Ang, Z. R.; Kong, I.; Lee, R. S. Y.; Kong, C.; Kakarla, A. B.; Chai, A. B.; Kong, W., Preparation of 3D graphene-carbon nanotube-magnetic hybrid aerogels for dye adsorption. *New Carbon Materials* 2022, *37* (2), 424–432.

67. Das, L.; Das, P.; Bhowal, A.; Bhattachariee, C., Synthesis of hybrid hydrogel nano-polymer composite using graphene oxide, chitosan and PVA and its application in waste water treatment. *Environmental Technology & Innovation* 2020, *18*, 100664.

68. Khedri, D.; Hassani, A. H.; Moniri, E.; Ahmad Panahi, H.; Khaleghian, M., Efficient removal of phenolic contaminants from wastewater samples using functionalized graphene oxide with thermo-sensitive polymer: Adsorption isotherms, kinetics, and thermodynamics studies. *Surfaces and Interfaces* 2022, *35*, 102439.

69. Liu, X.; Zhang, H.; Ma, Y.; Wu, X.; Meng, L.; Guo, Y.; Yu, G.; Liu, Y., Graphene-coated silica as a highly efficient sorbent for residual organophosphorus pesticides in water. *Journal of Materials Chemistry A* 2013, *1* (5), 1875–1884.

70. Zhang, C.; Zhang, R. Z.; Ma, Y. Q.; Guan, W. B.; Wu, X. L.; Liu, X.; Li, H.; Du, Y. L.; Pan, C. P., Preparation of cellulose/graphene composite and its applications for triazine pesticides adsorption from water. *ACS Sustainable Chemistry & Engineering* 2015, *3* (3), 396–405.

71. Queiroz, R. N.; Neves, T. d. F.; da Silva, M. G. C.; Mastelaro, V. R.; Vieira, M. G. A.; Prediger, P., Comparative efficiency of polycyclic aromatic hydrocarbon removal by novel graphene oxide composites prepared from conventional and green synthesis. *Journal of Cleaner Production* 2022, *361*, 132244.

72. Bibi, A.; Bibi, S.; Abu-Dieyeh, M.; Al-Ghouti, M. A., New material of polyacrylic acid-modified graphene oxide composite for phenol remediation from synthetic and real wastewater. *Environmental Technology & Innovation* 2022, *27*, 102795.

73. Shi, X.; Cheng, C.; Peng, F.; Hou, W.; Lin, X.; Wang, X., Adsorption properties of graphene materials for pesticides: Structure effect. *Journal of Molecular Liquid* 2022, *364*, 119967.

74. Al-Ghouti, M. A.; Sayma, J.; Munira, N.; Mohamed, D.; Da'na, D. A.; Qiblawey, H.; Alkhouzaam, A., Effective removal of phenol from wastewater using a hybrid process of graphene oxide adsorption and UV-irradiation. *Environmental Technology & Innovation* 2022, *27*, 102525.

75. Yang, Q.; Wang, J.; Zhang, W.; Liu, F.; Yue, X.; Liu, Y.; Yang, M.; Li, Z.; Wang, J., Interface engineering of metal organic framework on graphene oxide with enhanced adsorption capacity for organophosphorus pesticide. *Chemical Engineering Journal* 2017, *313*, 19–26.

76. Ren, H.; Cao, Z.-F.; Chen, Y.-Y.; Jiang, X.-Y.; Yu, J.-G., Graphene oxide-Bicine composite as a novel adsorbent for removal of various contaminants from aqueous solutions. *Journal of Environmental Chemical Engineering* 2021, *9* (6), 106769.

77. Rout, D. R.; Jena, H. M., Polyethylene glycol functionalized reduced graphene oxide coupled with zinc oxide composite adsorbent for removal of phenolic wastewater. *Environmental Research* 2022, *214*, 114044.

78. Zhao, R.; Li, Y.; Ji, J.; Wang, Q.; Li, G.; Wu, T.; Zhang, B., Efficient removal of phenol and p-nitrophenol using nitrogen-doped reduced graphene oxide. *Colloids and Surfaces A: Physicochemical and Engineering Aspects* 2021, *611*, 125866.

79. Boruah, P. K.; Sharma, B.; Hussain, N.; Das, M. R., Magnetically recoverable Fe3O4/graphene nanocomposite towards efficient removal of triazine pesticides from aqueous solution: Investigation of the adsorption phenomenon and specific ion effect. *Chemosphere* 2017, *168*, 1058–1067.

80. Maliyekkal, S. M.; Sreeprasad, T. S.; Krishnan, D.; Kouser, S.; Mishra, A. K.; Waghmare, U. V.; Pradeep, T., Graphene: A reusable substrate for unprecedented adsorption of pesticides. *Small* 2013, *9* (2), 273–283.

81. Du, H.; Lei, Y.; Chen, W.; Li, F.; Li, H.; Deng, W.; Jiang, G., Multifunctional magnetic bio-nanoporous carbon material based on zero-valent iron, angelicae dahuricae radix slag and graphene oxide: An efficient adsorbent of pesticides. *Arabian Journal of Chemistry* 2021, *14* (8), 103267.

82. Mahdavi, V.; Taghadosi, F.; Dashtestani, F.; Bahadorikhalili, S.; Farimani, M. M.; Ma'mani, L.; Mousavi Khaneghah, A., Aminoguanidine modified magnetic graphene oxide as a robust nanoadsorbent for efficient removal and extraction of chlorpyrifos residue from water. *Journal of Environmental Chemical Engineering* 2021, *9* (5), 106117.

83. Orozco, J.; Mercante, L. A.; Pol, R.; Merkoçi, A., Graphene-based Janus micromotors for the dynamic removal of pollutants. *Journal of Materials Chemistry A* 2016, *4* (9), 3371–3378.

84. Li, H.; Zhang, L.; Chen, J.; Lu, M.; Xie, J.; Wang, X.; Han, K.; Li, J.; Lu, J., Reduced graphene oxide based aerogels: Doped with ternary Prussian blue analogs and selective removal of Cs^+ from effluent. *Journal of Water Process Engineering* 2022, *47*, 102741.

85. Liu, X.; Wu, J.; Wang, J., Removal of Cs (I) from simulated radioactive wastewater by three forward osmosis membranes. *Chemical Engineering Journal* 2018, *344*, 353–362.

86. Chandra, V.; Park, J.; Chun, Y.; Lee, J. W.; Hwang, I.-C.; Kim, K. S., Water-dispersible magnetite-reduced graphene oxide composites for arsenic removal. *ACS Nano* 2010, *4* (7), 3979–3986.

87. Ren, Y.; Yan, N.; Wen, Q.; Fan, Z.; Wei, T.; Zhang, M.; Ma, J., Graphene/δ-MnO_2 composite as adsorbent for the removal of nickel ions from wastewater. *Chemical Engineering Journal* 2011, *175*, 1–7.

88. Li, K.; Xiong, T.; Liao, J.; Lei, Y.; Zhang, Y.; Zhu, W., Design of MXene/graphene oxide nanocomposites with micro-wrinkle structure for efficient separating of uranium(VI) from wastewater. *Chemical Engineering Journal* 2022, *433*, 134449.

89. Cao, M.; Chen, L.; Xu, W.; Gao, J.; Gui, Y.; Ma, F.; Liu, P.; Xue, Y.; Yan, Y., Preparation of graphene oxide composite nitrogen-doped carbon (GO@NCs) by one-step carbonization with enhanced electrosorption performance for U(VI). *Journal of Water Process Engineering* 2022, *48*, 102930.

90. Zhou, S.; Xie, Y.; Zhu, F.; Gao, Y.; Liu, Y.; Tang, Z.; Duan, Y., Amidoxime modified chitosan/graphene oxide composite for efficient adsorption of U(VI) from aqueous solutions. *Journal of Environmental Chemical Engineering* 2021, *9* (6), 106363.

91. Guo, B.; Kamura, Y.; Koilraj, P.; Sasaki, K., Co-sorption of Sr^{2+} and SeO_4^{2-} as the surrogate of radionuclide by alginate-encapsulated graphene oxide-layered double hydroxide beads. *Environmental Research* 2020, *187*, 109712.

92. Qi, H.; Liu, H.; Gao, Y., Removal of Sr(II) from aqueous solutions using polyacrylamide modified graphene oxide composites. *Journal of Molecular Liquids* 2015, *208*, 394–401.

93. Su, M.; Liu, Z.; Wu, Y.; Peng, H.; Ou, T.; Huang, S.; Song, G.; Kong, L.; Chen, N.; Chen, D., Graphene oxide functionalized with nano hydroxyapatite for the efficient removal of U(VI) from aqueous solution. *Environmental Pollution* 2021, *268*, 115786.

94. Huo, J.-B.; Yu, G.; Wang, J., Adsorptive removal of Sr(II) from aqueous solution by polyvinyl alcohol/graphene oxide aerogel. *Chemosphere* 2021, *278*, 130492.

95. Song, S.; Wang, K.; Zhang, Y.; Wang, Y.; Zhang, C.; Wang, X.; Zhang, R.; Chen, J.; Wen, T.; Wang, X., Self-assembly of graphene oxide/PEDOT:PSS nanocomposite as a novel adsorbent for uranium immobilization from wastewater. *Environmental Pollution* 2019, *250*, 196–205.

96. Liu, X.; Li, J.; Wang, X.; Chen, C.; Wang, X., High performance of phosphate-functionalized graphene oxide for the selective adsorption of U(VI) from acidic solution. *Journal of Nuclear Materials* 2015, *466*, 56–64.

97. Lingamdinne, L. P.; Choi, Y.-L.; Kim, I.-S.; Yang, J.-K.; Koduru, J. R.; Chang, Y.-Y., Preparation and characterization of porous reduced graphene oxide based inverse spinel nickel ferrite nanocomposite for adsorption removal of radionuclides. *Journal of Hazardous Materials* 2017, *326*, 145–156.

98. Zhong, X.; Liang, W.; Lu, Z.; Qiu, M.; Hu, B., Ultra-high capacity of graphene oxide conjugated covalent organic framework nanohybrid for U(VI) and Eu(III) adsorption removal. *Journal of Molecular Liquid* 2021, *323*, 114603.

99. Mu, W.; Yu, Q.; Hu, R.; Li, X.; Wei, H.; Jian, Y., Porous three-dimensional reduced graphene oxide merged with WO$_3$ for efficient removal of radioactive strontium. *Applied Surface Science* 2017, *423*, 1203–1211.

100. Sharma, M.; Laddha, H.; Yadav, P.; Jain, Y.; Sachdev, K.; Janu, V. C.; Gupta, R., Selective removal of uranium from an aqueous solution of mixed radionuclides of uranium, cesium, and strontium via a viable recyclable GO@chitosan based magnetic nanocomposite. *Materials Today Communications* 2022, *32*, 104020.

101. Zong, P.; Shao, M.; Cao, D.; Xu, X.; Wang, S.; Zhang, H., Synthesis of potential Ca–Mg–Al layered double hydroxides coated graphene oxide composites for simultaneous uptake of europium and fulvic acid from wastewater systems. *Environmental Research* 2021, *196*, 110375.

102. Sun, P.; Zheng, F.; Zhu, M.; Song, Z.; Wang, K.; Zhong, M.; Wu, D.; Little, R. B.; Xu, Z.; Zhu, H., Selective trans-membrane transport of alkali and alkaline earth cations through graphene oxide membranes based on cation–π interactions. *ACS Nano* 2014, *8* (1), 850–859.

103. Lingamdinne, L. P.; Koduru, J. R.; Chang, Y.-Y.; Karri, R. R., Process optimization and adsorption modeling of Pb(II) on nickel ferrite-reduced graphene oxide nano-composite. *Journal of Molecular Liquids* 2018, *250*, 202–211.

104. Dong, Z.; Zhang, Z.; Li, Z.; Feng, Y.; Dong, W.; Wang, T.; Cheng, Z.; Wang, Y.; Dai, Y.; Cao, X., 3D structure aerogels constructed by reduced graphene oxide and hollow TiO$_2$ spheres for efficient visible-light-driven photoreduction of U (VI) in air-equilibrated wastewater. *Environmental Science: Nano* 2021, *8* (8), 2372–2385.

7 Recyclability of Graphene-Based Adsorbent Materials

7.1 INTRODUCTION

Recyclability or regeneration of graphene-based adsorbents is the quick recycling or recovery of used adsorbents using practical, sustainable procedures. No single technique can sustain or improve the adsorption efficiency of all adsorbents on its own, particularly in the case of materials based on graphene. However, other methods have been found to work well with particular adsorbents. The stimulation of utilized adsorbents may also be accomplished by combining one or more options. The type and nature of the materials utilized, including elements like toxicity and radioactivity, are key considerations for regeneration processes. As a result, the chosen techniques must be effective, non-toxic, ecologically friendly, affordable, and simple to use. They must provide the option of reusing the used adsorbent after the water treatment.

Desorption studies are essential for designing an adsorption system, in addition to optimization and adsorption tests. The used adsorbents are recycled using thermal or chemical techniques in water and wastewater treatment facilities. In some instances, the used adsorbent is transported to another area for regeneration. As a result, this procedure significantly affects the operation's overall expenses. To create an adsorbent that performs better, it is crucial to develop affordable, sustainable regeneration techniques. The chance of secondary contamination is considerably diminished if the adsorbent material is simple to separate from the adsorbate.

It can be assumed that graphene derivatives regain the majority of their phenol and dye adsorption capacities after desorption-regeneration with a variety of potential eluents. Due to contaminants that bind firmly and do not desorb, some adsorbents do, however, experience some loss in adsorption capacity during each regeneration phase. A nice example is in the study, where the adsorbent still maintains a high adsorption efficiency after five cycles (98.76 percent, 97.32 percent, 95.60 percent, 93.45 percent, 90.62 percent removal efficiency, respectively) as figuratively presented in Figure 7.1.[1] However, as shown in Table 7.1 below, a retention capacity between 50 and 95 percent efficiency has apparently been reached. According to certain studies, the waste of adsorbents is the cause of the loss in efficiency. Due to the relatively weak attraction forces between contaminants and graphene adsorbents, simple desorption procedures were adequate to undergo rejuvenation and give a reuse capacity, with the majority of research using ethanol as well as DI water as eluents.

DOI: 10.1201/9781032621302-7

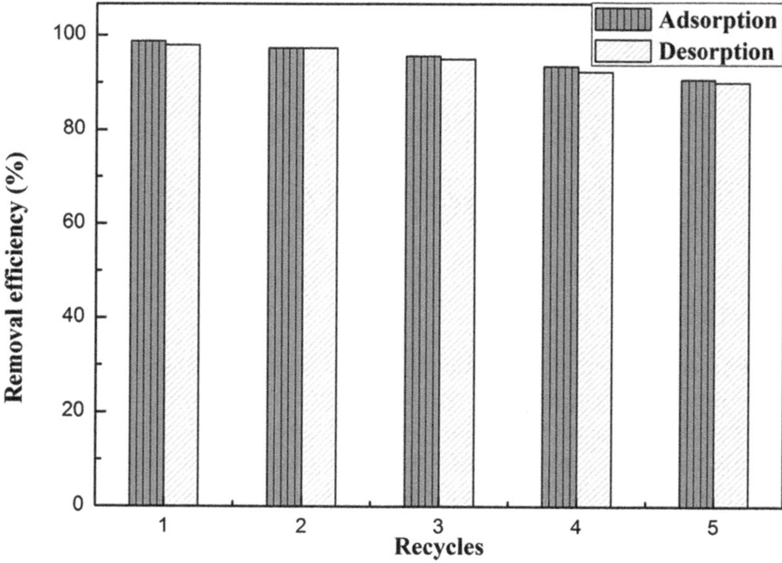

FIGURE 7.1 Adsorption-desorption recycles (m = 0.221 g, V = 100 mL, pH = 12, C_0 = 500 mg L^{-1}, 298 K). (Reproduced with permission from Liu et al.[1] Copyright 2018, Elsevier, Science Ltd.)

Although this is an important component of the "recovery and reuse" of adsorbate in a sustainable circular economy paradigm, it is obvious that not all researchers investigated the regenerative capacities of the GO adsorbents. This ought not to be so because, in the study of graphene-based adsorbent materials, information pertaining to their regeneration is of vital importance.

This chapter aims to discuss graphene-based adsorbent regeneration as per literature in order to guide emerging as well as existing researchers on the right approach and materials amalgamation towards advanced efficient graphene-based adsorbents.

7.2 GRAPHENE

The regeneration of graphene materials used as absorbents is of great importance, seeing it is both economical and eco-friendly. Most researchers have used the thermal approach of regeneration, which is the most used approach industrially and research-wise to regenerate graphene and its nanocomposites. Electrochemical regeneration techniques have also been greatly utilized by researchers to generate graphene and its composite materials.[2] The lack of functional groups on the surface of graphene platelets means its regeneration via an electrochemical approach presents no potential danger to its surface properties in comparison to GO, or rGO, whose surface functionalities are greatly affected by either chemical (depending on the reducing properties of the solvent used), electrochemical, thermal regeneration approaches.

TABLE 7.1

Graphene-Based Adsorbent Recyclability for Various Contaminants from Wastewater as per Literature

GBAM form	Formulation	Adsorbate	Efficiency (%)	Number of Cycles	Eluent	Ref.
Graphene						
	GO-EDA-CAC-PED	Cu(II)	99.5	10	Di water	8
		Ni(II)	99.5	10	Di water	8
		Co(II)	99.5	10	Di water	8
	Fe_3O_4/chitosan/graphene (FGC)	MB	94.81%	5	H_2SO_4 0.1 M and magnetic	7
	Porous graphene	Tenolol (ATL)	90	4	1 M NaOH, water and then ethanol	9
		Carbamazepine (CBZ),	90	4	1 M NaOH, water and then ethanol	9
		Ciprofloxacin (CIP),	90	4	1 M NaOH, water and then ethanol	9
		Diclofenac (DCF),	90	4	1 M NaOH, water and then ethanol	9
		Gemfibrozil (GEM)	90	4	1 M NaOH, water and then ethanol	9
		Ibuprofen (IBP)	90	4	1 M NaOH, water and then ethanol	9
	Graphene-TiO_2	MB	150	5	Electrochemical	10
	Fe_3O_4-G	MB	—	—	Magnetic	11
		CR	—	—	Magnetic	11
	G-Fe_3O_4	MB	—	5	Magnetic	12
Graphene Oxide						
	3D GO	BPA	90.3	5	n-hexane	13
	GO nanoplatelets	Ibuprofen	95.9	10	4.0 M HNO_3	3
	GO	Atenolol (ATL)	72	1	Di water, ethanol and isopropanol	14
	GO	Propranolol (PRO)	67	1	Di water, ethanol and isopropanol	14
	Acrylamide bonded sodium alginate (AM-SA)	Crystal violet (CV)	<80	10	1M HCL	15
	AM-GO-SA	CV	80	10	1M HCL	15

(Continued)

TABLE 7.1 (CONTINUED)

Graphene-Based Adsorbent Recyclability for Various Contaminants from Wastewater as per Literature

GBAM form	Formulation	Adsorbate	Efficiency (%)	Number of Cycles	Eluent	Ref.
	Xanthan gum-3D GO/TiO$_2$	Lead (Pb)	84.78	5	0.1 M HCL at 30 °C	16
	Nitrogen-GO	Nd^{3+}	—	10	1 M HCL	17
	GO-Fe$_3$O$_4$-Humic acid	MB	55	3	1 M HCl or 0.1 M NaOH	18
		Malachite Green MG	11.60	8	0.5 M HCl	19
		CV	5.63	8	0.5 M HCl	19
	GO	Eu(III)				20
	Colloidal GO	Gd(III)				21
	GO	Eu(III)				22
	RGO–Fe$_3$O$_4$	RhB,	97	5	Magnetic attraction	23
		R6G	97	5	Magnetic attraction	23
		AB92	97	5	Magnetic attraction	23
		OII	97	5	Magnetic attraction	23
		MG	97	5	Magnetic attraction	23
		NC	97	5	Magnetic attraction	23
Reduced Graphene Oxide						
	rGO	Tetracycline (TC)	—	6	Di water or methanol	24
	TrGO	Atenolol (ATL),	90	4	Di water + 1M NaOH	9
		Carbamazepine (CBZ)	90	4	Di water + 1M NaOH	9
		Ciprofloxacin (CIP)	90	4	Di water + 1M NaOH	9
		Diclofenac (DCF)	90	4	Di water + 1M NaOH	9
		Gemfibrozil (GEM)	90	4	Di water + 1M NaOH	9
		Ibuprofen (IBP)	90	4	Di water + 1M NaOHs	9

Graphene-based Composites and Hybrids

Ag_3PO_4/rGH	BPA	50%	5	—	25
CNC/GO	Basic blue 7 (BB7)	—	2	Water, EtOH, or acetone	26
	Reactive orange 122 (RO)	—	2	Water, EtOH, or acetone	26
	Rhodamine B (RhB)	—	2	Water, ethanol, or acetone	26
NiZrAl-GO-chitosan (NiZrAl-LDH-GO-CS NC))	Nalidixic acid (NA)	71	5	60% ultrapure water, 30% propanol and 10% HCL (36.5–38.0%)	27
Bi_2O_3@GO	RhB	80%	7	EtOH	28
PS/GO	MB	69.71%	5	EtOH	29
Ag_3PO_4/GO	MB	98%	8	3% H_2O_2	30
Mag/EC-Tyr/GRO	Phenol	56%	5	Water	31
AS/NZVI/GO	Methomyl	>70%	5	0.1 mol/L NaOH solution	32
	Isoprocarb	>70%	5	0.1 mol/L NaOH solution	32
	Carbaryl	>70%	5	0.1 mol/L NaOH solution	32
GO/PPy	Phenol	95%	6	EtOH	33
Cu/BDC@GrO	BPA	91%	5	EtOH	34
GH/AgBr@rGO	BPA	90%	5	—	35
GO/halloysite nanotube@polyaniline	Diclofenac sodium	~20	5	Ethanol (0.1 mol L^{-1})	36
PVA/polyacrylamide/GO semi-IPN	MB	70	5	Di water	37
GO/layered double hydroxide /poly acrylic acid nanocomposite (LDH-rGO-PAA NC)	Tetracycline (TC)	90	5		38
GO-Fe_3O_4-CMC	Chlorpyrifos	86.1	5	50 mM NaOH	39
Corn straw core (CSC)-5% GO (CSC-5GO)	MB	90.62	5	0.1 mol L^{-1} HCl	1
Lignosulfonate-GO	Cr(VI)	88	5	0.1 M NaOH	40

(Continued)

TABLE 7.1 (CONTINUED)
Graphene-Based Adsorbent Recyclability for Various Contaminants from Wastewater as per Literature

GBAM form	Formulation	Adsorbate	Efficiency (%)	Number of Cycles	Eluent	Ref.
	GO/montmorillonite	Cu^{2+}	83	7	1 M HNO_3	41
	GO@Fe_3O_4-phenopyridine (GFP)	Cr (VI)	15	5	Methanol	42
	GO@Fe_3O_4-2-mercaptobenzothiazole (GFM)	Cr (VI)	28	5	Methanol	42
	GO/polydopamine/β-cyclodextrin	MB	8.56	8	0.5 M HCl	19
	Poly(3-aminobenzoic acid/graphene oxide/ cobalt ferrite) nanocomposite (P3ABA/GO/ $CoFe_2O_4$)	CR	74.4	5	0.1 M NaOH	43

7.3 GRAPHENE OXIDE

The regeneration of graphene oxide base adsorbents have been reportedly based on chemical regeneration. This is because of the chemical and structural properties of graphene oxide, which are very complex and may be affected if a terminal or electrochemical approach is used for the regeneration. Therefore, a simple regeneration approach utilizing simple solvents, such as ethyl alcohol or dilute acids, are used to regenerate graphene oxide based as adsorbents.[3,4] Though graphene oxide nanocomposite containing magnetic iron oxide may be separated using a magnetic regeneration approach, GO has been shown to demonstrate excellent recyclability towards REEs, phenols, and dyes in a study on mixed contaminants.[5] Using 50.0 mg of GO-Bicine composite and 20.0 mL of PNP (10.0 mg/L), 20 mL of MB (10.0 mg/L), and 20 mL of Y^{3+} (10.0 mg/L), respectively, the reusability of GO-Bicine composite was investigated. GO-Bicine composite PNP, MB, and Y^{3+} desorption were investigated using 40.0 mL of HCl (0.1 mol/L) at 298 K with constant oscillation. Eight cycles of adsorption and desorption were completed. After each cycle, the GO-Bicine composite was recovered by suction filtration following adsorption or desorption. After eight adsorption-desorption cycles, as shown in Figure 7.2, the removal rate of the GO-Bicine composite was still above 90% toward PNP. At the same time, it decreased to 51% toward MB and 39% toward Y^{3+}, respectively. This indicates that it had relatively stronger interactions with positively charged substances like REE^{3+} and cationic dyes. Therefore, effective wastewater treatment could make use of the great reusability of GO-Bicine composite toward PNP.[5]

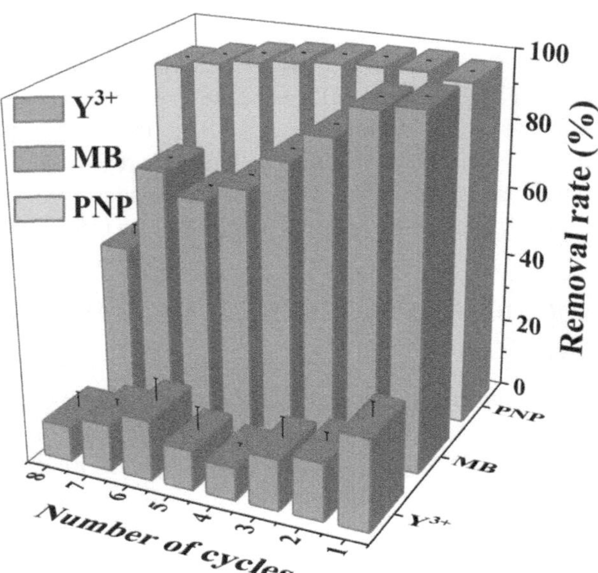

FIGURE 7.2 Reusability of GO-Bicine composite toward PNP, MB and Y^{3+}. (Reproduced with permission from Ren et al.[5] Copyright 2012, Elsevier Science Ltd.)

7.4   REDUCED GRAPHENE OXIDE

Reduced graphene oxide has been regenerated via diverse approaches, though the electrochemical regeneration approach has proven to be of great outstanding results, as reported by Sharif et al.[6]

A kind of GO called rGO has decreased its oxygen concentration through chemical, thermal, and other processes. Its qualities are similar to those of graphene. Additionally, it contains additional heteroatoms, as well as residual oxygen (rGOs). rGO is a desirable material that can be adequate in quality for a variety of applications at a lower cost and with simpler manufacturing procedures despite its less-than-ideal similarity to pristine graphene. The same uses for graphene, such as composite materials, sensors, and water treatment, also apply to rGO. In the water treatment process, the primary function of rGO (photocatalytic composite material) is to effectively increase electron mobility and permit the adsorption of organic contaminants, thereby significantly improving the removal performance.

7.5   GRAPHENE-BASED COMPOSITES AND HYBRIDS

Graphene hybrid materials as adsorbents have been regenerated by magnetic, chemical, thermal, hydrothermal, etc., approaches, though the regeneration approach selected is greatly dependent on the chemical and structural properties of the absorbent material concerned.

The removal of organic dyes from aqueous solutions, such as methylene blue (MB), methyl orange (MO), brilliant blue FCF (E133), eriochorome black T (EBT), and thymol blue (ThB), has been reported to be possible using a reusable nanosorbent made of Fe_3O_4/graphene/chitosan (FGC).[7] According to the data, the FGC adsorbent is more alluring to MB and EBT than the other colors under consideration. A maximum adsorption capacity (Q_{max}) of 94.16 mg/g of MB can be absorbed by FGC under optimal conditions, and the remaining adsorption efficiency for the fifth cycle was 74.81 percent when compared to the pristine FGC nanosorbent, where the regeneration was performed using both a magnetic and a chemical approach in combination. Additionally, the FGC nanocomposite was utilized to efficiently extract MB from actual wastewater.[7]

7.6   REGENERATION APPROACHES FOR GRAPHENE-BASED ADSORBENTS

The regeneration of graphene-based adsorbents is of great importance, as discussed earlier. For a competent adsorbent utilized in extracting heavy metals from real-world wastewater, low cost, high adsorptivity, quick regeneration, and outstanding reusability are strongly advised. It is important to differentiate these regeneration approaches with regard to graphene adsorbents. Therefore, we have decided to categorically discuss the adsorbent regeneration approaches individually for the purpose of simplicity. Diverse regeneration approaches like electrochemical,[2] sonication, chemical, hydrothermal, supercritical fluid extraction, thermal, photo-assisted, and biological regeneration have been reportedly used for the regeneration of graphene-based adsorbent materials as per literature.

7.6.1 Electrochemical Regeneration

The use of electrochemistry technology in the regeneration of graphene oxide has been of great interest to researchers within the last decade.[2,10] This approach entails the use of an electrolysis setup to regenerate graphene-based adsorbents. As per the literature, it is widely used by researchers and industrialists. Graphene-based adsorbent research has revealed that electrochemical regeneration suffers from low regeneration efficiency as a result of side reactions such as oxygen evolution and adsorbent oxidation.[2] Sharif et al.[6] have proposed this approach as an effective means to regenerate rGO-magnetic adsorbent after its utilization for the adsorption of MB dye. These authors reported an interesting regeneration efficiency of 100% achieved via electrochemical anodic oxidation though for five cycles, as depicted in Figure 7.3 below. The regeneration efficiency was directly proportional and largely influenced by the regeneration time, as per these authors' report. This regeneration approach has also been used by this same group (Sharif et al.[10]), where they regenerated a TiO_2-graphene nanocomposite used for the adsorptive removal of MB dye for over 5 cycles with high efficiency of adsorption.[10]

7.6.2 Regeneration via Sonication/Ultrasound Technology

This approach entails the use of sonication technology in the regeneration of the absorbent, though it may be very difficult to separate the absorbent from the adsorbate, especially in the case where the adsorbate released from the adsorbent does not sediment/settle. This approach is a very simple and clean technology, with regard to

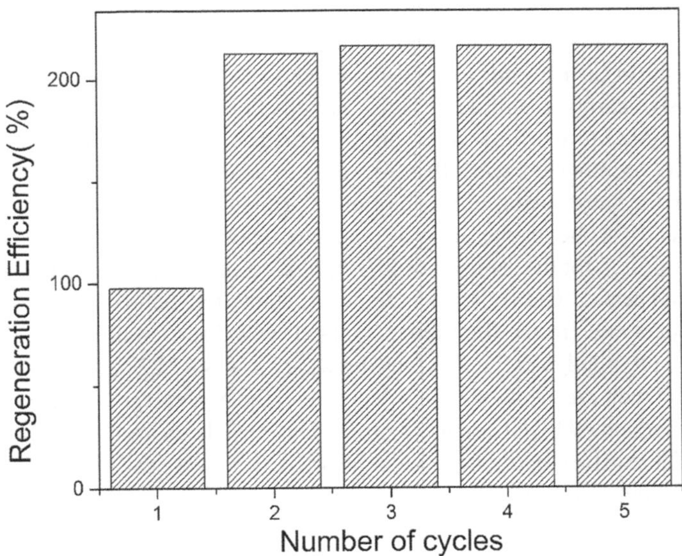

FIGURE 7.3 Regeneration efficiency over the number of adsorption and electrochemical regeneration cycles for MB adsorption on rGO/IO-60. (Reproduced with permission from Sharif et al.[6] Copyright 2017, Elsevier Science Ltd.)

the regeneration of graphene-based adsorbents. One major demerit of this approach is it may result in further breakdown of the graphene sheets into smaller units that may be difficult to recover after the regeneration process.

The utilization of ultrasound or sonication techniques to regenerate activated adsorbent materials have been majorly used for other carbon materials, especially MWCNTs,[44] though its adoption for the regeneration of G-based adsorbent materials is either rare or limited.

7.6.3 Chemical Regeneration

Chemical regeneration relies greatly on factors like solvent concentration, contaminant solubility, adsorbent surface charge, as well as the pH of the solution. This form of regeneration approach for graphene-based adsorbents is comparatively inexpensive and facile, though its demerits are alteration of adsorbent surface characteristics, the difficulty of handling chemicals, and may result in the generation of toxic or oxidized byproducts.

This approach has been reported in a recent study[45] where a one-step regeneration by means of organic reaction on rGO-PDTC/Fe_3O_4 resulted in outstanding adsorption-desorption regeneration cycles toward adsorption of (Cu(II), Cd(II), Pb(II) but also Hg(II)). According to these authors, the greatest adsorption capacities for Cu(II), Cd(II), Pb(II), and Hg(II) ions, respectively, were 113.64, 116.28, 147.06, and 181.82 mg/g. Metal-loaded rGO-PDTC/Fe_3O_4 composites were placed in a 0.1 M HCl solution (pH = 1) during the desorption process and stirred for 5 min. Due to the dithiocarbamate (DTC) functional group breaking down into the rGO-PEI/Fe_3O_4 composite and carbon disulfide (CS2) molecules in acidic environments, metal ions were rapidly removed from rGO-PDTC/Fe_3O_4 composites. 2 g of rGO-PEI/Fe_3O_4 residues were put into a 5 mol/L solution of NaOH (100 mL) in a flask while it was being stirred continuously in an ice bath to regenerate rGO-PDTC/Fe_3O_4 nanocomposite. After that excessive carbon disulfide (30 mL) and ethanol (50 mL) were thoroughly combined and added to the flask with constant stirring for 5 hours at 40 °C. The rGO-PDTC/Fe_3O_4 composites were then recycled to assess their ability to adsorb metal ions under the same adsorption conditions.[45] This technique has also been used by Zaman et al.,[4] for the regeneration of graphene oxide-nano cellulose nanocomposite adsorbent applied for the removal of MB dye. According to their findings, the adsorption effectiveness of GO-CNC was seen to be stable for the first seven cycles and to gradually decrease for the final three cycles. It was discovered that the percent MB elimination recorded over the previous three cycles was 85.22 percent, 84.14 percent, and 82.76 percent, respectively. This decrease in the percentage of MB removal from using regenerated GO-CNC as an adsorbent may have been brought on by changes in the GO-CNC surface caused by the acid or by the mass lost during each stage of retrieval and regeneration.[4] Banerjee et al. reported in their work using GO nanoplatelets in the adsorptive removal of ibuprofen drug that after ten cycles of adsorption-desorption studies, the exclusive removal of ibuprofen remained almost the same for eight regeneration cycles while suffering a slight decline for the ninth and tenth cycle whose percentage removal for the drug was 97.18 and 95.91%.[3] Their investigation was thus: the GONPs were treated with 4.0 M HNO_3 to perform desorption once the

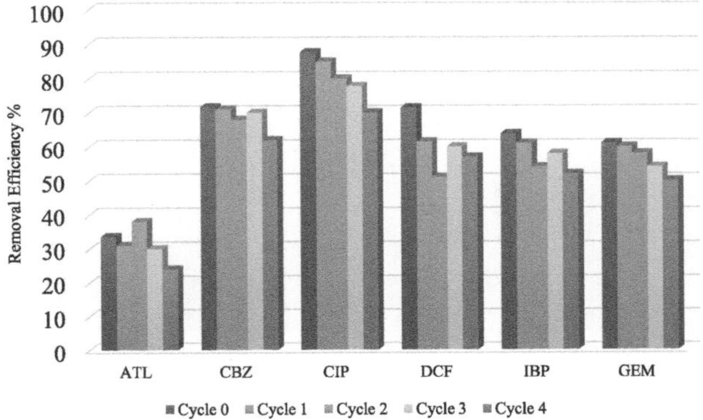

FIGURE 7.4 Recyclability of PG for the adsorption of ECs (Experimental conditions of adsorption test: 5 mg PG dosage, 20 mL EC of 10 mg/L, 2 h contact time, 200 rpm stirring speed). (Reproduced with copyright permission from Khalil et al.,[9] Elsevier 2020.)

sorbent had reached its maximum adsorption capability. Ten cycles of the process were completed.

Khalil et al. have shown in their work where they used porous graphene for the adsorptive removal of AATL, CBZ, CIP, DCF, IBP, and GEM, as shown in Figure 7.4 that the performance appeared to be enhanced after its regeneration with 1 M NaOH for four cycles. Though the removal efficiency was observed to be better in the second cycle for all the adsorbates.[9]

Chemical regeneration is the approach most adopted by researchers and industrialists, as per the available literature, though it presents the challenge of difficulty in chemical recovery.

7.6.4 HYDROTHERMAL REGENERATION

An alternate method is hydrothermal regeneration, which is supposed to be less expensive and easier to use than thermal regeneration. It is purportedly performed at temperatures between 120 and 260 °C.[46] The thermal approach, as is widely known, could use a regeneration temperature as high as 800 °C, which requires a significant amount of energy and is also known to lower the adsorption capacity significantly.

A group of researchers have adopted this approach where they did not only regenerate the spent adsorbent but modified it for better adsorption performance, along with the use of appropriate characterization techniques such as AFM and TEM to identify the graphene platelets within the C-GO regenerated adsorbent, as depicted in Figure 7.5 below.[46] The authors compared hydrothermal regeneration with thermal regeneration and found that the hydrothermal approach yielded a better-regenerated adsorbent.

The transformation mechanism is shown in Figure 7.5 below. When AC was subjected to hydrothermal treatment in step 1, two orbits of lactone lost two electrons, producing the lactone radical.[46] The authors reported that the EPR signal's power and spins increased as a result. Note that ideally, process (1) did not consume any lactone. The hydrolysis of anhydride groups, on the other hand, led to an increase in the

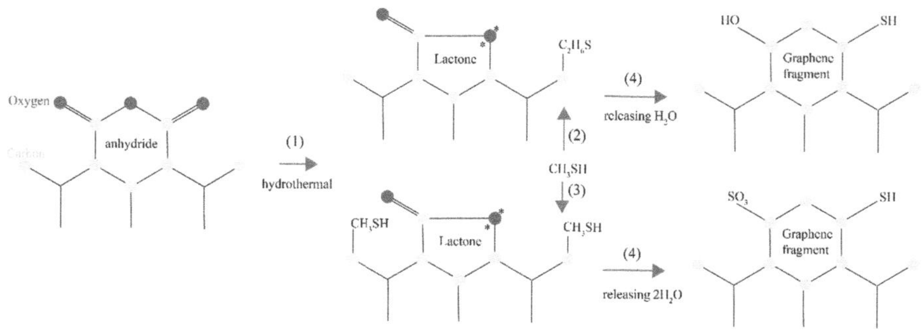

FIGURE 7.5   Mechanism for graphene-oxide formation. Processes (1)–(4). (Reproduced with permission from Liu et al.[46] Copyright 2021, Elsevier Science Ltd.)

number of lactone groups. In any case, the term "lactone radical" in this context refers to the lactone with a lone pair election. In steps 2 and 3, water was heated hydrothermally while the adsorbed CH_3SH was hydrolyzed into CH_3S^+, $C_2H_6S^+$, and H^+. TG-MS provided evidence for the existence of CH_3S^+ and $C_2H_6S^+$. In other words, AC contained 3.4 percent sulfur and obtained an adsorption quantity of 5.3 mg for CH_3SH. Comparatively, the elemental analysis showed that ACS and ACSH both contained 0.8% of S, while S was only found in 0.1 percent of ACS. In ACSH, S was fixed by hydrothermal treatment. Furthermore, the TG-MS supported this. Sulfur was still being resorbed to about 2.6%. When the autoclave was opened, no obnoxious gas was immediately detectable. The desorbed S was potentially still in the aqueous phase and might be employed again during a different regeneration procedure.

In step 4, the adsorbed $C_2H_6S^+/CH_3S^+$ reacted with the lactone radical, breaking and rearranging the hexagonal carbon layer and creating microdomains of graphitizable amorphous carbon, which resulted in the formation of tiny graphene-oxide fragments. The lactone radicals were also eaten at the same time. This process also involved the pyrone group changing into the phenol group, which was consistent with the formation of SO_3^{2-} due to oxidation. If the recycled carbon was employed for adsorption and regeneration once more, the produced S-containing groups may potentially be in favor of CH_3SH redox. The ratio of oxygen-centered radicals to carbon-centered radicals was reduced at the same time the lactone radical was consumed.

Actually, the production of nitrogen-doped graphene from activated carbon in an ammonia solution shows that the hydrothermal process caused the fundamental unit of amorphous AC to redistribute and produce lamellar graphene. The authors also proved that free radicals play in the production of graphene-oxide, which was an important finding.[46]

7.6.5 Supercritical Fluid Extraction

This approach involves the use of supercritical fluids for the regenerative recycling of graphene-based adsorbents. This approach greatly relies on the kind of supercritical fluid adopted, temperature, pressure, as well as solubility of the adsorbate. The advantage of this process is that it is fast. Though it's demerit is its operation at very high pressure, which may be destructive to the materials used in the construction of the instruments.

7.6.6 Thermal Regeneration

Thermal regeneration is the process of heating a sorbent to a specific temperature in order to break the chemical and physical bonds that hold sorbate with sorbent together. This process entails the use of heat to degrade the absorbed, absorbed on the surface of the adsorbents. It is a relatively cheap approach, though it may result in the generation of radioactive residues, which are dangerous. It is clear that this approach may be good or effective for graphene-based adsorbents that are made with pure graphene. However, in the case of graphene oxide or reduced graphene oxide, it may degrade the absorbent and completely transform the structure to another form of graphene material, which may not be the originally desired material as initially used. At the moment, this technique is used in both industrial and commercial settings to regenerate activated carbon.

The approach depends on factors such as the temperature of absorbing heating, as well as pollutants absorbed. The benefit of this approach is that it is effective on adsorbents that are loaded with diverse pollutants. However, its demerit is that it requires high temperature, which may be detrimental to some absorbance as it may degrade the adsorbent in the process of regeneration.

The adoption of thermal regeneration for effective regeneration of methyl orange saturated graphene has been explored. Though the authors revealed that the inorganic components of the dye, majorly NaCl was, left behind as residue, but also the porosity of the adsorbent was recovered at a temperature of 300 °C, even though the adsorption capacity of the adsorbent declined to 75–90%.[47] Yet in another instance, Labiadh and Kamali[48] have demonstrated the outstanding potential of thermal regeneration for graphene adsorbent used for the adsorptive removal of MO. The authors revealed that a facile thermal regeneration of graphene in the air at 300 °C showed great removal efficiency for the adsorbate exclusion, though with continued reduction in efficiency as the regeneration cycles went higher (Figure 7.6). As a result of the organic matter saturating the graphene during the adsorption process and obstructing its pores, efficiency may have been reduced. The material was heated to 300 °C in air, but this did not result in a complete recovery of the 3D graphene nanosheets.[48] Thermal regeneration has also been explored by Gao et al.,[49] for the regeneration of magnetic rGO-Fe_2O_3 nanocomposite, even as their report showed that the calcination of the adsorbent at 300 °C proved more effective than at 350 °C, due to partial oxidation of TrGO platelets resulting in declined adsorption capacity of the regenerated adsorbent.[49]

Thermal regeneration is the most convenient regeneration approach adopted for industrial purposes when compared to other regeneration approaches, seeing it is very effective for a wide variety of adsorbates such as oils and organic solvents.

7.6.7 Photo-Assisted Regeneration

This approach of regeneration of graphene-based absorbent is greatly encouraged in the sense that the same absorbent can be designed in such a way that it acts as an

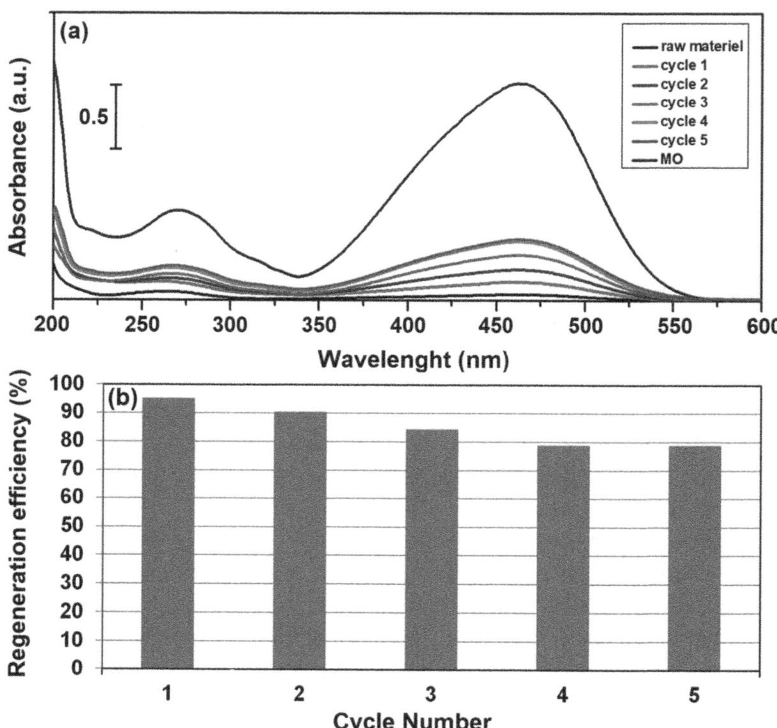

FIGURE 7.6   (a) UV–Vis spectra and (b) the corresponding thermal regeneration efficiency of the 3D graphene at different numbers of adsorption-desorption cycles. (Reproduced with permission from Labiadh and Kamali.[48] Copyright 2019, Elsevier Science Ltd.)

absorbent and a catalyst toward the degradation of the adsorbed adsorbate. This is a very efficient way of designing an absorbent and utilizing it without stressing about how to regenerate it. This approach is greatly reliant on the photocatalytic activity of the adsorbent, as well as the photosensitizers used. Its advantage is the quick extraction of the adsorbates. It is environmentally friendly. It transforms the pollutant into smaller particles, which will result in the complete regeneration of the adsorbent. And it is cheap, even though it also comes with its disadvantages, which include the creation of unwanted or toxic byproducts.

This approach has been reportedly utilized by a group of researchers for the adsorptive removal of naphthalene along with the adsorbent photo-assisted regeneration of ZnO/Ag/GO, even up to five cycles still showing more than 85% efficiency from an initial value of 92% as depicted in Figure 7.7a.[50] The adsorbent efficiency declined due to the loss of surface functionalities and pore-clogging.

Naphthalene was primarily adsorbed on the surface of the ZnO/Ag/GO nanocomposite via π-π interactions because both naphthalene and GO have aromatic properties in Step I of the overall adsorption mechanism, which was described by the authors (Figure 7.7b). In Step II, the adsorbed naphthalene was photocatalytically degraded in the presence of visible light. During photocatalysis, photogenerated holes and electrons

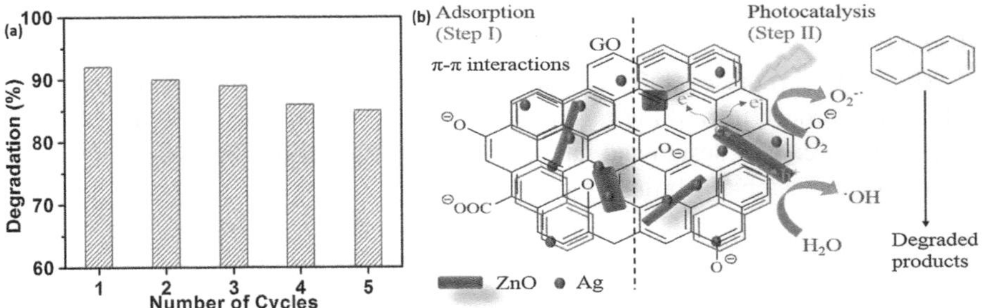

FIGURE 7.7 (a) Recyclability experiments for five cycles using ZnO/Ag/GO as photocatalyst and (b) Schematic showing the plausible adsorption-assisted mechanism. (Reproduced with permission from Mukwevho et al.[50] Copyright 2020, Elsevier Science Ltd.)

helped to produce active hydroxyl radicals and superoxide anion radicals, respectively. These radicals may consequently convert naphthalene into less harmful metabolites.[50]

In another work, Maswanganyi et al.[51] prepared Bismuth Molybdate@rGO (Bi_2MoO_6/rGO) adsorbent com-photocatalyst, which was used for the adsorptive removal of naphthalene followed by photocatalytic regenerative degradation of the adsorbate to regenerate the adsorbent.[51] They revealed that Bi_2MoO_6@rGO nanocomposite adsorbent demonstrated better performance in comparison to Bi_2MoO_6, though, with the nanocomposite having 2 wt.% rGO revealing better photocatalytic efficacy, which was large as a result of enhanced absorption of visible light and photogenerated charge carriers recombination rate arising from rGO inclusion.[51] The possible mechanism detailing the adsorption-com-photodegradative regeneration of naphthalene on the Bi_2MoO_6@rGO is presented in Figure 7.8. Although Bi_2MoO_6 has a narrow band gap that can form electron-hole pairs in the presence of visible light, its photocatalytic activity is constrained by the speedy recombination of charge carriers. The Bi_2MoO_6@rGO nanocomposite material's adsorption and photocatalytic effectiveness were significantly increased by the addition of rGO. Through π-π interactions, the rGO in the nanocomposite enhanced the adsorption of naphthalene

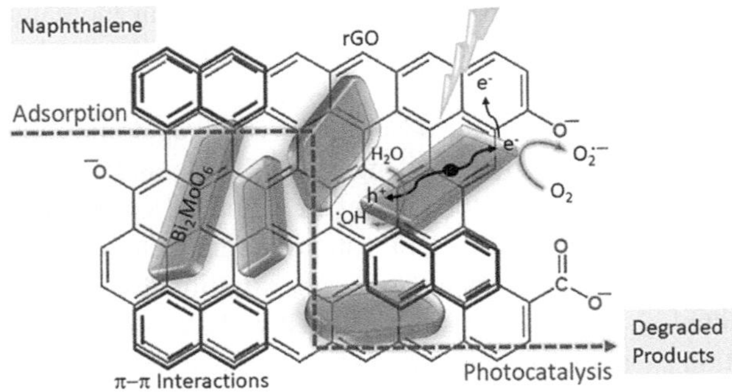

FIGURE 7.8 Plausible mechanism for the removal of naphthalene via adsorption–photodegradation. Reproduced with permission from Maswanganyi et al.[51] Copyright 2021, American Chemical Society.

molecules onto the surface of the photocatalyst, leading to effective naphthalene degradation.[51]

This approach presents itself as an eco-friendly, cost-effective, facile, as well as effective approach toward the regeneration of spent adsorbents, though such adsorbent requires careful design and fabrication in order to endow the adsorbent with the required photocatalytic ability.

7.6.8 BIOLOGICAL REGENERATION

This approach entails the use of biological or natural means to regenerate the graphene-based absorbent after use. The advantage of this approach is that it transforms the pollutants into smaller entities, which results in the complete regeneration of the absorbent in an environmentally friendly approach. Though this process has its own disadvantages also: it can only be used on biodegradable adsorbents, and it's not applicable with adsorbents that are altered/modified, the process is slow and may result in biofouling of the adsorbent porous structure. This approach greatly relies on the nature of the adsorbent, the concentration of the pollutants, the type of microorganism used, as well as conditions favorable to microorganism growth. It is mainly used where the adsorbate is a biological wastewater pollutant like viruses, insects, bacteria, plant pollen grains, microorganisms like bacteria, animal saliva creatures, etc.

In biological regeneration, the adsorbates are broken down by bacteria, using a biological process in both aerobic and anaerobic conditions. This approach generates less sludge overall, even though it is economically feasible and environmentally friendly. In fact, this method requires a long time for degradative regeneration and performs poorly on dyes having a long-lasting, high-molecular-weight polymeric structure. Wastewater treatment alone cannot completely get rid of toxic chemicals.

7.6.9 MAGNETIC REGENERATION

This approach involves the use of magnets to attract and separate the adsorbent from the adsorbate, which is a simple and cheap way to regenerate magnetic graphene-based adsorbents.[23] A unique method for magnetic separation involves adding magnetite nanoparticles based on iron oxide/compounds as an inorganic nanomaterial component. In this method, the nanosorbents are easily recovered and regenerable using an external magnet.

The use of this technique for the adsorption as well as desorption of mixed contaminants has been demonstrated by Geng et al.[23] These authors revealed that even after four cycles, the adsorption efficiency of the graphene-based adsorbent remained at 97%. This hybrid's adsorbability demonstrates a satisfactory level of tolerance for fluctuations in the pH environment and dye concentration. Even when subjected to a multi-dye cocktail, the hybrid may function well without impairing each dye's individual adsorption capability in comparison to that tested independently.

A group of researchers reported the adsorption efficiency of regenerated Fe_3O_4@ chitosan@graphene "FGC" to be 74.81% at the fifth cycle in comparison to the virgin FGC adsorbent used for adsorptive exclusion of MB dye.[7]

7.7 CONCLUSION

This chapter has covered the regeneration and reuse of graphene-based adsorbents within recent decades. We observe via our investigation of literature that researchers are tilting more toward composites/hybrid graphene-based materials, due to the enhancement in the overall performance of these systems. There is a need to design and engineer systems that will work well, especially in the selective adsorption of diverse adsorbates in a singular wastewater system. The development and use of regeneration approaches that are eco-friendly via the use of eco-friendly eluents should be used more by researchers and industrialists.

REFERENCES

1. Liu, S.; Ge, H.; Wang, C.; Zou, Y.; Liu, J., Agricultural waste/graphene oxide 3D bio-adsorbent for highly efficient removal of methylene blue from water pollution. *Science of the Total Environment* 2018, *628–629*, 959–968.
2. Sharif, F.; Roberts, E. P. L. Electrochemical oxidation of an organic dye adsorbed on tin oxide and antimony doped tin oxide graphene composites. *Catalysts* 2020, *10*(2), 263.
3. Banerjee, P.; Das, P.; Zaman, A.; Das, P., Application of graphene oxide nanoplatelets for adsorption of Ibuprofen from aqueous solutions: Evaluation of process kinetics and thermodynamics. *Process Safety and Environmental Protection* 2016, *101*, 45–53.
4. Zaman, A.; Orasugh, J. T.; Banerjee, P.; Dutta, S.; Ali, M. S.; Das, D.; Bhattacharya, A.; Chattopadhyay, D., Facile one-pot in-situ synthesis of novel graphene oxide-cellulose nanocomposite for enhanced azo dye adsorption at optimized conditions. *Carbohydrate Polymers* 2020, *246*, 116661.
5. Ren, H.; Cao, Z.-F.; Chen, Y.-Y.; Jiang, X.-Y.; Yu, J.-G., Graphene oxide-Bicine composite as a novel adsorbent for removal of various contaminants from aqueous solutions. *Journal of Environmental Chemical Engineering* 2021, *9* (6), 106769.
6. Sharif, F.; Gagnon, L. R.; Mulmi, S.; Roberts, E. P. L., Electrochemical regeneration of a reduced graphene oxide/magnetite composite adsorbent loaded with methylene blue. *Water Research* 2017, *114*, 237–245.
7. Tran, H. T. T.; Hoang, L. T.; Tran, H. V., Electrochemical synthesis of graphene from waste discharged battery electrodes and its applications to preparation of graphene/Fe$_3$O$_4$/chitosan-nanosorbent for organic dyes removal. *Zeitschrift für Anorganische und Allgemeine Chemie* 2022, *648* (3), e202100313.
8. Chaabane, L.; Beyou, E.; El Ghali, A.; Baouab, M. H. V., Comparative studies on the adsorption of metal ions from aqueous solutions using various functionalized graphene oxide sheets as supported adsorbents. *Journal of Hazardous Materials* 2020, *389*, 121839.
9. Khalil, A. M. E.; Memon, F. A.; Tabish, T. A.; Salmon, D.; Zhang, S.; Butler, D., Nanostructured porous graphene for efficient removal of emerging contaminants (pharmaceuticals) from water. *Chemical Engineering Journal* 2020, *398*, 125440.
10. Sharif, F.; Roberts, E. P. L., Anodic electrochemical regeneration of a graphene/titanium dioxide composite adsorbent loaded with an organic dye. *Chemosphere* 2020, *241*, 125020.
11. Yao, Y.; Miao, S.; Liu, S.; Ma, L. P.; Sun, H.; Wang, S., Synthesis, characterization, and adsorption properties of magnetic Fe$_3$O$_4$@graphene nanocomposite. *Chemical Engineering Journal* 2012, *184*, 326–332.

12. Ai, L.; Zhang, C.; Chen, Z., Removal of methylene blue from aqueous solution by a solvothermal-synthesized graphene/magnetite composite. *Journal of Hazardous Materials* 2011, *192* (3), 1515–1524.

13. Wang, W.; Gong, Q.; Chen, Z.; Wang, W. D.; Huang, Q.; Song, S.; Chen, J.; Wang, X., Adsorption and competition investigation of phenolic compounds on the solid-liquid interface of three-dimensional foam-like graphene oxide. *Chemical Engineering Journal* 2019, *378*, 122085.

14. Kyzas, G. Z.; Koltsakidou, A.; Nanaki, S. G.; Bikiaris, D. N.; Lambropoulou, D. A., Removal of beta-blockers from aqueous media by adsorption onto graphene oxide. *Science of the Total Environment* 2015, *537*, 411–420.

15. Pashaei-Fakhri, S.; Peighambardoust, S. J.; Foroutan, R.; Arsalani, N.; Ramavandi, B., Crystal violet dye sorption over acrylamide/graphene oxide bonded sodium alginate nanocomposite hydrogel. *Chemosphere* 2021, *270*, 129419.

16. Lai, K. C.; Lee, L. Y.; Hiew, B. Y. Z.; Thangalazhy-Gopakumar, S.; Gan, S., Facile synthesis of xanthan biopolymer integrated 3D hierarchical graphene oxide/titanium dioxide composite for adsorptive lead removal in wastewater. *Bioresource Technology* 2020, *309*, 123296.

17. Chen, Y.-Y.; Yu, J.-G., Nitrogen-containing graphene oxide composite with acid resistance and high adsorption selectivity for Nd^{3+}. *Journal of Environmental Chemical Engineering* 2022, *10* (5), 108348.

18. Li, D.; Hua, T.; Yuan, J.; Xu, F., Methylene blue adsorption from an aqueous solution by a magnetic graphene oxide/humic acid composite. *Colloids and Surfaces A: Physicochemical and Engineering Aspects* 2021, *627*, 127171.

19. Yan, J.; Li, K.; Yan, J.; Fang, Y.; Liu, B., A magnetically recyclable magnetic graphite oxide composite functionalized with polydopamine and β-cyclodextrin for cationic dyes wastewater remediation: Investigation on adsorption performance, reusability and adsorption mechanism. *Applied Surface Science* 2022, *602*, 154338.

20. Greaves, M.; Elderfield, H.; Klinkhammer, G., Determination of the rare earth elements in natural waters by isotope-dilution mass spectrometry. *Analytica Chimica Acta* 1989, *218*, 265–280.

21. Yusan, S.; Gok, C.; Erenturk, S.; Aytas, S., Adsorptive removal of thorium (IV) using calcined and flux calcined diatomite from Turkey: Evaluation of equilibrium, kinetic and thermodynamic data. *Applied Clay Science* 2012, *67*, 106–116.

22. Li, D.; Zhang, B.; Xuan, F., The sorption of Eu (III) from aqueous solutions by magnetic graphene oxides: A combined experimental and modeling studies. *Journal of Molecular Liquids* 2015, *211*, 203–209.

23. Geng, Z.; Lin, Y.; Yu, X.; Shen, Q.; Ma, L.; Li, Z.; Pan, N.; Wang, X., Highly efficient dye adsorption and removal: a functional hybrid of reduced graphene oxide–Fe_3O_4 nanoparticles as an easily regenerative adsorbent. *Journal of Materials Chemistry* 2012, *22* (8), 3527–3535.

24. Liu, Y.; Fu, J.; He, J.; Wang, B.; He, Y.; Luo, L.; Wang, L.; Chen, C.; Shen, F.; Zhang, Y., Synthesis of a superhydrophilic coral-like reduced graphene oxide aerogel and its application to pollutant capture in wastewater treatment. *Chemical Engineering Science* 2022, *260*, 117860.

25. Mu, C.; Zhang, Y.; Cui, W.; Liang, Y.; Zhu, Y., Removal of bisphenol A over a separation free 3D Ag_3PO_4-graphene hydrogel via an adsorption-photocatalysis synergy. *Applied Catalysis B: Environmental* 2017, *212*, 41–49.

26. da Silva, P. M. M.; Camparotto, N. G.; Figueiredo Neves, T.; Mastelaro, V. R.; Nunes, B.; Siqueira Franco Picone, C.; Prediger, P., Instantaneous adsorption and synergic effect in simultaneous removal of complex dyes through nanocellulose/graphene oxide nanocomposites: Batch, fixed-bed experiments and mechanism. *Environmental Nanotechnology, Monitoring & Management* 2021, *16*, 100584.

27. Radmehr, S.; Hosseini Sabzevari, M.; Ghaedi, M.; Ahmadi Azqhandi, M. H.; Marahel, F., Adsorption of nalidixic acid antibiotic using a renewable adsorbent based on Graphene oxide from simulated wastewater. *Journal of Environmental Chemical Engineering* 2021, *9* (5), 105975.

28. Das, T. R.; Patra, S.; Madhuri, R.; Sharma, P. K., Bismuth oxide decorated graphene oxide nanocomposites synthesized via sonochemical assisted hydrothermal method for adsorption of cationic organic dyes. *Journal of colloid and interface science* 2018, *509*, 82–93.

29. Qi, Y.; Yang, M.; Xu, W.; He, S.; Men, Y., Natural polysaccharides-modified graphene oxide for adsorption of organic dyes from aqueous solutions. *Journal of Colloid and Interface Science* 2017, *486*, 84–96.

30. Deng, M.; Huang, Y., RETRACTED: The phenomena and mechanism for the enhanced adsorption and photocatalytic decomposition of organic dyes with Ag_3PO_4/graphene oxide aerogel composites. *Ceramics International* 2020, *46* (2), 2565–2570.

31. Liu, N.; Liang, G.; Dong, X.; Qi, X.; Kim, J.; Piao, Y., Stabilized magnetic enzyme aggregates on graphene oxide for high performance phenol and bisphenol A removal. *Chemical Engineering Journal* 2016, *306*, 1026–1034.

32. Du, H.; Lei, Y.; Chen, W.; Li, F.; Li, H.; Deng, W.; Jiang, G., Multifunctional magnetic bio-nanoporous carbon material based on zero-valent iron, angelicae dahuricae radix slag and graphene oxide: An efficient adsorbent of pesticides. *Arabian Journal of Chemistry* 2021, *14* (8), 103267.

33. Hu, R.; Dai, S.; Shao, D.; Alsaedi, A.; Ahmad, B.; Wang, X., Efficient removal of phenol and aniline from aqueous solutions using graphene oxide/polypyrrole composites. *Journal of Molecular Liquids* 2015, *203*, 80–89.

34. Hugo, E. R.; Brandebourg, T. D.; Woo, J. G.; Loftus, J.; Alexander, J. W.; Ben-Jonathan, N., Bisphenol A at environmentally relevant doses inhibits adiponectin release from human adipose tissue explants and adipocytes. *Environmental Health Perspectives* 2008, *116* (12), 1642–1647.

35. Chen, F.; An, W.; Liu, L.; Liang, Y.; Cui, W., Highly efficient removal of bisphenol A by a three-dimensional graphene hydrogel-AgBr@ rGO exhibiting adsorption/photocatalysis synergy. *Applied Catalysis B: Environmental* 2017, *217*, 65–80.

36. Higgins, P.; Siddiqui, S. H.; Kumar, R., Design of novel graphene oxide/halloysite nanotube@ polyaniline nanohybrid for the removal of diclofenac sodium from aqueous solution. *Environmental Nanotechnology, Monitoring & Management* 2022, *17*, 100628.

37. Rahmatpour, A.; Soleimani, P.; Mirkani, A., Eco-friendly poly (vinyl alcohol)/partially hydrolyzed polyacrylamide/graphene oxide semi-IPN nanocomposite hydrogel as a reusable and efficient adsorbent of cationic dye methylene blue from water. *Reactive and Functional Polymers* 2022, *175*, 105290.

38. Omidi, M. H.; Azqhandi, M. H. A.; Ghalami-Choobar, B., Synthesis, characterization, and application of graphene oxide/layered double hydroxide /poly acrylic acid nanocomposite (LDH-rGO-PAA NC) for tetracycline removal: A comprehensive chemometric study. *Chemosphere* 2022, *308*, 136007.

39. Dolatabadi, M.; Naidu, H.; Ahmadzadeh, S., Adsorption characteristics in the removal of chlorpyrifos from groundwater using magnetic graphene oxide and carboxy methyl cellulose composite. *Separation and Purification Technology* 2022, *300*, 121919.

40. Sun, Y.; Liu, X.; Lv, X.; Wang, T.; Xue, B., Synthesis of novel lignosulfonate-modified graphene hydrogel for ultrahigh adsorption capacity of Cr(VI) from wastewater. *Journal of Cleaner Production* 2021, *295*, 126406.

41. Hao, X.; Yang, S.; Tao, E.; Liu, L.; Ma, D.; Li, Y., Graphene oxide/montmorillonite composite aerogel with slit-shaped pores: Selective removal of Cu^{2+} from wastewater. *Journal of Alloys and Compounds* 2022, *923*, 166335.

42. Sheikhmohammadi, A.; Hashemzadeh, B.; Alinejad, A.; Mohseni, S. M.; Sardar, M.; Sharafkhani, R.; Sarkhosh, M.; Asgari, E.; Bay, A., Application of graphene oxide modified with the phenopyridine and 2-mercaptobenzothiazole for the adsorption of Cr (VI) from wastewater: Optimization, kinetic, thermodynamic and equilibrium studies. *Journal of Molecular Liquids* 2019, *285*, 586–597.

43. Babakir, B. A.; Abd Ali, L. I.; Ismail, H. K., Rapid removal of anionic organic dye from contaminated water using a poly (3-aminobenzoic acid/graphene oxide/cobalt ferrite) nanocomposite low-cost adsorbent via adsorption techniques. *Arabian Journal of Chemistry* 2022, *15* (12), 104318.

44. Zhu, Q.; Wang, J.; Zhang, L.; Yan, D.; Yang, H.; Zhang, L.; Miao, X., Research progress of graphene for the adsorption, enrichment and analysis of environmental pollutants. *Bulletin of Materials Science* 2023, *46* (1), 17.

45. Fu, W.; Huang, Z., Magnetic dithiocarbamate functionalized reduced graphene oxide for the removal of Cu(II), Cd(II), Pb(II), and Hg(II) ions from aqueous solution: Synthesis, adsorption, and regeneration. *Chemosphere* 2018, *209*, 449–456.

46. Liu, J.; Li, C.; Kong, W.; Lu, Q.; Zhang, J.; Qian, G., Lactone radical transformed methyl mercaptan-adsorbed activated carbon into graphene oxide modified activated carbon. *Journal of Hazardous Materials* 2021, *413*, 124527.

47. Labiadh, L.; Kamali, A. R., Textural, structural and morphological evolution of mesoporous 3D graphene saturated with methyl orange dye during thermal regeneration. *Diamond and Related Materials* 2020, *103*, 107698.

48. Labiadh, L.; Kamali, A. R., 3D graphene nanoedges as efficient dye adsorbents with ultra-high thermal regeneration performance. *Applied Surface Science* 2019, *490*, 383–394.

49. Gao, Y.; Zhong, D.; Zhang, D.; Pu, X.; Shao, X.; Su, C.; Yao, X.; Li, S., Thermal regeneration of recyclable reduced graphene oxide/Fe3O4 composites with improved adsorption properties. *Journal of Chemical Technology & Biotechnology* 2014, *89* (12), 1859–1865.

50. Mukwevho, N.; Gusain, R.; Fosso-Kankeu, E.; Kumar, N.; Waanders, F.; Ray, S. S., Removal of naphthalene from simulated wastewater through adsorption-photodegradation by ZnO/Ag/GO nanocomposite. *Journal of Industrial and Engineering Chemistry* 2020, *81*, 393–404.

51. Maswanganyi, S.; Gusain, R.; Kumar, N.; Fosso-Kankeu, E.; Waanders, F. B.; Ray, S. S., Bismuth Molybdate Nanoplates Supported on Reduced Graphene Oxide: An Effective Nanocomposite for the Removal of Naphthalene via Adsorption–Photodegradation. *ACS Omega* 2021, *6* (26), 16783–16794.

8 Current Challenges, Conclusion, Future Prospects, and Cost-Effectiveness/ Comparison

8.1 CURRENT CHALLENGES

The understanding of (graphene G) as an adsorption material is advancing quickly; nevertheless, there are numerous obstacles as well as roadblocks that must be conquered. Despite the fact that nature abounds in the raw ingredients needed to make G, a big challenge remains for G researchers to reconfigure and/or advance ways of making extremely selective adsorbent materials with high adsorption characteristics. G-based systems having low aggregation but also high specific surface area possess a high adsorption capacity towards organic pollutants, particularly benzene-based compounds, in which the interaction between G and the adsorbate is important. As a result, one of the major issues encountered when using G adsorbent is G sheet aggregation. This aggregation between the G layers must be avoided since it affects accessibility and, as a result, the number of available adsorption sites for contaminants to bind. Aggregation can be avoided by adding oxygen groups, which enhance G's dispersion characteristics in solution and significantly boost G's capacity to remove impurities. In a different development, magnetic particles are added to the adsorbent to create a magnetic G composite for practical separation. The extra magnetic particles will be crucial in preventing G aggregation as well.

For large-scale G synthesis, Hummer's technique, which entails an oxidation and reduction procedure, is now in use. The fact that the oxidation process uses a lot of powerful acids and oxidants and produces a lot of acidic waste that needs to be expensively treated and disposed of is one of the main drawbacks of this technology. As a result, it could be an expensive and environmentally harmful practice. Additionally, the hydrazine employed in this procedure is potentially carcinogenic and harmful to the environment. Furthermore, a vigorous oxidation process could leave G with undesirable flaws, endangering its exceptional qualities. Hummers' approach is unable to produce G sheets with precise geometrical dimensions and forms. Therefore, it is crucial to investigate different techniques to handle the aforementioned problems. To some of the aforementioned problems, scientists have put forward a few remedies. Chemical/thermal oxidation categories have introduced and established a wide variety of G oxidation procedures to achieve desired outcomes.

DOI: 10.1201/9781032621302-8

Additionally, ascorbic acid (vitamin C), which is safe for the environment and non-toxic, can be used in place of hydrazine. A milder exfoliation procedure can be used with the aid of additional species, such as multi-pyrene tethered amphiphiles. Given that it will result in G sheets with adjustable size and internal conjugated structure, it has been mentioned that a synthetic direction may be offered. Despite the higher mechanical capabilities of a single G sheet, weak inter-sheet interactions usually degrade the mechanical performance of pure collective G products or G composites. Consequently, it is currently exceedingly difficult to find ways to improve the interaction between neighboring G sheets or the interaction between G sheets and their modifier functionalities.

Given that G can be altered through covalent or non-covalent bonding, including stacking, hydrogen bonding, and other processes, a multifunctional inter-G "welding" molecule or metallic nanoparticles may be a practical choice for enhancing inter-layer electrical contact. Because there is always a need to research more straightforward, reliable, and effective techniques to prepare GO, G, and their composites, the manufacturing of GO and rGO is challenging. In spite of the fact that research studies examining the chemistry of GO and G are currently developing at a quick rate, according to various sources and our own experience, substantial studies are still needed for a thorough grasp of G structure. The lack of synthetic procedures that provide repeatable and regulated ways is another crucial issue that needs to be addressed; for now, studies in this field are critically needed.

8.2 GENERAL CONCLUSIONS

It is clear that researchers are in the phase of advanced G-based adsorbent for effective remediation of wastewater as well as the exclusion of toxic materials from the human system, as recently focused on by some researchers. This book has provided up-to-date information on G-based adsorbent materials, the kinetic, thermodynamic, and isotherms utilized in this niche by diverse researchers. It also oversees the application of these systems, the current challenges faced by researchers, and the future prospects thereof. Finally, this book presents a conclusion with regard to its contents.

8.3 FUTURE PROSPECTS

The potency of using G nanostructures to remove different chemical and environmental contaminants is enormous. G's exceptional physicochemical characteristics will play a critical part in environmental toxic waste remediation as well as management at present and in the near future. For starters, G composite demonstrated a high potential for environmental application in the detection of environmental pollutants. Due to the unresolved issue of G's toxicity for both short- and long-term exposure to the biosphere as well as the human body, it is still too early to adopt large-scale applications of these nanocomposites in environmental monitoring and cleanup. Despite all odds, G composites continue to be one of the fascinating research platforms for the environment as well as energy.

Furthermore, the production of nanomaterials through the use of "chemical" processing routes, such as G oxidation followed by reduction of the GO platelet obtained

TABLE 8.1
Pros and Cons of GBAM

GBAM	Pros	Cons	Ref.
Functionalized using organic molecules.	High surface area; excellent colloidal stability; enhanced quantity of $-NH_2$, and $-OH$ functionalities.	The modification approach affects how stable the loaded molecules are.	1
Functionalized using polymeric as well as copolymeric molecules.	Excellent structural characteristics within acid media; improved adsorption efficiency.	Poor structural solubility, as well as lack of proper processability.	2
Functionalized using inorganic substances	Simple isolation used for recycling; "increased reductivity and adsorptivity;" as well as strengthening of the structural system.	The colloidal stability is decreased by co-reduction of GO during the attachment of the particles.	1,3
Functionalized using inorganic acid	Acid higher adsorption capacity at acidic pH levels; decrease in self-aggregation; and increased the aqueous solution's dispersion.	The challenging separation from the solution raises the cost of industrial application and/or results in re-pollution of the remediated water.	4,5

through exfoliation, may be capable of producing relatively large amounts of cost-effective "G." The chemical details, such as the mechanism of oxidation or reduction, but also detailed chemical structure, are currently lacking. Fewer G-based adsorbent materials have been generated and studied in comparison to other well-known nano-materials; hence, future research should assess whether it is feasible to produce additional G-based materials with further developments in the manufacturing of nanostructured materials. Furthermore, advancements in foundational physical sciences, as well as techniques that can be applied, will enable revolutionary applications based on G and its composites to broaden the horizons of G-based adsorbent materials, as well as open new doors in the exclusion of pollutants from our environment and its restoration. GBAMs still present some pros and cons that require the necessary solutions from emerging researchers as presented in Table 8.1 above.

8.4   COST EFFECTIVENESS/COMPARISON

In comparison to other competing methods, a significant aspect determining an adsorbent's suitability for use in wastewater treatment is the cost of preparation and utilization. In order to determine the cost of an adsorbent, factors such as the price of raw materials, discounted cash flow, cost indices, the cost of the adsorbent per gram of adsorbate removed, annual capital expenditures (CAPEX), operating expenditures (OPEX), and the cost of the adsorbent application in an adsorption operation can all be taken into account. This subsection is required because of the necessity to harmonize the numerous conceptions of adsorbent cost proposed by different researchers and to assess the impact of process efficiency on process cost. The cost of employing

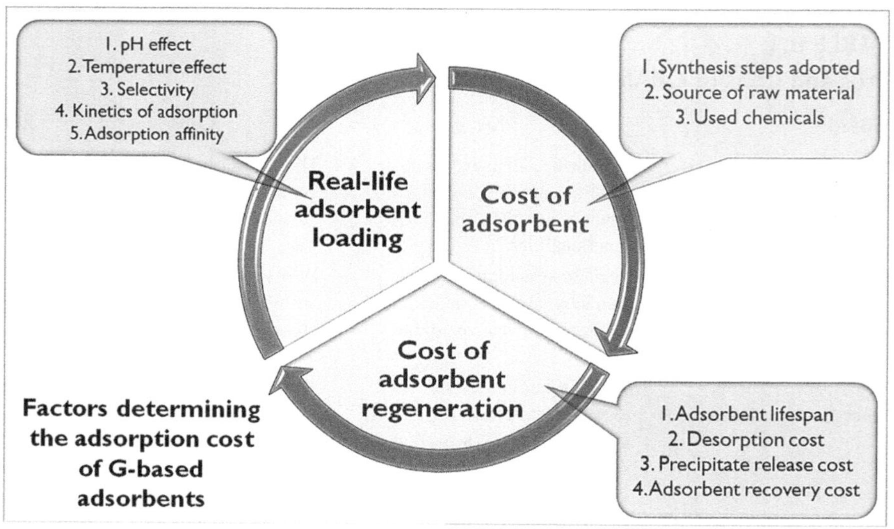

FIGURE 8.1   Factors determining the adsorption cost of/using G-based adsorbents.

various adsorbents for wastewater treatment is summarized in this subsection. The "adsorbent cost performance, "denoted as (and calculated in \$/mol), was also established as a quantitative indicator for comparing adsorbent costs. It was explained as the monetary cost of making and using 1 g of an adsorbent for the removal of 1 mole of a chemical species in the aqueous phase, when the chemical species was considered at its theoretical point of maximum absorption. Adsorbent cost performance varies greatly, although most lie between \$1 and \$200/mol. For the intended use, adsorbents priced under \$1/mol can be thought of as being very affordable, while those priced above \$200/mol can be thought of as being highly expensive. To increase practical applicability, more research into adsorbent cost analysis is needed, particularly in routine adsorption studies.

It is important to remember that the adsorption of diverse adsorbates using G-based adsorbents is primarily governed by the cost of adsorbent, the amount of adsorbent practically used/loaded, and the cost of regeneration as pictorially presented in Figure 8.1.

8.4.1 Factors Determining the Adsorption Cost of/Using G-Based Adsorbents

8.4.1.1 Cost of Adsorbent

In order to drastically lower the cost of the adsorbents overall, an ideal adsorbent should not only have higher adsorption capacity but also improved reusability properties. As a result, economic viability is crucial. Along with the type of pollutant adsorbed, graphene adsorbents undergo a wide range of desorption and regeneration processes. But, choosing the right eluent is crucial for effective desorption, which typically depends on the adsorption mechanism, as well as the characteristics of both the adsorbates and the adsorbents.

Many different G-based adsorbents have been created and studied, as per the literature that is currently available.[6–9] These G-based adsorbents may be synthesized using waste materials to the use of fabricated composites/hybrids. However, information on the cost of these adsorbents is often not provided, making it hard to deduce. The cost of G-based adsorbents largely depends on the raw material source/cost as well as the cost of chemicals used for its synthesis.

Also, the number of steps necessary in synthesizing/modifying the adsorbent has a great influence on the cost of the adsorbent and, consequently, the adsorption cost.[10] For instance, composite adsorbents with active sites immobilized or impregnated on a backbone need more chemicals and processes to create or change the adsorbent; as a result, these methods are often linked to more expensive adsorbents.[11] Granular iron oxide-based adsorbents fall within the group of synthetic adsorbents that have an intermediate cost and are mostly composed of metal oxides without any extra backbones.[12] As a result, adsorbents are divided into three cost groups. (a) Low cost – These include biomass waste and cost much less than \$1 per kilogram. (b) Granular porous metal oxides, for example, typically cost between \$3 and \$6 per kilogram (based on the information provided for the granular ferric hydroxide GEH and FerroSorp). Hybrid ion exchange resins like BioPhree®, which cost between \$15 and \$20 per kg, fall under this category of (c) High Cost. Adsorbents can be categorized into these many groups to aid in cost analysis for various scenarios. Studies describing the creation of novel adsorbents must calculate the cost of the chemicals used, indicating which of these three cost categories it will fall under.

8.4.2 Real-Life Adsorbent Loading

The amount of adsorbate that can be removed per mass of an adsorbent is how the term "adsorption capacity" signifies an adsorbent's removal capacity. The most variable adsorbent property is also the one that has been the subject of the most research. Adsorbate adsorption capacity is dependent on the adsorbent characteristics, such as surface area, surface charge, and surface functionality, as well as the previously mentioned physicochemical characteristics of the solution, such as adsorbate concentration, temperature, pH, and the presence of other ions or molecules. It is exceedingly challenging to compare the adsorption capabilities of various adsorbents because different investigations are carried out under different conditions. Adsorption capacity alone can suggest either the greatest adsorption capacity or the adsorption capacity at equilibrium, which can be confusing and misleading. Here, the term "practical loading" refers to the adsorbate adsorption capacity that will actually be realized under a specific set of circumstances. This phrase, for instance, can be used to distinguish between the adsorption capacity attained at brief contact durations and the adsorption capacity at equilibrium conditions or at various adsorbate concentrations.[13] In this vein, Kumar and his colleagues[13] considered LM adsorption models over a period of five years and found out that there exists a great discrepancy between the max adsorption capacity and the achievable adsorption at lower adsorbate concentrations. Their research reveals that the maximum adsorption capacity is unimportant for the use of wastewater cleanup. When creating G-based wastewater adsorbents for this application, it is more crucial to consider the affinity (for instance, expressed as the constant in the Langmuir equation).

8.4.2.1   Affinity

An adsorption isotherm describes the relationship between the variation in adsorbate adsorption capacity and the equilibrium adsorbate concentration. Since it can be modeled, the adsorption isotherm is a particularly useful characterization for determining the adsorption capacity at various concentrations. The most popular models for characterizing adsorption are the Langmuir and Freundlich isotherms. The Langmuir adsorption model is predicated on the notion of chemisorption, which frequently occurs when the adsorbate is adsorbed onto adsorbent (hydr)oxides. The Langmuir model also enables comparison of the maximal adsorption capacity with the adsorption capacity at different adsorbate concentrations. Chapter 4 of this book, for instance, has the Langmuir equation. Even at lower equilibrium adsorbent concentrations (C_e), having a high equilibrium adsorption capacity (q_e) is beneficial for wastewater cleanup: the affinity of the adsorbent is the name given to this characteristic. This is dependent on both the Langmuir isotherm constant and the maximal adsorption capacity. However, studies typically only consider the adsorbent's maximum adsorption capacity. When used in the context of wastewater remediation, such capacities are frequently reported at equilibrium concentrations much higher than 10 mg adsorbent/L. Adsorbent affinity varies with the kind of adsorbent used for a specific adsorbate. For instance, the affinity of diverse G-based adsorbents differs with respect to their physical and chemical characteristics/properties as discussed earlier in the preceding chapters.

8.4.2.2   Kinetics of Adsorption

Adsorption isotherms are used to evaluate the adsorbate adsorption capabilities under equilibrium conditions. The time it takes to attain equilibrium, however, might vary from a few minutes to a few days, with certain adsorbents potentially needing a few weeks. Studies have shown that the time required to attain 90% (t_{90}) of equilibrium adsorption capacity based on PSOM for data considered for 5 years is 90 hours. The duration of the majority of these studies is also taken into consideration, probably because any change in adsorption after this point is only observed very slowly experimentally. The pseudo-second-order model's mathematical formula, which states that it takes 11 times longer to attain equilibrium adsorption capacity (t_{99}) than t_{90}, also explains this.[13] Even though t_{90} is thought of as the adsorption kinetics indicator, in many instances, it occurs over a period of several hours to days. Time is money in the real world, thus an adsorbent with superior kinetics is tremendously advantageous. The contact times, particularly when run in column mode, are typically in the range of a few minutes to less than an hour (measured as empty bed contact time or space velocity).[13] When used in column mode, porous adsorbents have a large enough particle size and high surface area to prevent pressure drop issues. For example, nano-perforated graphene (NPG) is a more desirable option for the removal of emerging contaminants because of its greater surface area, superhydrophobicity, straightforward water separation, recyclable materials, and chemical stability, especially when it is produced using a low-cost, scalable synthesis technique.[14]

8.4.2.3   pH Effect

The effect of pH on the adsorbent performance toward adsorptive exclusion of contaminants from wastewater has been discussed in Chapter 5. We understand that adjusting the pH of the medium can change the electrostatic interaction between the adsorbents with the adsorbates; hence, pH variation is frequently considered while improving the adsorption of contaminants in wastewater. The engineering of cheap G-based adsorbents that perform excellently at diverse pH will go a long way in augmenting the adsorption cost, ease of regeneration, as well as its ease of application in real life.

8.4.2.4 Temperature Effect

An adsorbent's thermodynamic characteristics control how temperature affects its performance. At higher temperatures, an endothermic process will result in better adsorption; an exothermic reaction will result in the opposite. With different adsorbents, the impact of temperature varies in strength. Although some of these studies employ temperatures above those typically found in practical applications, these are used as examples to highlight potential variations in the thermodynamic properties of these adsorbents.

It's crucial to take into account how temperature affects adsorption kinetics. This is particularly pertinent to performance in continuous modes with brief contact times. The majority of the time, the kinetic constants for various adsorbents increased as the temperature rose. This may be the result of enhanced adsorbates/contaminants diffusion at higher temperatures, which suggests that realistic loading for such adsorbents will decline at lower temperatures. This phenomenon is discussed in depth in Chapter 5.

8.4.2.5 Selectivity

Selectivity is the capacity of an adsorbent, such as a G-based adsorbent, to remove pollutants from wastewater in preference to competing ions. The competing ion's kind of contact with the adsorbent surface determines the selectivity of the adsorbent. In general, ions like chloride and nitrate exhibit little to no competition, but arsenate and silicate exhibit intense competition.[13,15] Tetrahedral ions like arsenate, phosphate, and silicate combine with G-adsorbent-containing metal (hydr)oxides like iron oxides to generate inner-sphere complexes. A weaker contact is created when nitrate and chloride combine to generate outer-sphere complexes. Sulphate and carbonate, on the other hand, have different competing effects, while carbonate frequently has a greater competitive effect.[13] On metal oxide surfaces, sulphate can exist as inner- or outer-sphere complexes.[16] Calcium carbonate surface precipitates can also occur when carbonate binds through electrostatic attraction.

With a look at the ultimate cost of adsorbent, which determines the final cost of adsorption, it is important to design/engineer adsorbents that have high performance along with excellent selectivity for diverse contaminants from wastewater.

8.4.3 Cost of Adsorbent Regeneration

The adsorbent concentrated flow could be recovered in the form of new chemicals: recovery of these chemicals requires additional use of other chemicals along with the adjustment of pH.

The amount of chemicals needed to precipitate the adsorbate will vary depending on the makeup of the stream that was regenerated. For instance, the molar ratio of magnesium:ammonium: phosphate necessary to create struvite in the regenerate stream was 1.5:1.5:1 following phosphate adsorption from secondary wastewater effluent.[17,18] Due to competing parallel processes, the ratio was larger than the stoichiometric value of 1:1:1. Likewise, optimal calcium phosphate production needed a molar ratio of between 2 and 2.5 of Ca to P.[18] This is greater than the molar ratios of 1.5 and 1.67 needed for the stoichiometric formation of tricalcium phosphate and hydroxyapatite, correspondingly.[19]

The conditions during regeneration and the selectivity of the absorbent will therefore affect how many chemicals are used to recover the adsorbent. More research is also required to demonstrate the regenerate solution's potential for reuse, particularly the impact that the build-up of adsorbent in the regenerate has on desorption. In order to effectively reuse the regenerate solution, this will provide crucial information on how often and when the adsorbate needs to be retrieved.

8.4.3.1 Adsorbent Lifespan

Reusing the G-based adsorbent more than once lengthens its useful life. According to our analysis of the literature, which is shown in Table 7.1 of Chapter 7, numerous research studies have assessed the adsorbent's reusability between 1 and 10 cycles.[20,21] Reusability is frequently impacted by adsorbent attrition during the adsorption or regeneration process as earlier discussed in Chapter 7. However, there are studies where the adsorbent removal rate comparatively remains unchanged, like in the report by Ren et al.[7] Diverse G-based adsorbents behave differently with regard to their regeneration due to their structural and chemical properties.

Practically speaking, the adsorbent lifetime should be much greater than 1 to 10 reuse cycles. However, studies should concentrate on creating a better understanding of the variables that affect reusability rather than performing endless regeneration cycles. Thus, the most effective regeneration techniques can be created. If it can be demonstrated using such techniques that the adsorbent properties do not change after a specific number of reuse cycles, the extrapolation of the adsorbent lifetime to larger reuse cycles will be more accurate.

The design and engineering of G-based adsorbents having infinite recyclability is of extreme demand and should, therefore, be the focus of researchers in this dispensation.

8.4.3.2 Adsorbent Recovery

The release of the molecules bound to the loaded adsorbent occurs through regeneration. This could involve both the adsorbate and other rival ions. Regeneration is carried out for two purposes: (a) To retrieve the separated molecule of interest or adsorbate(s). (b) To restore the adsorbent's active sites so that it can be reused. This adsorbent is G-based. The procedure is made to be economical and environmentally beneficial by recycling the adsorbent and any extra chemicals from the regeneration. The elements influencing the costs of chemicals for regeneration are the cost of releasing adsorbed complexes, surface adsorbed precipitates, and neutralizing acids(s)/bases used in many cases.

Knowing how much chemical is used to desorb one or more adsorbates from the surface of a G-based adsorbent is crucial from an economic perspective for adsorbate, in this example, Cs^{+1}.[22] As an inner sphere complex, adsorption occurs as a monodentate or bidentate complex. Accordingly, only one or two molecules of hydroxide ion should be used to desorb each adsorbate molecule, such as a phosphate molecule. However, in order to provide a driving force, too many hydroxide ions are needed. As a result, high NaOH concentrations, typically between 0.1 and 1 M, are utilized for phosphate desorption.[17] However, fewer hydroxide ions will actually be consumed, and any extra ions in the regenerate solution can be recycled. In pilot research, for instance, the NaOH solution was refilled and used 60 times to desorb phosphate.[17]

The actual consumption of the hydroxide ion during desorption depends on the other ions that bind via the same mechanism when the adsorbent is utilized in a water matrix made up of numerous competing ions. A selective adsorbent will reduce the overall hydroxide ion consumption per mole of adsorbate desorbed. But, having a highly selective adsorbent also means that it will have a high affinity for adsorbate, and the binding may be too strong to allow for simple desorption.

Accordingly, the conditions during regeneration, as well as the selectivity of the adsorbent, will determine the amount of chemicals used for the adsorbent recovery. Additionally, more research is required to demonstrate the possibility of recycling the regenerate solution, particularly the impact on desorption brought on by the build-up of adsorbent in the regenerate. For the regenerate solution to be effectively reused, this will provide crucial information on how often and when the adsorbate needs to be recovered.

8.4.4 Diverse Techniques Used for the Analysis of Adsorbents

This subsection discusses the many methods that various researchers have used to conduct a cost analysis of the adsorbent preparation used for the treatment of contaminated water.

8.4.4.1 Raw Material As Well As Preparation Cost Usage to Decide Adsorbent Cost

The cost per unit of the mass of the raw materials utilized in the synthesis process is used to determine if an adsorbent production is economically viable.[23] Depending on the activation technique utilized, the cost of the adsorbent material may change, with chemical activation being more expensive than physical activation.[23] Based on the price of the raw materials or industrial-grade precursors and the chemicals used to process them, the cost of adsorbent preparation can be calculated.[23] Adsorption capacity, selectivity, operational costs, and adsorbent degradation rate are further cost considerations. Ahmad and his colleagues[24] evaluated the prices of feedstock transportation, chemicals (hydroxides and hydrochloric acid), CO_2 gas, and the power necessary to produce GWAC to determine the price of gas waste activated carbon (GWAC). By taking into account the costs of the precursor, the activating agent, and the operation of relevant equipment, Bello and co.[25] used this method to estimate the cost of the synthesis of acid-modified kola nut husk (KNHA) for the removal of ibuprofen from aqueous media. When using materials that are common agricultural or industrial wastes, the cost of the precursor material might not always be taken into account. However, other operational costs and the cost of preparation will be taken into account.[23]

8.4.4.2 Excluded Adsorbate Comparison with Adsorbent Per Gram Used

If the operational and regeneration costs are comparable, it is possible to compare the cost of the manufactured adsorbent per gram of adsorbate removed with that of other well-known adsorbents in order to assess its economic viability.[26] Equation (8.1) illustrates how to calculate the cost of the adsorption operation using the cost of an adsorbent per gram of adsorbate.[23,27]

$$\text{Adsorption cost}\left(\text{USD}\middle/ g_{adsorbate}\right) = \frac{\left(\text{Cost of purchased chemical}\left(\dfrac{\text{USD}}{g}\right) + \text{Energy cost}\left(\dfrac{\text{USD}}{g}\right)\text{kWh}\right)}{\text{Adsorption capacity}\left(\text{mg/g}\right)\times 10^{-3}\,\text{g/mg}} \tag{8.1}$$

Herewith, adsorption capacity is measured in mg g^{-1} and represents the quantity of adsorbate extracted per adsorbent utilized. For extensive industrial uses, this can be scaled up.

8.4.4.3 Using Discounted Cash Flow to Estimate the Price of Creating an Adsorbent

The life cycle of a unit mass of adsorbent is tracked through production, deployment as well as elution cycles, and disposal in a cost analysis of the textile adsorbent method for recovering uranium from saltwater.[28] This method often relies on raw material and chemical costs. Precursor and chemical prices can be determined using historical data, vendor quotes, and a time-value of money calculation.

8.4.4.4 Application Cost for Adsorbents in Adsorption Operations

This scenario accounts for extra operations such as drying, energy production, and other reaction apparatus in addition to the cost of the chemical used to activate the carbon precursor material (for chemical processes) or the cost of the precursor material alone (for physical procedures).[23] TCI (total capital investment), which includes WCC (working capital investment) and FCE (fixed capital estimation), is required to cover the costs of creating an adsorbent and operating an adsorption plant.[29] The FCE includes all equipment acquisition costs, process pipes, equipment installation, electrical systems, instrumentation and controls, buildings, construction, and yard upgrades, as well as WCC, which could account for up to 6.5 percent of the FCE. If the price is expressed in dollars per m^3 of treated water, this cost can be calculated in relation to the amount of water that needs to be treated.[30]

$$\text{TCI} = \text{WCC} + \text{FCE} \tag{8.2}$$

The total capital investment (TCI), working capital investment (WCC), and fixed capital estimation (FCE) are all capitalized terms. The cost of raw materials (C_{RM}), the cost of managing waste created during the operation (C_{WG}), the cost of utilities (C_U), and additional cost (C_E) can all be used to compute the annual operation cost (AOC) for the preparation of adsorbents, as shown in equation (8.3).[29]

$$\text{ACC} = C_{RM} + C_{WG} + C_U + C_E \tag{8.3}$$

The annual adsorbent production, E_P, and the cost per unit of the manufactured adsorbent, C_E, can be used to compute the cost as per equation (8.4).

$$C_E = \text{AOC} / E_P \tag{8.4}$$

8.4.5 Using Cost Indices as the Price of Creating an Adsorbent

The adsorption potential, given in grams or milligrams, is the amount of adsorbate that can be removed per unit of money.[23] The relationship in equation (8.5) provides a clearer understanding of this.

The following Equation illustrates how the AOC and annual adsorbent production, E_P, can be used to determine the cost per unit of the produced adsorbent, C_E.

This can be understood to be the adsorption potential, which is the amount of adsorbate that can be removed per dollar and is expressed in grams or milligrams.[31] From the relationship in equation (8.5) below, this can be understood more clearly.

$$\text{Cost Index} = \frac{\text{Adsorption capacity}}{\text{Unit cost of adsorbent}} = \frac{\left(\dfrac{\text{Adsorbate mg}}{\text{Adsorbent mg}}\right)}{\left(\text{USD per mg adsorbent}\right)} \tag{8.5}$$

The procedure is more practical and vice versa a higher cost index. The cost of treatment may be established after the adsorbent's price, calculated adsorption capacity, and treated volume are known.[32]

8.4.6 Using Annual Capital Expenditures (CAPEX) and Operational Costs, the Cost of an Adsorption Operation (OPEX)

The sum of the CAPEX and the OPEX will give the entire cost of the adsorption operation.[33] By dividing the CAPEX by the annuity factor (AF), as given in Equation, one may determine the annual expense (equation (8.6)).

$$\text{Annual capital expenditure}\left(\text{USD per year}\right) = \text{CAPEX/AF} \tag{8.6}$$

AF is expressed in terms of equations and is based on the anticipated lifespan of the medical system and the interest rate as per equation (8.7).

$$\text{AF} = \frac{1 - \left(1 + r\right)^{-n}}{r} \tag{8.7}$$

The interest rate is indicated by r, and the operation's lifespan is indicated by n. You may determine the OPEX by taking into account the following factors: coagulant and adsorbent materials, the energy required to run the system, maintenance costs, and labor costs. The sum of OPEX and CAPEX, as given in equation (8.8), can be used to calculate the price of an adsorbent.[34]

$$C_{\text{adsorbent}} = C_c + C_o + C_{fh} + C_{ft} \tag{8.8}$$

Here C_{fh} is the total yearly cost of harvesting feedstock, C_{ft} is the total annual cost of transporting feedstock, while C_c is the annualized CAPEX of the treatment facility, C_o is the annualized OPEX. Another important component of the cost, which may not directly affect the price of the adsorbent material, is maintenance expenditures, in addition to CAPEX and OPEX.[35]

The exact experimental results from the adsorbent cost analysis need to be discussed; herewith, key papers concerning this aspect are discussed with regard to publications obtainable in the literature. The price of treating wastewater with various adsorbent classes, as reported in published studies, is shown in Table 8.2. The adsorbent class reported in Tables 8.2 and 8.3 is based on the classification of Iwuozor et al.,[36] which is class A for biosorbents, class B for activated carbon, class C for biochar, class D for clays and minerals, class E for polymers and resins, class F for nanoparticles. In contrast, class G is for (nano)composite adsorbents.

A group of researchers have proven that the price of G-based adsorbents can be even lowered to around 0.2 USD/g in their work.[40] We are aware that G-based adsorbents are high performance adsorbents and though they are high-performance adsorbents, they can be found for a reasonable price. The good news is that it is important to highlight that GO's price dropped significantly in recent decades, going from 50 USD/g to 2 USD/g.[40]

With due consideration to the G-based adsorbents cost per g or Kg aimed at wastewater treatment as presented in Table 8.2, one can say G-based adsorbents are comparatively cheap when compared on a holistic ground with a look at their lifetime usage. Also, the adsorbent class G may be comparatively more expensive, but in most instances, they offer greater performance than their counterparts, as shown in Table 8.3.

8.5   MARKET TRENDS FOR ADSORBENTS

With an array of varied as well as distinctive applications, the global adsorbent market has been expanding at a healthy rate. Adsorbents are employed to remove pollutants and impurities from a variety of substances, rendering them pure and safe for consumption. The market is being propelled by the rising demand coming from sectors including water treatment, petrochemicals, food, pharmaceuticals, and beverage. The size of the global adsorbent market was USD 4.6 billion in 2022 and is projected to increase by USD 8.2 billion by 2030, increasing at a CAGR of 6.8%.[49]

Adsorbents can be found in a number of different materials, such as activated carbon, silica gel, zeolites, graphene, and clay. One of the most popular kinds is activated carbon. Due to their excellent selectivity and effectiveness, other types of adsorbents,

TABLE 8.2

The Price of Utilizing Various Adsorbent Classes to Treat the Pollutants in Wastewater

Formulated Adsorbent	Class of Adsorbent	Excluded Contaminant	Adsorption Capacity (mg/g)	Cost ($/g)	Ref.
Durian shell	A	Basic brown 16	—	0.0002	[29]
Poly(3-aminobenzoic acid/graphene oxide/cobalt ferrite) (P3ABA/GO/ CoFe$_2$O$_4$)	G	Congo red (CR)	153.92	2.96	[10]
Tiger nut residue	A	Methylene blue	146	0.00001	[37]
Chitosan/GLA/GIO	G	Hg^{+2}	187	€15.60 ($17.49)	[38]
Tomato seeds	A	Acid red 14	125.00	0.0340	[39]
Carboxylated GO/ chitosan/cellulose	G	Cu^{+2}	22.4	0.2	[40]
AC from sugarcane bagasse	B	Methylene blue	—	0.0021	[41]
Rice husk biochar	C	Basic red 09	44	0.1504	[34]
Modified steel converter slag	D	Methylene blue	41.62	0.0002	[42]
Fe$_3$O$_4$@GO-EDA	G	As^{+3}	13.3	$1753.19/kg ($1.753/g)	[43]
Alkali-activated steel slag	D	Cu^{+2}	161.29	0.0001	[44]
Pristine steel slag	D	Cu^{+2}	109.89	0.00003	[44]
Nano-zeolite	F	Nitrophenol	156.6	0.0300	[27]
Nanoscale zero-valent iron	F	Phosphate	0.312	3.15/m^3	[30]
Copper impregnated tufa	G	As^{+5}	104.62	0.0066	[26]
Microporous biomass-polyphenol	G	U	0.0278	0.275	[45]
n-layer amino-functionalized graphene oxide (nGO-(NH)R)	F	MB	3036.43	$1900/kg ($1.9/g)	[46]
nGO-(NH)R		RB	335.86	$1900/kg	[46]
Titania/graphene oxide	G	Pb^{+2}	228.311	0.0460	[47]
Titania/graphene oxide	G	Cu^{+2}	168.067	0.0460	[47]
Titania/graphene oxide	G	Phenol	24.0	0.1875	[47]
Titania/graphene oxide	G	Pb^{+2}	228.311	0.0460	[47]
Titania/graphene oxide	G	Cu^{+2}	168.067	0.0460	[47]
3D GBMs	F	MB	31.94	0.1	[48]

TABLE 8.3

Calculated Adsorbent Cost Performance for Different Adsorbent Classes

Adsorbent Class	Target Pollutant	Adsorption Capacity (mg/g)	Adsorption Capacity (g/g)	Molar Mass (g/mol)	Adsorption Capacity (mol/g)	Cost per Treatment ($/g)	$\hat{C}$ ($/mol)
A	Acid blue 92	36.23	0.03623	656.2	5.52118 05	0.118	2137.223
A	Methylene blue	146	0.146	319.85	0.000456464	0.00001	0.021908
A	Acid red 14	125	0.125	502.431	0.00024879	0.034	136.6612
A	Crystal violet	20.84	0.02084	407.979	5.10811 E-05	0.0009	17.61905
A	Ibuprofen	39.22	0.03922	206.29	0.000190121	0.0425	223.5422
B	Malachite green	226.06	0.22606	364.911	0.000619494	0.0002	0.322844
C	Methylene blue	340.3	0.3403	319.85	0.001063936	0.0009	0.845915
C	Phosphate	96.9	0.0969	94.9714	0.001020307	0.0009	0.882087
C	Pb(II)	94.48	0.09448	207	0.000456425	0.0419	91.80038
C	Cd(II)	92.42	0.09242	112.41	0.000822169	0.0419	50.96277
C	Ni(II)	66.22	0.06622	58.693	0.001128244	0.0419	37.13737
C	Zn(II)	35.71	0.03571	65.4	0.000546024	0.0026	4.761691
C	Basic red 09	10	0.01	237.258	4.21482 E-05	0.7037	16,695.85
C	Basic red 09	46.3	0.0463	237.258	0.000195146	0.1411	723.0476
C	Basic red 09	44	0.044	237.258	0.000185452	0.1504	810.991
D	Methylene blue	41.62	0.04162	319.85	0.000130123	0.0002	1.537001
D	Cu(II)	161.29	0.16129	67.58	0.002386653	0.0001	0.0419
D	Cu(II)	109.89	0.10989	67.58	0.001626073	0.00003	0.018449
D	U(VI)	820.7	0.8207	450.745	0.001820763	0.0678	37.23713
E	Ni(II)	63	0.063	58.693	0.001073382	0.000006	0.00559
F	Nitrophenol	156.6	0.1566	139.11	0.001125728	0.03	26.64943

G	As(V)	104.62	0.10462	74.92159	0.001396393	0.0066	4.726462
G	U(VI)	0.0278	0.0000278	450.745	6.16757 E-08	0.275	4,458,808
G	Pb(II)	228.311	0.228311	207	0.001102952	0.046	41.70627
G	Cu(II)	168.067	0.168067	67.58	0.002486934	0.046	18.49667
G	Phenol	24	0.024	94.11	0.000255021	0.1875	735.2344
G	Phosphorus	165.9	0.1659	30.97	0.005356797	0.6495	121.2478
G	Cr(VI)	217.8	0.2178	51.996	0.004188784	0.62	148.0143
G	Fluoride	28.57	0.02857	18.998	0.001503843	0.0258	17.15605
G	Cr(VI)	54.42	0.05442	51.996	0.001046619	0.0895	85.51345
G	Phosphorus	21.36	0.02136	30.97	0.0006897	0.2646	383.6452

Reproduced with permission from Ighalo et al.,[23] Copyright 2022, Elsevier Science Ltd.

including covalent organic frameworks (COFs) and metal-organic frameworks (MOFs), are becoming more and more popular.

With a worldwide market share, the water treatment industry is the biggest consumer of adsorbents. Adsorbents are used to clean water of pollutants, including suspended particles, organic matter, and other contaminants so that it is suitable for consumption. Adsorbents are utilized in the pharmaceutical industry, another significant consumer of them, for the purification and separation of numerous medicines and pharmaceutical intermediates.

Gas filtration, air separation, and industrial waste treatment are some more uses for adsorbents. Cosmetics, pet food, and personal care items are also made using adsorbents.

The greatest market for adsorbents is in the Asia Pacific area, driven by rising demand from numerous end-use sectors in nations like China, India, and Japan. Cabot Corporation, Clariant AG, BASF SE, Grace & Co., Arkema S.A., W.R. as well as Zeochem AG are the main players in the adsorbent market.[49] The following paragraphs present the recent trends within the adsorbent materials niche.

- Utilization of graphene-based adsorbents: High surface area, variable pore size, and outstanding mechanical qualities make graphene-based adsorbents a desirable material for adsorption applications.

Demand for eco-friendly adsorbents is on the rise, and more people are choosing to use renewable, non-toxic, as well as biodegradable adsorbents.

- An increase in the usage of magnetic adsorbents: After adsorption and regeneration, magnetic adsorbents are simple to remove from the medium, which has led to their growing utilization.
- Use of adsorbents in CO_2 capture: The oil and gas industry, in particular, uses adsorbents to absorb and store CO_2. Adsorbents are being created for applications involving water desalination, particularly in areas where there is a shortage of freshwater.
- Adsorbents utilization in energy storage: Adsorbents are utilized in energy storage applications, such as the creation of cutting-edge batteries.

Adsorbents for air purification are in higher demand, notably in the healthcare/hospitality niches. Adsorbents are used to purify the air in indoor environments.

- Adsorbents utilization in drug delivery: Adsorbents are utilized in drug delivery applications, especially in the creation of formulations for prolonged release.

8.6 CONCLUSIONS

G, as well as its derivatives, have been used to exclude heavy metals as well as dyes, as discussed in this chapter. Numerous applications require large-scale, cost-effective, as well as environmentally friendly production of high-purity G and its derivatives.

Low pH values usually favor anionic contaminants in electrostatic interactions. Anions are thought to be adsorbed via both specific and non-specific adsorption. Nonetheless, cation, as well as anion adsorption is based on three adsorption processes: electrostatic interaction, ion exchange, and complex formation. The LM and the PSOM can be used to describe the adsorption isotherm and kinetics. For G-based materials, the adsorption process is spontaneous, endothermic, as well as feasible.

REFERENCES

1. Perreault, F.; De Faria, A. F.; Elimelech, M., Environmental applications of graphene-based nanomaterials. *Chemical Society Reviews* 2015, *44* (16), 5861–5896.

2. Sun, Y.; Shao, D.; Chen, C.; Yang, S.; Wang, X., Highly efficient enrichment of radionuclides on graphene oxide-supported polyaniline. *Environmental Science & Technology* 2013, *47* (17), 9904–9910.

3. Upadhyay, R. K.; Soin, N.; Roy, S. S., Role of graphene/metal oxide composites as photocatalysts, adsorbents and disinfectants in water treatment: A review. *RSC Advances* 2014, *4* (8), 3823–3851.

4. Liu, X.; Li, J.; Wang, X.; Chen, C.; Wang, X., High performance of phosphate-functionalized graphene oxide for the selective adsorption of U (VI) from acidic solution. *Journal of Nuclear Materials* 2015, *466*, 56–64.

5. Hu, X. -J.; Liu, Y.-G.; Wang, H.; Chen, A.-W.; Zeng, G.-M.; Liu, S.-M.; Guo, Y.-M.; Hu, X.; Li, T.-T.; Wang, Y.-Q., Removal of Cu (II) ions from aqueous solution using sulfonated magnetic graphene oxide composite. *Separation and Purification Technology* 2013, *108*, 189–195.

6. Ranjan Rout, D.; Mohan Jena, H., Synthesis of novel reduced graphene oxide decorated β-cyclodextrin epichlorohydrin composite and its application for Cr(VI) removal: Batch and fixed-bed studies. *Separation and Purification Technology* 2021, *278*, 119630.

7. Ren, H.; Cao, Z.-F.; Chen, Y.-Y.; Jiang, X.-Y.; Yu, J.-G., Graphene oxide-Bicine composite as a novel adsorbent for removal of various contaminants from aqueous solutions. *Journal of Environmental Chemical Engineering* 2021, *9* (6), 106769.

8. Sari Yilmaz, M., Graphene oxide/hollow mesoporous silica composite for selective adsorption of methylene blue. *Microporous and Mesoporous Materials* 2022, *330*, 111570.

9. Sheikhmohammadi, A.; Hashemzadeh, B.; Alinejad, A.; Mohseni, S. M.; Sardar, M.; Sharafkhani, R.; Sarkhosh, M.; Asgari, E.; Bay, A., Application of graphene oxide modified with the phenopyridine and 2-mercaptobenzothiazole for the adsorption of Cr (VI) from wastewater: Optimization, kinetic, thermodynamic and equilibrium studies. *Journal of Molecular Liquids* 2019, *285*, 586–597.

10. Babakir, B. A. M.; Abd Ali, L. I.; Ismail, H. K., Rapid removal of anionic organic dye from contaminated water using a poly(3-aminobenzoic acid/graphene oxide/cobalt ferrite) nanocomposite low-cost adsorbent via adsorption techniques. *Arabian Journal of Chemistry* 2022, *15* (12), 104318.

11. Xing, H. T.; Chen, J. H.; Sun, X.; Huang, Y. H.; Su, Z. B.; Hu, S. R.; Weng, W.; Li, S. X.; Guo, H. X.; Wu, W. B.; He, Y. S.; Li, F. M.; Huang, Y., NH2-rich polymer/graphene oxide use as a novel adsorbent for removal of Cu(II) from aqueous solution. *Chemical Engineering Journal* 2015, *263*, 280–289.

12. Genz, A.; Kornmüller, A.; Jekel, M., Advanced phosphorus removal from membrane filtrates by adsorption on activated aluminium oxide and granulated ferric hydroxide. *Water Research* 2004, *38* (16), 3523–3530.

13. Kumar, P. S.; Korving, L.; van Loosdrecht, M. C. M.; Witkamp, G.-J., Adsorption as a technology to achieve ultra-low concentrations of phosphate: Research gaps and economic analysis. *Water Research X* 2019, *4*, 100029.

14. Khalil, A. M. E.; Memon, F. A.; Tabish, T. A.; Salmon, D.; Zhang, S.; Butler, D., Nanostructured porous graphene for efficient removal of emerging contaminants (pharmaceuticals) from water. *Chemical Engineering Journal* 2020, *398*, 125440.

15. Temane, L. T.; Orasugh, J. T.; Ray, S. S., Adsorptive removal of pollutants using graphene-based materials for water purification. In Kumar, N.; Gusain, R.; Ray, S. S., *Two-Dimensional Materials for Environmental Applications*; Springer, 2023; pp. 179–244.

16. Wijnja, H.; Schulthess, C. P., Vibrational spectroscopy study of selenate and sulfate adsorption mechanisms on Fe and Al (hydr) oxide surfaces. *Journal of Colloid and Interface Science* 2000, *229* (1), 286–297.

17. Drenkova-Tuhtan, A.; Schneider, M.; Franzreb, M.; Meyer, C.; Gellermann, C.; Sextl, G.; Mandel, K.; Steinmetz, H., Pilot-scale removal and recovery of dissolved phosphate from secondary wastewater effluents with reusable ZnFeZr adsorbent@Fe_3O_4/SiO_2 particles with magnetic harvesting. *Water Research* 2017, *109*, 77–87.

18. Kalaitzidou, K.; Mitrakas, M.; Raptopoulou, C.; Tolkou, A.; Palasantza, P.-A.; Zouboulis, A., Pilot-scale phosphate recovery from secondary wastewater effluents. *Environmental Processes* 2016, *3*, 5–22.

19. Song, Y.; Hahn, H. H.; Hoffmann, E., Effects of solution conditions on the precipitation of phosphate for recovery: A thermodynamic evaluation. *Chemosphere* 2002, *48* (10), 1029–1034.

20. Liu, Y.; Fu, J.; He, J.; Wang, B.; He, Y.; Luo, L.; Wang, L.; Chen, C.; Shen, F.; Zhang, Y., Synthesis of a superhydrophilic coral-like reduced graphene oxide aerogel and its application to pollutant capture in wastewater treatment. *Chemical Engineering Science* 2022, *260*, 117860.

21. Sharif, F.; Roberts, E. P. L. Electrochemical oxidation of an organic dye adsorbed on tin oxide and antimony doped tin oxide graphene composites *Catalysts* [Online], 2020, *10*, 263.

22. Tan, L.; Wang, S.; Du, W.; Hu, T., Effect of water chemistries on adsorption of Cs(I) onto graphene oxide investigated by batch and modeling techniques. *Chemical Engineering Journal* 2016, *292*, 92–97.

23. Ighalo, J. O.; Omoarukhe, F. O.; Ojukwu, V. E.; Iwuozor, K. O.; Igwegbe, C. A., Cost of adsorbent preparation and usage in wastewater treatment: A review. *Cleaner Chemical Engineering* 2022, *3*, 100042.

24. Ahmad, A. A.; Ahmad, M. A.; Yahaya, N. K. E.; Din, A. M.; Yaakub, A. R. W., Honeycomb-like porous-activated carbon derived from gasification waste for malachite green adsorption: Equilibrium, kinetic, thermodynamic and fixed-bed column analysis. *DWT* 2020, *196*, 329–347.

25. Bello, O. S.; Alao, O. C.; Alagbada, T. C.; Agboola, O. S.; Omotoba, O. T.; Abikoye, O. R., A renewable, sustainable and low-cost adsorbent for ibuprofen removal. *Water Science and Technology* 2021, *83* (1), 111–122.

26. Bajić, Z. J.; Veličković, Z. S.; Djokić, V. R.; Perić-Grujić, A. A.; Ersen, O.; Uskoković, P. S.; Marinković, A. D., Adsorption study of arsenic removal by novel hybrid copper impregnated tufa adsorbents in a batch system. *Clean–Soil, Air, Water* 2016, *44* (11), 1477–1488.

27. Pham, T.-H.; Lee, B.-K.; Kim, J., Improved adsorption properties of a nano zeolite adsorbent toward toxic nitrophenols. *Process Safety and Environmental Protection* 2016, *104*, 314–322.

28. Das, S.; Liao, W.-P.; Flicker Byers, M.; Tsouris, C.; Janke, C. J.; Mayes, R. T.; Schneider, E.; Kuo, L.-J.; Wood, J. R.; Gill, G. A., Alternative alkaline conditioning of amidoxime based adsorbent for uranium extraction from seawater. *Industrial & Engineering Chemistry Research* 2016, *55* (15), 4303–4312.

29. Gopalakrishnan, Y.; Al-Gheethi, A.; Abdul Malek, M.; Marisa Azlan, M.; Al-Sahari, M.; Radin Mohamed, R. M. S.; Alkhadher, S.; Noman, E., Removal of basic brown 16 from aqueous solution using durian shell adsorbent, optimization and techno-economic analysis. *Sustainability* 2020, *12* (21), 8928.

30. Mahmoud, A. S.; Mostafa, M. K.; Nasr, M., Regression model, artificial intelligence, and cost estimation for phosphate adsorption using encapsulated nanoscale zero-valent iron. *Separation Science and Technology* 2019, *54* (1), 13–26.

31. Li, S.; Huang, X.; Liu, J.; Lu, L.; Peng, K.; Bhattarai, R., PVA/PEI crosslinked electrospun nanofibers with embedded La(OH)$_3$ nanorod for selective adsorption of high flux low concentration phosphorus. *Journal of Hazardous Materials* 2020, *384*, 121457.

32. Pap, S.; Kirk, C.; Bremner, B.; Sekulic, M. T.; Shearer, L.; Gibb, S. W.; Taggart, M. A., Low-cost chitosan-calcite adsorbent development for potential phosphate removal and recovery from wastewater effluent. *Water Research* 2020, *173*, 115573.

33. Mahmoud, A. S.; Mostafa, M. K.; Peters, R. W., A prototype of textile wastewater treatment using coagulation and adsorption by Fe/Cu nanoparticles: Techno-economic and scaling-up studies. *Nanomaterials and Nanotechnology* 2021, *11*, 18479804211041181.

34. Praveen, S.; Gokulan, R.; Pushpa, T. B.; Jegan, J., Techno-economic feasibility of biochar as biosorbent for basic dye sequestration. *Journal of the Indian Chemical Society* 2021, *98* (8), 100107.

35. Youssef, P. G.; Mahmoud, S. M.; Al-Dadah, R. K., Numerical simulation of combined adsorption desalination and cooling cycles with integrated evaporator/condenser. *Desalination* 2016, *392*, 14–24.

36. Iwuozor, K. O.; Ighalo, J. O.; Ogunfowora, L. A.; Adeniyi, A. G.; Igwegbe, C. A., An empirical literature analysis of adsorbent performance for methylene blue uptake from aqueous media. *Journal of Environmental Chemical Engineering* 2021, *9* (4), 105658.

37. Kani, A. N.; Dovi, E.; Mpatani, F. M.; Li, Z.; Han, R.; Qu, L., Tiger nut residue as a renewable adsorbent for methylene blue removal from solution: Adsorption kinetics, isotherm, and thermodynamic studies. *Desalination and Water Treatment* 2020, *191*, 426–437.

38. Kyzas, G. Z.; Travlou, N. A.; Deliyanni, E. A., The role of chitosan as nanofiller of graphite oxide for the removal of toxic mercury ions. *Colloids and Surfaces B: Biointerfaces* 2014, *113*, 467–476.

39. Najafi, H.; Pajootan, E.; Ebrahimi, A.; Arami, M., The potential application of tomato seeds as low-cost industrial waste in the adsorption of organic dye molecules from colored effluents. *Desalination and Water Treatment* 2016, *57* (32), 15026–15036.

40. Zhao, L.; Yang, S.; Yilihamu, A.; Ma, Q.; Shi, M.; Ouyang, B.; Zhang, Q.; Guan, X.; Yang, S.-T., Adsorptive decontamination of Cu2+-contaminated water and soil by carboxylated graphene oxide/chitosan/cellulose composite beads. *Environmental Research* 2019, *179*, 108779.

41. Fingolo, A. C.; Klein, B. C.; Rezende, M. C.; Silva e Souza, C. A.; Yuan, J.; Yin, G.; Bonomi, A.; Martinez, D. S.; Strauss, M., Techno-economic assessment and critical properties tuning of activated carbons from pyrolyzed sugarcane bagasse. *Waste and Biomass Valorization* 2020, *11*, 1–13.

42. Cheng, M.; Zeng, G.; Huang, D.; Lai, C.; Liu, Y.; Zhang, C.; Wang, R.; Qin, L.; Xue, W.; Song, B., High adsorption of methylene blue by salicylic acid–methanol modified steel converter slag and evaluation of its mechanism. *Journal of Colloid and Interface Science* 2018, *515*, 232–239.

43. Tabatabaiee Bafrooee, A. A.; Moniri, E.; Ahmad Panahi, H.; Miralinaghi, M.; Hasani, A. H., Ethylenediamine functionalized magnetic graphene oxide (Fe_3O_4@GO-EDA) as an efficient adsorbent in Arsenic(III) decontamination from aqueous solution. *Research on Chemical Intermediates* 2021, *47* (4), 1397–1428.

44. Nikolić, I.; Đurović, D.; Tadić, M.; Radmilović, V. V.; Radmilović, V. R., Adsorption kinetics, equilibrium, and thermodynamics of Cu^{2+} on pristine and alkali activated steel slag. *Chemical Engineering Communications* 2020, *207* (9), 1278–1297.

45. Luo, W.; Xiao, G.; Tian, F.; Richardson, J. J.; Wang, Y.; Zhou, J.; Guo, J.; Liao, X.; Shi, B., Engineering robust metal–phenolic network membranes for uranium extraction from seawater. *Energy & Environmental Science* 2019, *12* (2), 607–614.

46. Fraga, T. J. M.; de Souza, Z. S. B.; Marques Fraga, D. M. d. S.; Carvalho, M. N.; de Luna Freire, E. M. P.; Ghislandi, M. G.; da Motta Sobrinho, M. A., Comparative approach towards the adsorption of Reactive Black 5 and methylene blue by n-layer graphene oxide and its amino-functionalized derivative. *Adsorption* 2020, *26* (2), 283–301.

47. Fu, C.-C.; Juang, R.-S.; Huq, M. M.; Hsieh, C.-T., Enhanced adsorption and photodegradation of phenol in aqueous suspensions of titania/graphene oxide composite catalysts. *Journal of the Taiwan Institute of Chemical Engineers* 2016, *67*, 338–345.

48. Tewari, C.; Tatrari, G.; Kumar, S.; Pandey, S.; Rana, A.; Pal, M.; Sahoo, N. G., Green and cost-effective synthesis of 2D and 3D graphene-based nanomaterials from Drepanostachyum falcatum for bio-imaging and water purification applications. *Chemical Engineering Journal Advances* 2022, *10*, 100265.

49. Jonathan Tersur Orasugh, C. P.; Ali, Mir Sahidul; Chattopadhyay, Dipankar., 8 – Electromagnetic interference shielding property of polymer-graphene composites. In *Polymer Nanocomposites Containing Graphene*, Mostafizur Rahaman, L. N., ... Das, Narayan Chandra. Eds.; Woodhead Publishing, 2022; pp. 211–243.

Index

Pages in *italics* refer to figures and pages in **bold** refer to tables.